Internal Combustion Engines:
Performance, Fuel Economy and Emissions

Organising Committee

Professor Paul Shayler *(Committee Chairman)*	Nottingham University
Dr Roy Horrocks	Ford Motor Company
Humphrey Niven	Consultant
Hugh Blaxill	MAHLE Powertrain Ltd
Dr Martyn Twigg	Johnson Matthey plc
Professor Choongsik Bae	KAIST
Professor Colin Garner	Loughborough University
Steven Whelan	Clean Air Power Ltd
Lt Cdr Jacques Oliver	UK MoD
Brian Cooper	Ricardo UK Ltd
Dr Matthias Wellers	AVL Powertrain UK Ltd

Internal Combustion Engines: Performance, Fuel Economy and Emissions

IMechE, London
11–12 December 2007

Chandos Publishing (Oxford) Limited
TBAC Business Centre
Avenue 4
Station Lane
Witney
Oxford OX28 4BN
UK
Tel: +44 (0) 1993 848726 Fax: +44 (0) 1865 884448
Email: info@chandospublishing.com
www.chandospublishing.com

First published in Great Britain 2008

ISBN: 978-1-84334-451-3 (1-84334-451-3)

Copyright ©2008: The Institution of Mechanical Engineers, unless otherwise stated.

British Library Cataloguing-in-Publication Data.
A catalogue record for this book is available from the British Library.

All rights reserved. No part of this publication may be reproduced, stored in or introduced into a retrieval system, or transmitted, in any form, or by any means, (electronic, mechanical, photocopying, recording or otherwise) without the prior written permission of the Publishers. This publication may not be lent, resold, hired out or otherwise disposed of by way of trade in any form of binding or cover other than that in which it is published without the prior consent of the Publishers. Any person who does any unauthorised act in relation to this publication may be liable to criminal prosecution and civil claims for damages.

The Publishers make no representation, express or implied, with regard to the accuracy of the information contained in this publication and cannot accept any legal responsibility or liability for any errors or omissions.

The material contained in this publication constitutes general guidelines only and does not represent to be advice on any particular matter. No reader or purchaser should act on the basis of material contained in this publication without first taking professional advice appropriate to their particular circumstances.

Designations used by companies to distinguish their products are often claimed as trademarks. All brand names and product names used in this book are trade names, service marks, trademarks or registered trademarks of their respective owners. The Publisher is not associated with any product or vendor mentioned in this book.

Produced from electronic copy supplied by authors.
Printed in the UK and USA.

Cover photograph by kind permission of MAHLE Powertrain Ltd.

CONTENTS

COMBUSTION DIAGNOSTICS

C660/024/07   Optical studies of spray development in a quiescent chamber and in a direct-injection spark-ignition engine   3
P.G. Aleiferis, J. Serras-Pereira, Z. van Romunde, University College London, UK; D. Richardson, S. Wallace, Jaguar Advanced Powertrain Engineering, UK; R.F. Cracknell, H.L. Walmsley, Shell Global Solutions (UK) Ltd., UK

C660/017/07   Full bore imaging of combustion and quantitative AFR PLIF with a multi-component fuel and co-evaporating tracers   15
B. Williams, X. Wang, P. Ewart, R. Stone, University of Oxford, UK; H. Ma, H.L. Walmsley, R.F. Cracknell, Shell Global Solutions (UK) Ltd., UK; R. Stevens, S. Wallace, D. Richardson, Jaguar Cars, UK

C660/050/07   Experimental investigation of combustion and heat transfer in a direct injection spark ignition (DISI) engine through instantaneous combustion chamber surface temperature measurements   27
K. Cho, D. Assanis, Z. Filipi, University of Michigan, USA; G. Szekely, P. Najt, R. Rask, General Motors R&D, USA

DIESEL

C660/013/07   The potential of downsizing diesel engines considering performance and emissions challenges   49
T. Körfer, M. Lamping, A. Kolbeck, T. Genz, FEV Motorentechnik GmbH, Germany; S. Pischinger, H. Busch, D. Adolph, Institute for Internal Combustion Engines (VKA), Germany

C660/015/07   Delphi's 2000 bar common rail development for the Multec™ diesel common rail system   61
R.W. Jorach, D. Schoeppe, R.T. Nevard, I.R. Thornthwaite, N.D. Wilson, Delphi Diesel Systems Ltd., UK

C660/040/07   Application of JCB Dieselmax high performance technology to future off-highway diesel engines   71
A. Banks, M. Beasley, A. Skipton Carter, Ricardo Consulting Engineers Ltd., UK; A. Tolley, T. Leverton, JCB Power Systems Ltd., UK

GASOLINE

C660/014/07	Controlled Auto Ignition based on GDI – strategies derived from experiment and simulation A. Kulzer, J.-P. Hathout, C. Sauer, A. Christ, Robert Bosch GmbH, Germany	87
C660/020/07	A comparison study of different NVO strategies in a diesel HCCI engine using 3D CFD M. Jia, Z. Peng, University of Sussex, UK	99
C660/007/07	Cyclic combustion variability in gasoline engines R. Tily, C.J. Brace, University of Bath, UK	111

EMISSIONS

C660/043/07	Characteristics of catalytic oxidation for carbon black simulating diesel particulate matter over promoted Pt/Al_2O_3 catalysts assembled in two stage J.-W. Jeong, B. Choi, M.-T. Lim, Chonnam National University, Korea	123
C660/045/07	Phenomenological NO model for conventional heavy-duty diesel engine combustion R.S.G. Baert, X.L.J. Seykens, Eindhoven University of Technology, The Netherlands	131
C660/047/07	The porous medium approach applied to CFD modelling of SCR in an automotive exhaust with injection of urea droplets S.F. Benjamin, C.A. Roberts, Coventry University, UK	143
C660/002/07	Homogeneous diesel combustion challenges under vehicle compliant operating conditions R. Otte, J. Müller, F. Weberbauer, B. Guggenberger, Robert Bosch GmbH, Germany	161
C660/006/07	Investigation of fuel injection strategies on a low emission heavy-duty diesel engine with high EGR rates A.J. Nicol, C. Such, Ricardo UK Ltd., UK; U. Sarnbratt, Volvo Technology Corp., Sweden	173
C660/051/07	Recent developments in the control of particulate emissions from diesel vehicles M.V. Twigg, Johnson Matthey plc, UK	185
C660/044/07	Feasibility study of emission control of hydrogen fueled S.I. engine T. Ohira, Suzuki Motor Corporation, Japan; K. Nakagawa, K. Yamane, Musashi Institute of Technology, Japan; H. Kawanabe, M. Shioji, Kyoto University, Japan	199

FUELS AND LUBRICANTS

C660/011/07   Fuel dilution effects in a direct injection of natural gas engine   213
G.P. McTaggart-Cowan, Loughborough University, UK; S.N. Rogak, P.G. Hill, W.K. Bushe, University of British Columbia, Canada; S.R. Munshi, Westport Power Inc., Canada

C660/022/07   A study of alcohol blended fuels in a new optical spark-ignition engine   223
J.S. Malcolm, P.G. Aleiferis, University College London, UK; A.R. Todd, A. Cairns, A. Hume, H. Blaxill, MAHLE Powertrain Ltd., UK; H. Hoffman, J. Rueckauf, MAHLE International GmbH, Germany

C660/042/07   Influence of gasoline engine lubricant on tribological performance, fuel economy and emissions   235
P.M. Lee, M. Priest, University of Leeds, UK

C660/010/07   Determining the effect of lubricating oil properties on diesel engine fuel economy   247
C.D. Bannister, J.G. Hawley, C.J. Brace, University of Bath, UK; I. Pegg, Ford Motor Company Limited, UK; J.C. Dumenil, A. Brown, BP plc, UK

CALIBRATION

C660/009/07   A novel approach to investigating advanced boosting strategies of future diesel engines   261
S. Akehurst, M. Piddock, University of Bath, UK

C660/035/07   Smart calibration – turbocharger speed limitation – an example   277
M. Wellers, B. Carnochan, F. Ewen, U. Genc, E. Martini, AVL Powertrain UK Ltd., UK

C660/018/07   The effect of multiple fuel-injections on emissions of NOx and smoke with partially-premixed diesel combustion in a common-rail diesel engine   285
P. Eastwood, T. Morris, K. Tufail, T. Winstanley, Ford Motor Company Limited., UK; Y. Hardalupas, A.M.K.P. Taylor, Imperial College London, UK

C660/041/07   Effects of extended exhaust and intake duration on CAI combustion in a multi-cylinder DI gasoline engine   303
N. Kalian, H. Zhao, Brunel University, UK

AUTHOR INDEX

COMBUSTION DIAGNOSTICS

Optical studies of spray development in a quiescent chamber and in a direct-injection spark-ignition engine

P.G. Aleiferis, J. Serras-Pereira, Z. van Romunde
Department of Mechanical Engineering, University College London, UK

D. Richardson, S. Wallace
Jaguar Advanced Powertrain Engineering, UK

R.F. Cracknell, H.L. Walmsley
Shell Global Solutions (UK) Ltd., UK

ABSTRACT

The effects of fuel type and in-cylinder flow on spray formation from a multi-hole injector were studied by high-speed imaging techniques in a quiescent injection chamber and in a single-cylinder Direct-Injection Spark-Ignition (DISI) engine. To examine the effect of fuel volatility on spray formation, the injector was heated from 20 °C to 120 °C in the chamber for *iso*-octane and gasoline. The injection chamber was operated at 0.5 and 1.0 bar to mimic in-cylinder pressures for early injection strategies. Droplet sizing was also employed in the chamber using Phase Doppler Anemometry (PDA). Fuel-type and temperature effects were studied in-cylinder by operating the engine at 20 °C and 90 °C head temperature at 1500 RPM. For both sets of experiments, the study was carried out for two orthogonal views, relating to the tumble and swirl planes of in-cylinder flow motion. Spray formation was observed to be different for the two fuels, especially at high injector temperatures. Wetted footprint spray areas were calculated for both experimental setups.

1 INTRODUCTION

Experimental investigations of spray development are crucial for effective optimisation of different engine combustion systems. However, many investigations into this subject are carried out using single-component fuels due to ease of handling and known, consistent properties (1–4). The 'success' of different combustion systems in practice however, is dependent on real-world, multi-component gasoline behaviour. To address this issue, recent work by the current authors has examined the effects of different fuel properties on multi-hole injector spray development, considering a variety of single-component and multi-component fuels in a quiescent chamber (5, 6). In addition, the authors have examined the mixture preparation and combustion in an optical engine using a multi-hole injector for different injection timings with single and split-injection strategies (7). In this paper, the spray development in the quiescent chamber is directly compared to that in the engine to de-couple the effects of in-cylinder flow on spray development through both observation and quantitative comparison of the projected spray areas as viewed from the engine piston crown and a replicated view in the chamber. Spray droplet sizes are also presented, as measured 25 mm below the injector tip for a range of gas pressures and injector temperatures.

2   EXPERIMENTAL SETUP

2.1   Injection system

For both the chamber and engine work the fuel injection system comprised of a pneumatic pump which provided a fuel pressure of 150 bar. The injector used is a prototype with a 6-hole nozzle designed for vertical close spacing arrangement with the centrally located spark plug in the engine. The nozzle essentially accommodates two groups of three holes producing the spray pattern illustrated in Figure 1, where plumes 1 and 6 pass around the spark plug. Due to confidentiality agreements no further details regarding the injector manufacturer or the operating mechanism can be given. However, further details of the fuel supply and injection system, including driver and associated delays, have been quantified and presented in (5).

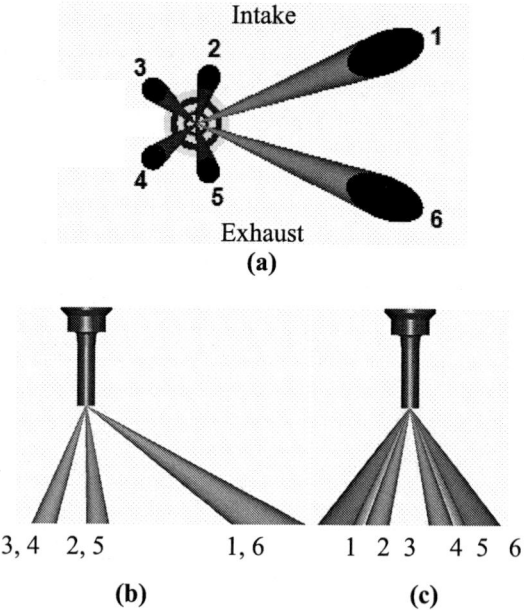

Figure 1. Injector configuration and orientation of spray plumes.

2.2 Fuel properties

The fuels used in this study were *iso*-octane and a pump-grade gasoline that has been commercially available in continental Europe. Figure 2 illustrates the distillation curves for the fuels and highlights the single boiling point of *iso*-octane at ~99 °C.

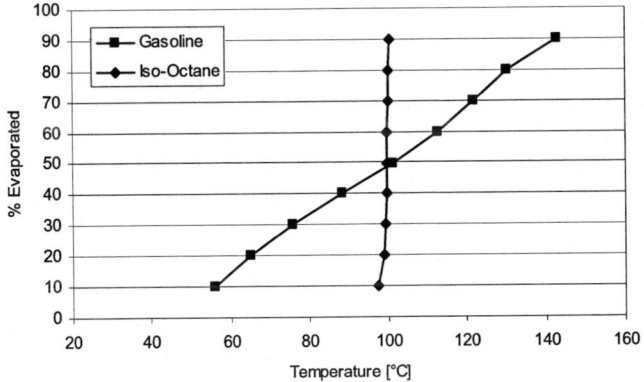

Figure 2. Fuel distillation curves.

2.3 Optical pressure chamber

Imaging and drop sizing of the fuel sprays was carried out in a pressure chamber that has been designed with an octagonal geometry for application of simultaneous optical diagnostic techniques, as shown in Figure 3(a). In addition to side windows, the chamber was fitted with a quartz window in its base to replicate the piston crown view of the optical engine. The injector was mounted at the top of the chamber at an angle of 19° to allow optimised imaging of all three injector spray plume pairs, as will be shown in the results section. Further details pertaining to the chamber and associated instrumentation can be found in (5, 6). The injector body was heated by an Omega MB1 150W band heater and temperature control feedback was provided via a K-type thermocouple located as near to the injector tip as physically possible with closed loop control. As such, the assumption has been made that given a heat soak time of at least half an hour, the fuel temperature inside the injector nozzle had risen to that of the injector body. The spray was either back-lit or side-lit by a Multiblitz Variolite 500 photographic flashgun and was imaged on 640×480 pixel-resolution frames using a high-speed CMOS camera (Photron APX-RS) set at a frame rate of 9 kHz, corresponding to 1° CA between frames for an engine running at 1500 RPM. Each imaging test batch consisted of 100 spray events per condition. Droplet sizing was carried out using a TSI Phase Doppler Anemometry (PDA) system consisting of a Coherent Innova 70C Argon-Ion laser coupled to a TSI beam splitter and Bragg cell. Both the transmitter and receiver had an optical focal length of 250mm. The PDA technique is well documented in the literature and numerous authors have examined the influence of both hardware and software settings on the measured values (8–10), so no further details will be provided here. The droplet size measurements were taken 25 mm downstream from the injector tip along the chamber central axis in the central region of plume 2. Due to the complex geometry of the nozzles, a plate was placed near the injector tip which allowed the unobstructed passage of plume 2 for all conditions, whilst deflecting the other plumes away from the PDA measurement volume. 200 injections were measured at each test point. Injector and instrumentation triggering was provided by an AVL 427 Engine Timing Unit. The chamber was purged after every 20 injections of 2 ms duration each for both imaging and droplet sizing techniques to prevent the build up of fuel vapour and obscuration of measurements.

 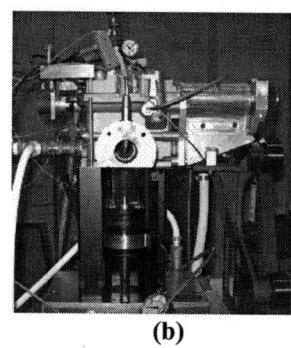

(a) (b)

Figure 3. Optical pressure chamber and optical single-cylinder engine.

2.4 Optical single-cylinder engine

The single-cylinder engine used for this work is based on a modular Ford (US) design using a prototype Jaguar DISI engine head, as shown in Figure 3(b). The engine has 2 intake and 2 exhaust valves, a bore of 89 mm and a stroke of 90 mm. A number of optical configurations are possible depending on the level of in-cylinder access necessary. In-cylinder pressure was measured with a Kistler 6041 water-cooled piezo-electric pressure transducer. A piezo-resistive absolute pressure transducer was installed in the inlet plenum to set the engine load by adjusting the throttle. Further details on the engine and test bed can be found in (7). The engine was motored at 1500 RPM under part-load (0.5 bar intake pressure) and at full-load (1 bar intake pressure) conditions. Injection was set early in the intake stroke to promote mixing and evaporation during the intake and compression strokes. In total 100 cycles were recorded for every condition. Images were recorded with the Photron APX-RS CMOS camera at a frame rate of 9 kHz through the piston crown (640×480 pixel resolution), to give a temporal resolution of 1° CA at 1500 RPM, and at 5 kHz for the images through the quartz liner (512×1024 pixels), giving 1.8° CA resolution at 1500 RPM. Imaging was performed with illumination from a high-repetition rate Nd:YLF laser, using either flood or laser-sheet lighting. Laser-sheet imaging was performed on two orthogonal planes. A vertical plane, termed 'tumble plane', 'sliced' the injector tip through the centre thereby allowing mainly the imaging of spray plumes 2 and 5, see Figure 1(a, c). A horizontal plane allowed imaging of the 'footprint' of all the spray plumes as these penetrated down into the cylinder bore at a vertical location 1 mm above the head gasket (*i.e.* 1 mm into the pent-roof). This plane will be referred to as the 'head gasket plane'. The influence of temperature was observed by controlling the engine coolant from 20–90 °C.

3   RESULTS AND DISCUSSION

Instantaneous images of typical sprays taken for gasoline and *iso*-octane in the quiescent chamber at 0.5 bar are presented in Figure 4 for two timings After the Start Of Injection (ASOI, time interval after the start of the trigger pulse sent to injector driver) and for two fuel temperatures (20 °C and 120 °C). As mentioned earlier, the injector for these experiments was mounted at an angle of 19° and so reference should be made to the schematic diagram shown at the top of this figure, defining the imaged views termed 'side', 'angle', 'end' and 'base'. The effect of increasing the temperature at this low pressure condition is clearly illustrated in the images of Figure 4, namely the contraction and combination of the closely spaced plumes (termed spray 'collapse') for the multi-component gasoline. *Iso*-octane can be observed not to exhibit the same extent of 'collapse' at the same condition.

The spray was also imaged in-cylinder through the piston crown with the engine not running (*i.e.* 'static') so that a baseline image of the spray at atmospheric conditions was available for comparison with the injector rig and the motoring engine. The top row in Figure 5 shows the spray development for gasoline at 'static' engine conditions on the head gasket plane. This is similar in pattern to that observed in the chamber for the same 'ambient' conditions. The line of symmetry between the top and bottom pair of nozzles is clearly seen, as expected from the design of the injector that has been presented schematically in Figure 1. In contrast, the second row of images in Figure 5 shows the spray development with the engine motored at 0.5 bar intake pressure and with an injection timing of 80° CA After intake Top Dead Centre (ATDC). The injection duration was set to 0.78 ms (~7° CA), corresponding to stoichiometric conditions for a firing engine at that load. The effect of flow on spray development is evident as the symmetry of the spray plumes is destroyed and, in particular, spray plumes 1 and 6 are deflected towards the exhaust side by the intake flow. Spray deformation and break-up is also demonstrated by the increased area of light scattering around the spray cores relative to the static condition. Although there are reports in the literature suggesting that the engine intake flow has a relatively small effect on the in-cylinder spray development for pressure-swirl atomisers (11, 12), such interactions are highly sensitive to both engine operating conditions and engine/injection system used. In the current study of a multi-hole injector with a complex layout, the levels of interaction between the spray and air motion will be different for each plume and will be governed by the magnitude, duration and incident angle of the flow over the intake valves that forms an intense 'valve jet'. The intake flow has been characterised by Particle Image Velocimetry (PIV) in an engine of identical configuration (13) and it has been shown that the valve jet reaches its peak magnitude at ~80°–90° CA ATDC which coincides with the injection event in the current study. Maintaining the same injection timing and increasing the engine load (1 bar intake pressure) has an even more dramatic effect on spray formation during and after the injection event, as shown in the third row of Figure 5. For comparison with the part-load operation, the injection duration for stoichiometric firing of the engine at full-load was 1.6 ms (~14.5° CA).

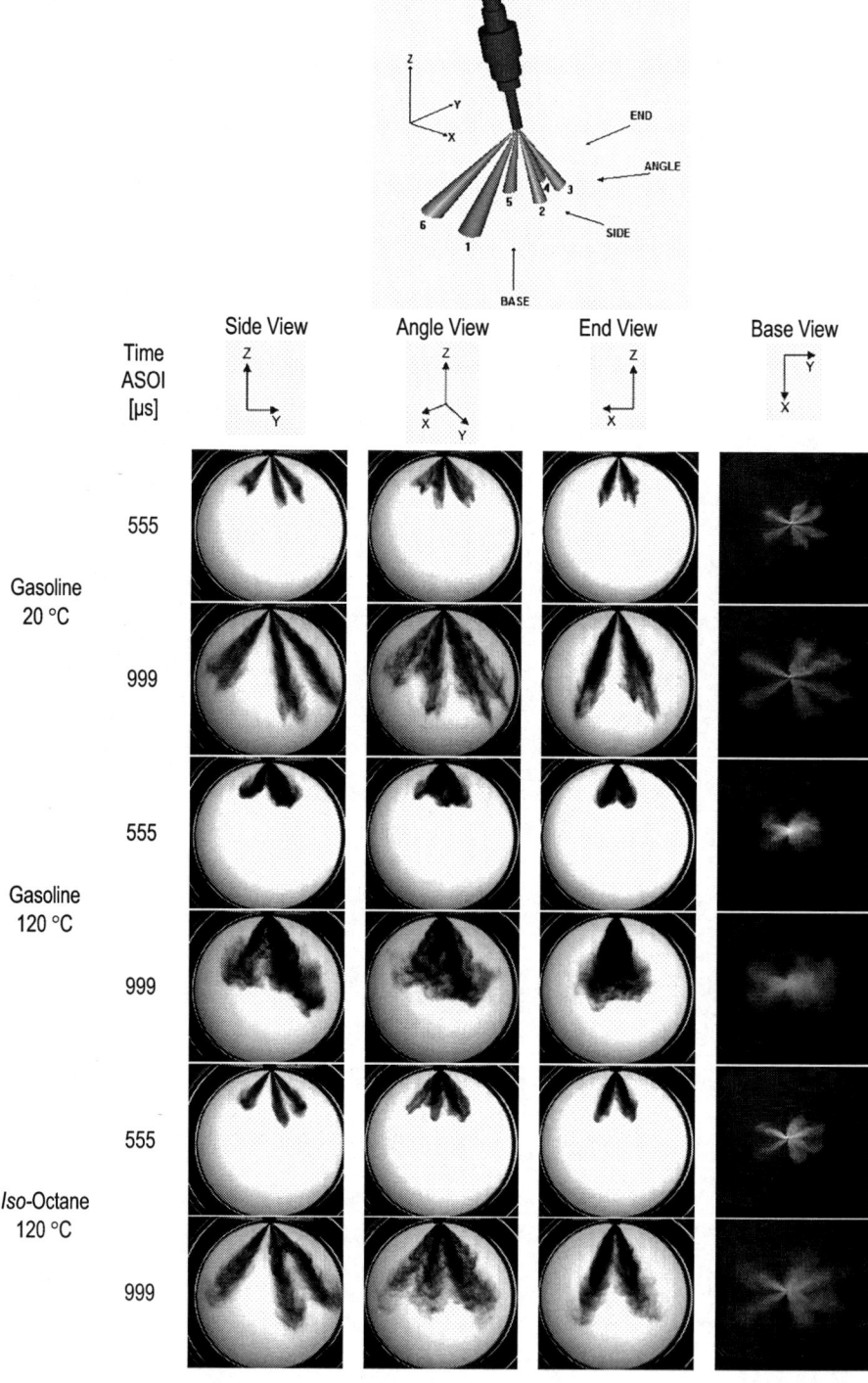

Figure 4. Spray development in the pressure chamber (gasoline and *iso*-octane).

The in-cylinder spray development for the part-load operating condition (0.5 bar intake pressure) is shown in Figure 6 for gasoline fuel and for single representative cycles at 20° C and 90 °C engine head temperature (SOI 80 °CA ATDC). The left column shows the spray formation on the tumble plane using laser-sheet imaging, the central column shows the spray formation on the head gasket plane and the right column shows the spray formation through the piston crown using flood illumination for direct comparison with the chamber 'base' view spray images. Laser-sheet imaging allowed clearer determination of penetration variability and spray plume targeting on planes similar to those used for the PIV measurements in (13), whereas global imaging provided insights into the 3-D nature of the air-fuel interactions over the whole injection duration. The effects of engine temperature on spray development are evident in Figure 6 and similar in nature to those presented earlier in the chamber in Figure 4. However, the flow field effects in the engine do alter the spray characteristics. The primary point of note is that in the engine a 'normal' spray formation is observed at low engine head temperatures, Figure 6(a), whilst a 'collapsed' spray is seen at 90 °C engine head temperature, Figure 6(b). In the chamber the spray is only partially collapsed at 90 °C, hence comparison at high temperatures is best made between the chamber sprays at 120 °C injector temperature and the in-cylinder sprays at 90 °C engine head temperature.

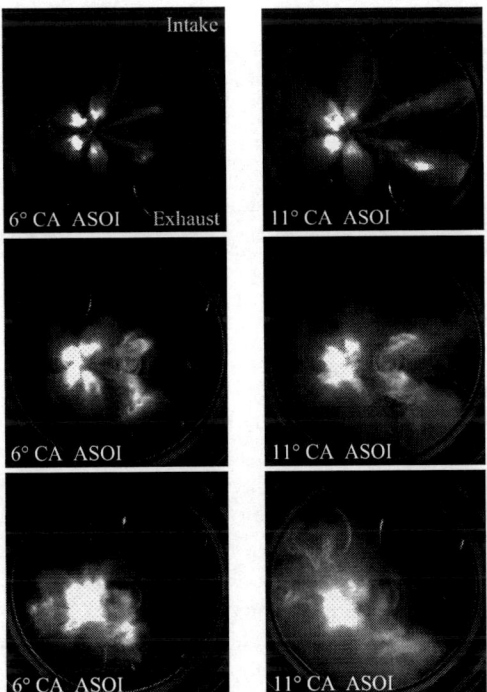

Figure 5. In-cylinder gasoline spray development. Top row: static engine; Middle row: motored engine 0.5 bar intake pressure; Bottom row: motored engine 1.0 bar intake pressure (engine head 20 °C).

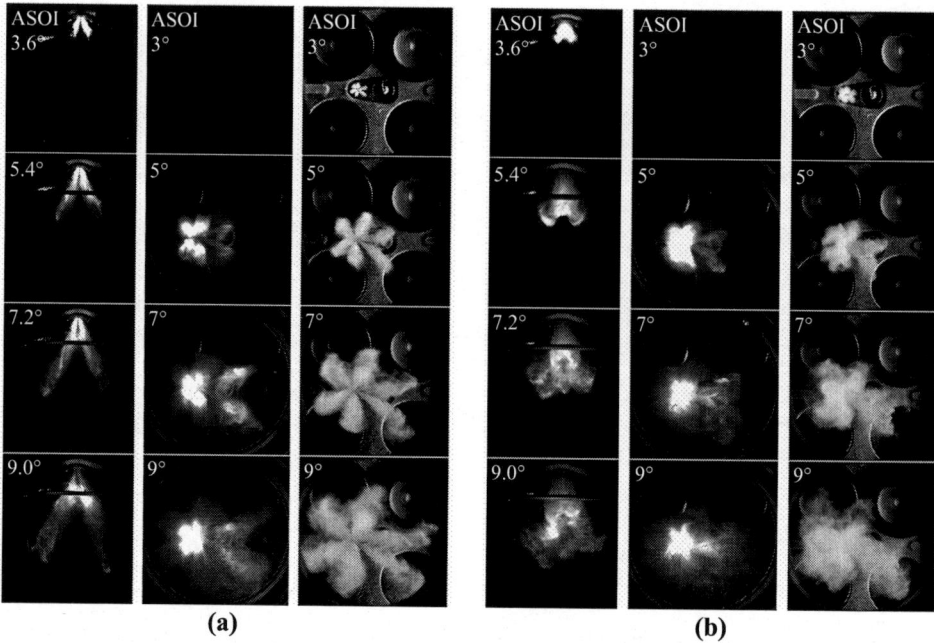

Figure 6. In-cylinder gasoline spray development. Engine head (a) 20 °C, (b) 90 °C. Left: tumble plane; Centre: head gasket plane; Right: flood imaging.

To illustrate the effect of the in-cylinder air flow on spray formation the spray areas as viewed through the piston crown (*i.e.* the flood-illumination images in Figure 6) are compared to the equivalent spray images in the chamber (*i.e.* the 'base' view in Figure 4). Spray images from both experimental set-ups were binarised using appropriate thresholding and these were used to calculate the 'wetted footprint' spray area. The spray images acquired in the chamber were corrected for the 19° injector inclination angle. Each calculated area was then normalised by the area of the engine piston crown window that was 65 mm in diameter. Figure 7 shows the normalised areas for 0.5 bar chamber and intake pressures. As can be seen from this graph, the overall trends concerning the differences in spray development for the two fuels are replicated in both the engine and chamber. These observations mirror the comments made earlier when comparing the respective spray images. The air motion effect on the in-cylinder measurements can be observed from the larger areas calculated for the engine sprays due to increased degree of mixing between the air and the liquid fuel. In both engine and chamber, the rate of growth of area is initially larger at collapsed spray conditions and then reduces to below that of the non-collapsed spray pattern. This is probably due to the greater constant radial component of the outwardly growing plumes for the non-collapsed case. In both experimental setups, the spray area viewed for *iso*-octane for all conditions is smaller than gasoline's, except at 20 °C in the chamber where both fuels show nearly identical calculated spray areas. This difference is thought to be due to the multi-component nature of the evaporation process of the gasoline fuel, in particular the vaporization of the light fractions 'swelling' the spray plumes. After the end of injection in the engine (~1200 µs) the spray area of *iso*-octane reduces at a faster rate in comparison to gasoline.

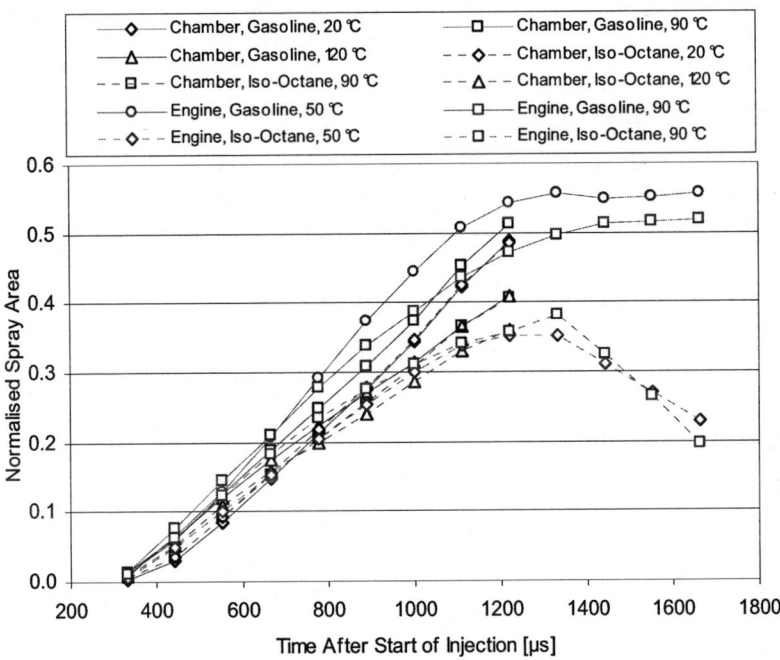

Figure 7. Spray areas through the base chamber view and the piston crown.

In order to investigate further the link between the spray development and atomisation, including the observed differences in calculated spray areas in the chamber and in the engine, droplet sizing was carried out in the chamber for a range of gas pressures and injector temperatures. Figure 8 shows the Sauter Mean Diameter (SMD, $D_{3,2}$) measured 25 mm below the injector tip for plume 2 using gasoline. This shows that droplet sizes reduce with both a decrease in gas pressure (for a given injector temperature) and an increase in injector temperature (for a given gas pressure), ranging from ~17 to 9 µm. Spray images at 777 µs ASOI have been superimposed on the graph to aid interpretation. Initial convergence of the far right plume pair can be observed to occur from the images when the measured SMD falls below ~12 µm. A similar trend between droplet size and onset of collapse was observed to occur in measurements taken for single-component fuels, but not presented here. At the same distance from the injector tip (25 mm) and at a fuel temperature of 90 °C using a "multi-component petroleum product" van der Wege and Hochgreb (14) measured an SMD of approximately 16–19 µm in a square bore engine for an intake pressure of 0.6 bar and an injection pressure of 50 bar, the exact value of the SMD being a function of the radial distance from the injector axis for the pressure swirl injector tested. These droplets are slightly larger than those measured for the multi-hole injector under investigation here; the increased fuel pressure and the alternative form of atomisation utilised would both lend themselves to the production of smaller droplets.

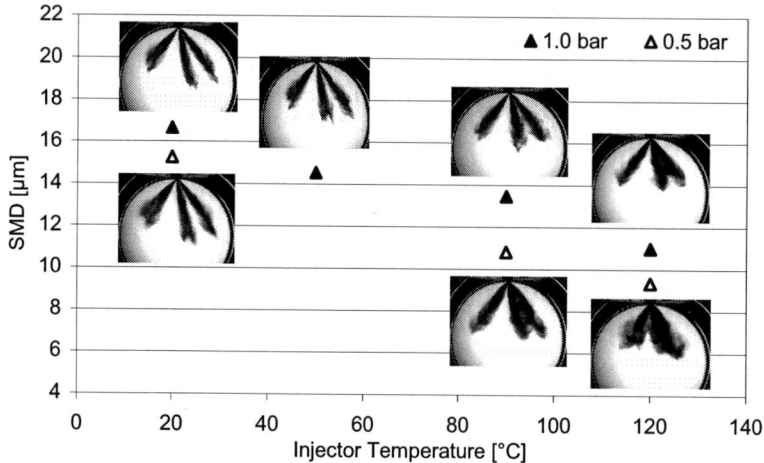

Figure 8. Droplet sizes for gasoline sprays.

4 CONCLUSIONS

Spray development from a multi-hole DISI engine injector was compared in a quiescent chamber and in a single-cylinder optical engine. High-speed imaging and droplet sizing techniques were employed to study the effect of fuel type, injector temperature and in-cylinder pressure. The following conclusions can be drawn:

- The development of sprays produced using *iso*-octane and pump-grade gasoline fuels vary considerably at increased fuel temperature and reduced gas pressure which has implications on the use of *iso*-octane for the diagnostics of injection systems and engines in general.
- Onset of spray collapse was observed for gasoline in the quiescent chamber at conditions of 0.5 bar gas pressure and 120 °C injector temperature. In the engine a similar degree of collapse was observed for 0.5 bar intake pressure and 90 °C engine head temperature.
- The wetted footprint of the fuel spray increases initially for both gasoline and *iso*-octane as injector temperature is raised from 20 °C to 90 ° C in the chamber. Similar observations were made for the in-cylinder spray formation when the engine head temperature was raised from 50 °C to 90 °C.
- Further increase in injector temperature for the quiescent chamber sprays leads to a reduction in wetted footprint as the spray is drawn into the area directly below the injector tip, concentrating along the chamber axis as it 'collapses'.
- The area of the wetted footprint of the spray is smaller for *iso*-octane than for gasoline due to the single boiling point of *iso*-octane and its uniform propensity to evaporate at any temperature.
- At the onset of spray collapse the SMD of the spray was measured to be below 12 µm.

ACKNOWLEDGEMENTS

Support from the EPSRC (grant GR/S58850/01), Jaguar Cars and Shell Global Solutions is gratefully acknowledged.

REFERENCE LIST

(1) Gavaises, M., Abo-Serie, E. and Acroumanis, C., "Nozzle Hole Film Formation and its Link to Spray Characteristics in Swirl-Pressure Atomisers for Direct-Injection Gasoline Engines", SAE Paper 2002-01-1136, 2002.
(2) Alexander, P., Begg S., Heikal, M., Li G. and Gold, M., "Airflow and Fuel Spray Interaction in a Gasoline DI Engine", SAE Paper 2005-01-2104, 2005.
(3) Hayakawa M., Takada S., Yonesige K., Nagaoka M. and Takeda K., "Fuel Spray Simulation of Slit Nozzle Injector for Direct-Injection Gasoline Engine", SAE Paper 2002-01-1135, 2002.
(4) Mitroglou, M., Nouri, J.M., Yan, Y., Gavaises, M. and Arcoumanis, C., "Spray Structure Generated by Multi-Hole Injectors for Gasoline Direct-Injection Engines", SAE Paper 2007-01-1417, 2007.
(5) van Romunde, Z. and Aleiferis, P.G., "Effect of Operating Conditions and Fuel Volatility on Development and Variability of Sprays from Gasoline Direct-Injection Multi-Hole Injectors", Atomization and Sprays, in Press, 2007.
(6) van Romunde, Z., Aleiferis, P.G., Cracknell, R.F. and Walmsley, H.L., "Effect of Fuel Properties on Spray Development from a Multi-Hole DISI Engine Injector", SAE 2007 Powertrain & Fluid Systems Conference and Exhibition, October 29–31, Chicago, USA, SAE Paper 2007-01-4032, 2007.
(7) Serras-Pereira, J., Aleiferis, P.G., Richardson, D. and Wallace, S., "Mixture Preparation and Combustion Variability in a Spray-Guided DISI Engine", SAE 2007 Powertrain & Fluid Systems Conference and Exhibition, October 29–31, Chicago, USA, SAE Paper 2007-01-4033, 2007.
(8) Albrecht, H.-E., Borys, M., Damaschke, N. and Tropea, C., "Laser Doppler and Phase Doppler Measurement Techniques", Springer-Verlag, 2003.
(9) Araneo, L. and Tropea, C. "Improving Phase Doppler Measurements in a Diesel Spray", SAE Paper 2000-01-2047, 2000.
(10) Wigley, G., Goodwin, M., Pitcher, G. and Blondel, D., "Imaging and PDA Analysis of a GDI Spray in the Near-Nozzle Region", Experiments in Fluids, Vol. 36, pp. 565–574, 2004.
(11) Kashdan, J.T., Shrimpton, J.S. and Arcoumanis, C., "Dynamic Structure of Direct-Injection Gasoline Engine Sprays: Air Flow and Density Effects", Atomization and Sprays, Vol. 12, pp. 539–557, 2002.
(12) Davy, M.H., "Effect of Impinging Airflow on the Near Nozzle Characteristics of a Gasoline Spray from a Pressure-Swirl Atomiser", SAE Paper 2006-01-3343, 2006.
(13) Justham, T., "Cyclic Variability of Engine Intake and In-Cylinder Flows", PhD Thesis, Loughborough University, UK, in Preparation, 2007.
(14) van der Wege, B.A. and Hochgreb, S., "Effects of Fuel Volatility and Operating Conditions on Fuel Sprays in DISI Engines: (2) PDPA Investigation", SAE Paper 2000-01-0536, 2000.

Full bore imaging of combustion and quantitative AFR PLIF with a multi-component fuel and co-evaporating tracers

Ben Williams, Xiaowei Wang, Paul Ewart, Richard Stone
University of Oxford, UK

Hongrui Ma, Harold Walmsley, Roger Cracknell
Shell Global Solutions (UK) Ltd, UK

Robert Stevens, Stan Wallace, David Richardson
Jaguar Cars, UK

ABSTRACT

A technique has been developed for full bore imaging, and a high speed video camera used to obtain images that have been processed for soot temperature and loading, and flame growth. The soot measurements have been made with a colour ratio technique that avoids the need for an in-situ absolute radiation calibration. Instead, the spectral response of the camera has to be determined for each of the red, green and blue channels.

Planar Laser-Induced Fluorescence (PLIF) has been extensively used for visualizing species of combustion and reacting flows. However, the relationship between PLIF signal strength and species concentration is complicated by other dependencies, so careful calibration is essential. In order to study the effects of evaporation of a multi-component fuel in an IC engine, a gasoline-like component fuel has been devised which has three components of low, medium and high-volatility. Each component is made from two non-fluorescing constituents chosen so that the component (as a whole) will co-evaporate with one of three selected tracers: namely acetone, toluene or 1,2,4-trimethylbenzene (TMB). Calibration results are presented along with fuel concentration distributions derived from the PLIF images that have been analyzed on a cycle-by-cycle basis.

1 INTRODUCTION

The Bowditch piston imaging system (1) has been used for many decades, but conventionally the piston window is about half the bore area. Because of the density ratio across the flame front, then the field of views is filled once about 10% of the

© IMechE 2007

charge is combusted. By using a transparent piston with a flat crown and concave under surface, then the whole bore can be imaged (2). The optical diagnostic techniques that can be used with the piston are determined by its durability and optical transmissivity. Natural light images will be presented here that have been analysed for flame front position, and soot temperature and loading.

The two-colour method (3) is a well established technique for determining soot temperature and loading, but the variant reported here is rarely used. In the three-colour method, the spectral filters incorporated into a high speed video camera are used to give ratios of the red/green and red/blue intensities. This eliminates the need (with the two-colour method) for an absolute calibration of the detector sensitivity at each of the two wavelengths being used. The background to this technique is reported in detail elsewhere (4), and it can be argued that as the dynamic range of video cameras increases, then the use of ratios (as opposed to absolute values) become much more attractive.

The final technique to be considered here is Planar Laser Induced Fluorescence (PLIF) for the 2-D spatial determination of air fuel ratio (AFR). A sheet of laser light is used to make a tracer fluoresce, and the fluorescence is detected by an Intensified Charge Coupled Device (ICCD) camera. This sounds simple: given a suitable laser, engine, optics, camera, triggering and data acquisition system. However, making quantitative measurements (Q-PLIF) is a major challenge, and some of the challenges and their solutions are presented here.

All of the work reported here used LabView to control the engine operation and imaging, and provide integrated high and low speed data acquisition. The key unit was an Integrated Timing Control System (ITCS) that was a combination of digital hardware and an 8 channel counter-timer card. This controlled the injection timing and duration, the ignition coil on and off times with provision for various modes of skip firing or skip injection. Pulses were generated for image capture and a 'demand' to the laser and ICCD system for the PLIF measurements; there was a facility for incrementing the image capture after a batch had been recorded at a particular crank angle.

2 FULL BORE IMAGING

Figure 1 shows the Perspex piston used in the experiments reported here. The major dimensions are: od 88 mm, id 58 mm, crown thickness (min) 20 mm, radius of curvature 34 mm.

The piston could be operated for up to 50 cycles at part load before reverting to motoring. Images were recorded with a Phantom v7.1 CMOS colour camera, that has a resolution of 800 by 600 pixels (adjustable in 16 x 8 pixel increments) and

Fig. 1   Perspex piston with concave inner surface

a dynamic range of 8 bits, with a recording speed of up to 4800 fps for full frame pictures.

The plano-concave piston distorts the image, and the distortion depends on the piston position. A reference grid was placed on the cylinder head gasket plane, and this was imaged with the piston being incremented in 30°ca increments. The correction was carried out using control point selection and a local weighted mean (LWM) spatial transformation in MatLab[1] ®. All corrected grid images showed an accuracy in circle diameter of ~1%. Because the distance between the grid and the piston crown changed in a quasi-sinusoidal manner, cubic interpolation (as opposed to linear) was used on the coordinates of the selected control points at known crank angles to derive the transformations for other crank positions in the range 0–120°ca.

Flame images were smoothed using nonlinear diffusion filtering before applying an edge detecting algorithm. The MatLab ® edge detecting functions were applied on the filtered flame images to track the flame front. The routine took intensity images as input and returned binary images of the same size as the input images. The Sobel edge detector was adopted for the derivative calculation which returned edges at those points where the gradient of intensity was a maximum. Occasionally, the edge detection routine generated 'holes' within the detected boundary or 'spots' outside it. These were due to extremely bright pixel clusters such as a cloud of soot. Median filtering and region filling were then applied in MatLab ® to remove these false features and make the boundary a simple closed contour. The median filtering process also helped to smooth the edge and eliminate redundant shape features.

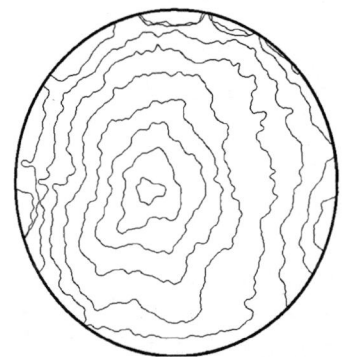

Fig. 2 Flame front images from 25°btdc to 15°atdc in steps of 5°ca, with toluene and λ = 1.2

Figure 2 illustrates the development of the flame at an engine speed of 1500 rpm and imep of 3.6 bar. It must be remembered that the camera image is from a line of sight integration, and that parts of the image will be out of focus. None the less, these images can be compared with the burn rate analysis from pressure records, by considering the enflamed area and its perimeter. It is also possible to track the centroid of the enflamed area. It is also possible to interpolate between the measured flame positions so as to estimate intermediate flame positions, but this is probably of limited utility.

---

[1] MatLab is a registered trademark of MathWorks, Inc, Natick MA.

3   SOOT PYROMETRY

The spectral intensity of radiation at a particular wavelength is given by:

$$I_\lambda = \varepsilon_\lambda I_{\lambda b}(\lambda, T_b) = \varepsilon_\lambda \frac{C_1}{\lambda^5 \left(e^{C_2/\lambda T_t} - 1\right)} \quad (1$$

where: $I_{\lambda b}$ is the spectral intensity emitted by a black body, C_1 and C_2 are known as Planck's radiation constants, having values of $3.742 \cdot 10^{-16} W \cdot m^2$ and $1.439 \cdot 10^{-2} m \cdot K$ respectively.

The emissivity of the soot can be described by the empirical correlation of Hottel and Broughton (5)

$$\varepsilon_\lambda = 1 - e^{-KL/\lambda^\alpha} \quad (2$$

where: λ is the wavelength and KL represents the soot concentration over the optical path length L; α is 1.39 in the visible range.

If the absolute intensity is known at two different wavelengths, then Eqs 1 & 2 can be solved simultaneously to yield the temperature and KL. However, to obtain absolute measurements, then a calibrated source needs to be placed in the combustion chamber of the engine, so that the detectors can be calibrated. The alternative is to take three measurements and use these to give two independent ratios.

Normally narrow band filters are used (so as to avoid wavelengths where there is chemiluminescence), but this then requires the use of ICCD cameras. The alternative adopted here is to use the broad band filters that are an integral part of colour video cameras, and to then rely on algorithms to avoid using data that have been affected by chemiluminescence. A database of colour ratios (usually R/G and/or R/B) is created by using the Planck and Hottel and Broughton equations and the spectral properties of the camera which are camera specific.

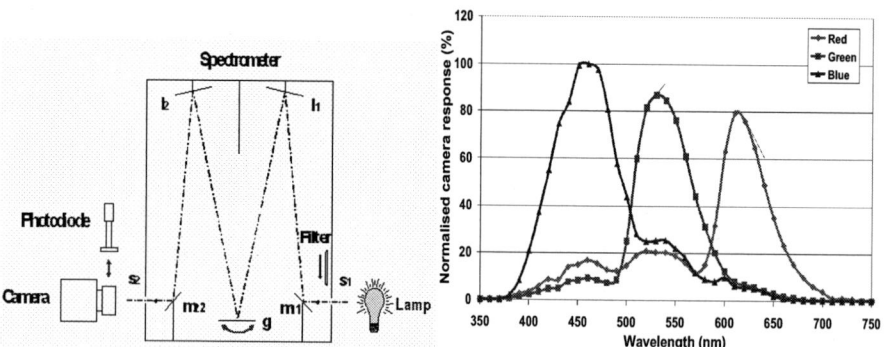

Fig. 3   Spectral Calibration of the video camera

The calibration (4) was achieved by shining light at different wavelengths onto the camera. Then the digital outputs of each of the three channels were extracted from the images using the image processing functions in MatLab ®. Light was focused onto the grating and then focused onto the camera chip (no objective lens) or photodiode. Both slits s_1 and s_2 were set to 2 mm, giving an output bandwidth of 8 nm. To calibrate the emission spectrum of the lamp, a photodiode of known spectral sensitivity was used to measure the relative intensities when extracting the spectral response curves.

The spectrometer could potentially introduce sources of error:
- because the bulb emitted much less in the blue range than red, light scatter of the red constituents could bias the measurements in the blue range;
- the second order diffraction effect of the grating could cause some blue radiation (300~400 nm) to appear in the signal in the red range (600~800 nm).

To attenuate these sources of noise, four filters were selected and put behind slit s_1 according to the wavelength range being measured. For example, when a blue filter was used, most of the radiation above 500 nm would have been heavily attenuated, hence eliminating the cross talk between wavelength ranges. The transmission characteristics of these filters and the camera lens were obtained from a spectrophotometer (Perkin-Elmer Lambda 9, usable range 185~3200nm).

Although a video camera will output red, green and blue values for each pixel, in reality there is usually only a single detector for each pixel. Above the detector is a colour filter array that is aligned with the pixels. During the calibration it is necessary to identify which pixel values are true ones, as opposed to those that have been created by interpolation ('colour de-mosaicing'). This is so that there is control over the de-mosaicing.

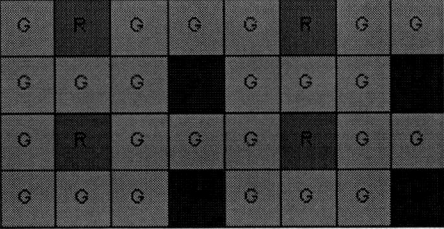

Fig 4 Colour Filter Array

Figure 5 shows some of the frames from a typical video sequence.

4 QUANTITATIVE PLANAR LASER INDUCED FLUORESCENCE Q-PLIF

The underlying theory of PLIF can be found in several references, for example (6 or 7); some energy is absorbed from light at a particular wavelength, and some of this energy is then released as light at a longer wavelength. PLIF might appear to be a comparatively simple technique, but even qualitative measurements can be a challenge.

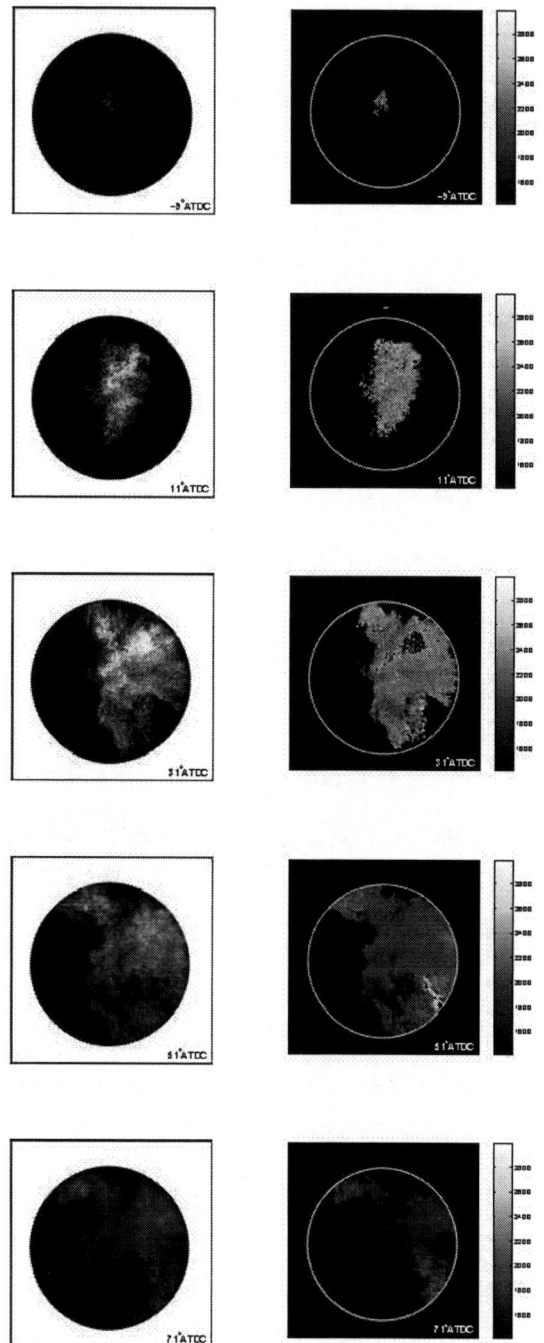

Figure 5: Example results of high-speed colour ratio pyrometry from a combustion sequence (starts from 9°BTDC with images every 20°CA) with toluene and $\lambda = 0.9$ — corrected flame image (left column), temperature K (right column), see www.eng.ox.ac.uk/ice for the corresponding videos.

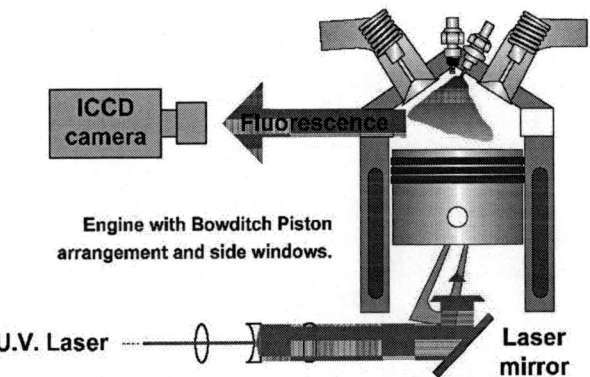

Fig 6 Typical PLIF set up for fuel distribution measurements in an engine with optical access; in reality the ICCD camera would be imaging in a direction normal to the page.

To make quantitative PLIF measurements it is necessary to have:

- A good signal to noise ratio
- A tracer that co-evaporates with the fuel
- A calibration that relates fluorescence yield to the tracer concentration
- A cycle resolved measurement of the laser sheet intensity profile

Each of these will be discussed in turn.

4.1 Signal to noise ratio (SNR)
The ICCD camera has a good SNR because the detector chip is cooled. Scattered light from the laser can be minimised by time gating of the intensifier, so that imaging only starts after the laser pulse has decayed; a filter can also be used, since the fluorescence is at a longer wavelength (c290 nm for toluene) than the excitation (248 nm). Stray fluorescence also has to be eliminated. The cylinder head was grease lubricated, to avoid oil that passes through the valve stem seals entering the combustion chamber and fluorescing. The cylinder head material (an aluminium alloy that had been resin impregnated) also fluoresced, so a metallic insert was used as a target for where the laser sheet illuminated the cylinder head. Finally, the base fuel must be non-fluorescing.

4.2 Co-evaporation of the tracer and fuel
Often iso-octane is used to represent unleaded gasoline, and then a tracer with a high yield is chosen on the basis of what can be excited by an available laser. This, in general, will not lead to co-evaporation of the fuel and the tracer, so that it is the tracer concentration that is being measured, not the AFR. Furthermore, a single component fuel is not representative of a gasoline which might have a boiling point range of 40-180°C. The fuel used here has 6 constituents that have been matched in pairs to co-evaporate with tracers; these high, medium and low volatility components then enable mimicking of the distillation characteristics of a full boiling point range gasoline (6).

Table 1 Multicomponent fuel and its tracers

Species	Tb [°C]	Volume fraction	Tracer	Tb [°C]
Butanes	~0	0.304	2-propanone (Acetone)	56
Isopentane	28			
Isooctane	99	0.550	Toluene (methyl-benzene)	111
n-Octane	126			
Iso-dodecane	175-185	0.146	1,2,4-trimethylbenzene	168
n-Decane	174			

Although the acetone has a higher boiling point than the corresponding hydrocarbons, its polar nature increases its vapour pressure when mixed with hydrocarbons, so that it will in fact co-evaporate with the light fraction. Polar tracers have non-ideal thermodynamic behaviour, so modelling (and matching) their co-evaporation with hydrocarbons is more difficult, but there are no alternative tracers. Also, a different excitation wavelength needs to be used to increase the fluorescence yield and make the variation in yield less sensitive to temperature. A Raman cell was filled with hydrogen (40 bar), and the 1st Stokes harmonic (275nm) was separated and used to excite the acetone. Very high purity is needed in the base fuel, since even small quantities of aromatic compounds would lead to an increase in the background fluorescence and reduce the SNR.

4.3 Calibration

Because the absorption and fluorescence characteristics are an unknown function of temperature and pressure, and because quenching by oxygen needs to be considered, then extensive *in situ* calibration is required. The approach to calibration is similar to [6], and uses a recirculating system with a motored engine. The loop can be heated or cooled and pressurized, so that the compression process of the target engine operating point can be replicated. The recirculation loop has provision for adding both fuel and air to make-up for any leakage. The air-fuel ratio (AFR) is determined from a HC analyzer (utilizing a FID) that has been calibrated on a mixture of methane and nitrogen that has been prepared using partial pressures - allowance being made for the relative sensitivity of the FID to different hydrocarbon species. The FID is fuelled on a H_2/He mix so as to minimize the sensitivity of the FID to the presence or absence of oxygen in the sample. The sampling system has been modified to reduce the sample flow. This is so that when the flammable air fuel mixtures are extracted from the engine, they have less impact on the FID flame temperature. The calibration exercise leads to a data set for each tracer, that show fluorescence as a function of pressure and lambda.

4.4 Laser sheet energy measurement

The intensity of the laser sheet is attenuated by absorption from the tracers, but when low tracer concentrations are used (as is the case here), then the attenuation can be determined from Beer's Law. However, there are also shot-by-shot variations in laser energy, and the intensity variation across the sheet. A beam splitter separates some of the laser sheet and uses this to excite rhodamine dye in methanol, that is then imaged by a line camera. The image from the line camera can then be mapped to give the laser sheet intensity distribution, and this has been validated by imaging a homogeneous mixture in the combustion chamber. Figure 7 illustrates the variation in laser sheet energy.

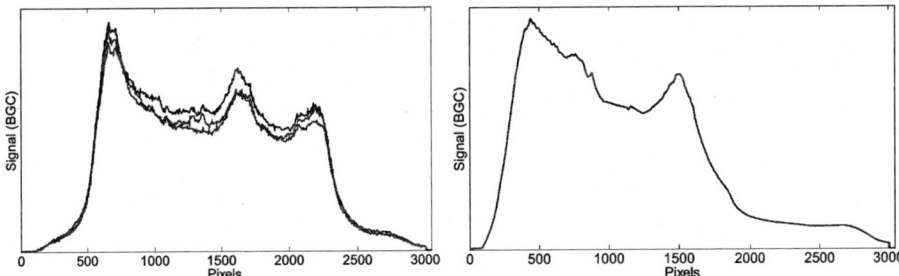

Fig. 7 Variations in laser sheet intensity obtained with the line camera; shot-to-shot variations, and day-to-day variation

4.5 PLIF results

The aim in the work reported here is not just to obtain quantitative results, but to do so on a cycle resolved basis. This depends on a rigorous calibration and a high signal to noise ratio, and evidence that this has been achieved is in Table 2. Port injection was used with an overall stoichiometric air fuel ratio, and 2% toluene has been used as a tracer for the mid-volatility component. Firstly, it is very encouraging that the mean values are close to unity during compression; after ignition (35°btdc) the mean fuel concentration falls and the standard deviation rises because of combustion. If the mixture was perfectly homogeneous, and there was no noise in the signal, then the standard deviation would be zero. The results here indicate that an absolute calibration was achieved, and that under favourable circumstances, then cycle resolved measurements are possible. Figure 8 shows the first of the toluene images from an animation using an average of 32 images taken at 10°ca increments.

Table 2 Fraction of mid-component vaporized for port injection – tracer is 2% toluene

CAD btdc	Mean	Std Dev
90	0.98	0.25
80	0.99	0.27
70	1.01	0.24
60	1.00	0.25
50	0.95	0.22
40	0.98	0.22
30	0.98	0.23
20	0.95	0.24
10	0.88	0.29

5 CONCLUSIONS

Full-bore imaging is applicable to many optical techniques. PLIF is a very difficult technique to use in a quantitative manner, and cycle resolved measurements are at the limit of what is achievable.

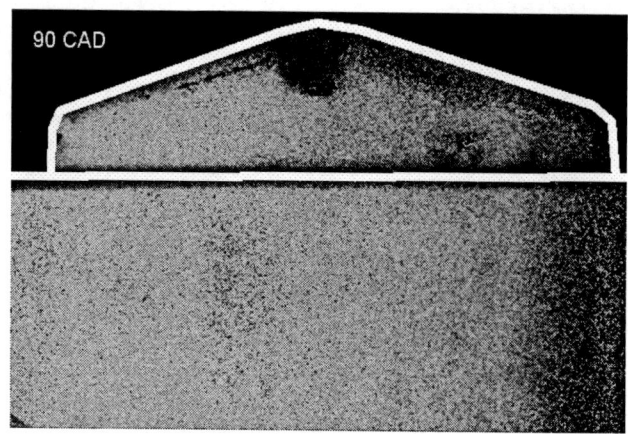

Fig. 8 A toluene image 90°btdc from an animation using an average of 32 images taken at 10°ca increments

ACKNOWLEDGEMENT

Support from EPSRC (Grant GR/S58829/01), Jaguar Cars, and Shell Global Solutions is most gratefully acknowledged.

REFERENCE LIST

1. FW Bowditch, 'A New Tool for Combustion Research--A Quartz Piston Engine', *SAE Transactions*, 69, 1961, pp. 17-23.

2. H. Ma, S. Marshall, R. E. Stevens and C. R. Stone, 'Full Bore Crank-Angle Resolved Imaging of Combustion in a four-stroke GDI Engine', Accepted for publication *Proc IMechE PartD, J. Automobile Engineering*.

3. H. Zhao, and N. Ladommatos 'Optical Diagnostics for Soot and Temperature Measurement in Diesel Engines', *Progress in Energy and Combustion Science*, Vol. 24, No. 3, pp. 221-255, 1998.

4. Hongrui Ma, Robert Stevens, and Richard Stone 'In-Cylinder Temperature Estimation from an Optical Spray-Guided DISI Engine with Three-Color Pyrometry', *SAE 2006*. Paper 2006-01-1198. Also in SP-2015 *Combustion and Flow Diagnostics and Fundamental Advances in Thermal Fluid Sciences*. 2006.

5. H. C. Hottel and F. P. Broughton 'Determination of True Temperature and Total Radiation from Luminous Gas Flames', *Industrial and Engineering Chemistry: Analytical Edition*, **4**(2), pp. 166–175, 1932.

6. R. E. Stevens, H. Ma, C. R. Stone, H. L. Walmsley and R. Cracknell, 'On Planar Laser-Induced Fluorescence with Multi-component fuel and tracer design for Quantitative Determination of Fuel Concentration in IC Engines' *Proc IMechE Pt D J. Automobile Engineering*.

7. C. Schulz and V. Sick, 'Tracer-LIF Diagnostics: Quantitative Measurement of Fuel Concentration, Temperature and Fuel/Air Ratio in Practical Combustion Systems', *Progress in Energy and Combustion Science* **31**, pp75-121, 2005

8. G. de Sercey, M. Heikal, M. Gold, S. Begg, R. Wood, G. Awcock and O. Laguitton, 'On the use of laser-induced fluorescence for the measurement of in-cylinder air-fuel ratios', *Proc. Inst. Mech. Eng. Part C-J. Eng. Mech. Eng. Sci.* 216 (2002), no. 10, 1017-1029.

Experimental investigation of combustion and heat transfer in a direct injection spark ignition (DISI) engine through instantaneous combustion chamber surface temperature measurements

Kukwon Cho[1], Dennis Assanis, Zoran Filipi
Department of Mechanical Engineering, University of Michigan, USA

Gerald Szekely, Paul Najt, Rod Rask
General Motors R&D, USA

[1] currently affiliated at Oak Ridge National Laboratory

ABSTRACT

An experimental study was performed to provide the combustion and in-cylinder heat transfer characteristics resulting from different injection strategies in a direct injection spark ignition (DISI) engine. Fast response thermocouples were embedded in the piston top and cylinder head surface to measure instantaneous combustion chamber surface temperature and heat flux, thus providing critical information about the combustion characteristics and a thorough understanding of the heat transfer process.

Two distinctive operating modes, homogeneous and stratified, were considered and their effect on combustion and heat transfer in a DISI engine was investigated. The results show the stratified operating mode yielded higher spatial variations of heat flux than the homogenous operating mode. This behavior is directly caused by the main features of stratified combustion, i.e. vigorous burning of a close-to-stoichiometric mixture near the spark plug, and cool, extremely lean mixture at the periphery. The cooling effect of the spray impinging on the piston bowl surface when the fuel is injected late in compression could be detected, too.

Comparison between the calculated global heat fluxes and measured local heat fluxes were performed in order to assess the behavior of classic heat transfer models. Comparisons between the global and local heat fluxes provide additional insight into spatial variations, as well as indications about the suitability of different classic models for investigations of the heat transfer aspect of DISI engines. The general observation indicates that special consideration is required when applying classic heat transfer correlations to stratified DISI operation.

© *University of Michigan 2007*

NOTATION

ABDC	after bottom dead center	MAP	manifold absolute pressure
ATDC	after top dead center	m_{fuel}	mass of fuel
AWG	American wire gage standard	N (or n)	harmonic number
		NMEP	net indicated mean effective pressure
A/F	air to fuel ratio		
A_n	Fourier coefficient	p	cylinder pressure
A_s	cylinder surface area	PCP	peak cylinder pressure
BBDC	before bottom dead center	PFI	port fuel injection
BSFC	brake specific fuel consumption	PHF	peak heat flux
		PID	proportional-integral-derivative
BTDC	before top dead center		
B_n	Fourier coefficient	Q_{ch}	chemical heat release
COV	coefficient of variance	Q_{loss}	heat loss
c_p	specific heat at constant pressure	Q_n	net heat release
		SCV	swirl control valve
DI	direct injection	SI	spark ignition
DI-CI	direct injection compression ignition	T	gas temperature
		TDC	top dead center
DISI	direct injection spark ignition	THC	total hydrocarbon
		$T_{coolant}$	coolant temperature
EI	emission index	$T_{exhaust}$	exhaust gas temperature
EGR	exhaust gas recirculation	T_{intake}	intake charge temperature
EOI	end of injection	T_{oil}	oil temperature
EVC	exhaust valve closing	T_m	time-averaged temperature
EVO	exhaust valve opening	T_w	wall temperature
E_{fuel}	energy content in fuel	$(T_w)_{avg}$	average wall temperature
FEM	finite element method	$(T_w)_{swing}$	wall temperature swing
h	convective heat transfer coefficient	UEGO	universal exhaust gas oxygen sensor
H/C	hydrogen to carbon ratio	V	cylinder volume
IVC	intake valve closing	α	thermal diffusivity
IVO	intake valve opening	β	heat loss scaling factor
k	thermal conductivity	γ	specific heat ratio
LFE	laminar flow element	δ	distance
LHV	lower heating value	η_c	combustion efficiency
LPHF	location of peak heat flux	ρ	density
LPP	location of peak cylinder pressure	Φ	equivalence ratio
		ω	angular frequency

1 INTRODUCTION

The increasing emphasis on achieving substantial improvements in vehicle fuel economy calls for new engine concepts providing improved efficiency, while complying with future stringent emission requirements. The fuel efficiency of direct injection compression ignition (CI), or diesel engine is superior to that of the port fuel injection (PFI) spark ignition (SI) engine, mainly due to the use of a significantly higher compression ratio and overall lean combustion, coupled with un-throttled operation. However, sturdy structure, sophisticated high-pressure injection system and complex aftertreatment for removing nitric oxides (NO_x) and particulate matters (PM) from exhaust contribute to the high cost of a diesel. This stimulates work on more efficient gasoline engine technologies that could provide a better tradeoff for light passenger car vehicles. In short, intense research and development efforts are under way to develop an internal combustion engine that combines the best features of SI and DI diesel engines. A critical step on the path towards increasing the competitiveness of a gasoline engine is introduction of direct injection and development of the stratified DISI engine capable of part-load lean and unthrottled operation.

The progress has been remarkable, but there are still significant challenges limiting a wide-spread introduction of DISI engines. The range of the stratified DISI operation is limited. The upper load limit is typically established based on the NO_x emission constraint, and the practical stratified range of operation depends on a complex interplay between spray dynamics, including impingement, evaporation, and gas motion. The robustness of the mixing process is the key to achieving desired combustion and emission formation characteristics. The heat transfer has a significant impact on all of the key processes, particularly impingement, and near-wall mixing and combustion, but it has to be fully characterized under stratified SI conditions. Even though many investigations have addressed the development of the heat flux measurement methodology [1, 2, 3, 4] and applied it to PFI [5, 6, 7, 8, 9, 10, 11, 12, 13, 14, 15] and diesel engines [16, 17, 18, 19], only a couple studies have been published that involve combustion chamber surface temperature measurements in DISI engines [20, 21]. Steeper at al. [20] performed piston surface temperature measurements for a DISI optical engine, while Kato et al. [21] measured average piston surface temperatures using thermistors. Even though the latter provided valuable information for finite element method (FEM) studies of piston design, it is impossible to draw any information about instantaneous heat transfer and combustion processes. Systematic measurements of the instantaneous combustion chamber surface temperatures and heat flux are needed for analyzing the impact of the different in-cylinder conditions on heat transfer, and establishing quantitative and qualitative insight required for development and evaluation of new engine concepts.

Therefore, the objective of this work is to advance the understanding of the combustion and heat transfer phenomena in a DISI engine. The analysis is based on detailed measurements of the instantaneous wall temperature using fast response thermocouples. The temperature histories are processed to yield heat flux profiles. Analysis of measurements obtained simultaneously at different locations on the cylinder head and the piston allows assessments of the impact of stratification on spatial variations of heat transfer. Comparison of results obtained under DISI stratified conditions and homogeneous, stoichiometric conditions offers qualitative insight into differences between two combustion processes, and an illustration of quantitative differences in rates of heat transfer. Finally, local heat flux measurements are used to evaluate the validity of classic heat transfer correlations when applied to the DISI engine. In addition to direct application to DISI combustion/emission development, the insight can be invaluable in providing guidance and boundary conditions for modeling of heat transfer within the framework of one-dimensional or Computational Fluid Dynamics simulations.

The paper is organized as follows. The experimental engine setup is described in the next section. This includes detailed discussion of the installation of fast response thermocouples, development of the telemetry linkage for transferring signals from probes on the piston top, and processing of the signals to determine crank-angle resolved heat fluxes. The section on results begins with the assessment of main features of instantaneous temperature and heat flux profiles measured on the piston top and the cylinder head. The spatial variations obtained in the homogenous mode are contrasted to those observed during stratified operation, with very late fuel injection. Conclusions are drawn about the degree of stratification and the possible effect of impingement on local heat flux. Next, the local heat flux profiles are compared to traces predicted using classic heat transfer correlations, such as Woschni [22] and Hohenberg [23]. This yields further insight in to spatial variations, as well as guidance regarding the suitability of correlations for simulating DISI stratified conditions. The paper ends with conclusions.

2 EXPERIMENTAL SETUP

2.1 Engine

The experiments were performed with a single cylinder engine configured for DISI mixture preparation and combustion. The engine specifications are summarized in Table 1. The engine has a pent-roof combustion chamber and 4 valves per cylinder. The spark plug is located at the center of the cylinder head and the electrode length has been extended to 11 mm in order to facilitate stable combustion during stratified operation. A gasoline direct injection fuel injector is located at side between two intake valves. A deep-bowl piston was used to obtain an 11:1 compression ratio and assist in mixture preparation. The spray targeting and combustion chamber shape are typical of a spray-guided system with swirl charge motion. The intake ports are split and one of the ports has a swirl control valve (SCV) that is used to vary the amount of intake generated swirl. The levels of in-cylinder swirl could be varied from 0.55 to 3.35.

Table 1. Engine specifications

Bore	86.0 mm
Stroke	94.6 mm
Displacement	550 cm^3
Compression Ratio	11.0:1
IVO	28 °BTDC
IVC	80 °ABDC
EVO	56 °BBDC
EVC	26 °ATDC

A hollow-cone swirl type injector with a 53 ° cone angle was used for all of the experiments in this study. A bladder-type high pressure fuel delivery system was designed for this study and used to maintain a fuel pressure of 8.5 MPa. The air flow to the engine was measured with a laminar flow element (LFE), and a plenum was used to damp the pressure pulsations in the intake. The plenum was equipped with heaters so that the intake charge could be pre-heated and controlled to maintain the temperature of 95 ± 2 °C. Engine oil temperature and coolant temperature were kept at 90 ± 2 °C. The exhaust gas was sampled at the exhaust plenum (~55 cm from the exhaust valves) and analyzed to determine the total hydrocarbon (THC), nitric oxides (NO_x), carbon monoxide (CO), carbon dioxide (CO_2), and oxygen (O_2) concentrations.

Cylinder pressure was measured with a piezoelectric transducer located at the rear of the combustion chamber along the axis of the pent-roof. A multi-slot flame trap was used minimize effects of thermal shock on the pressure measurements. Crankshaft position was measured with a resolution of 0.5 degrees crank angle and a TDC reference was obtained a thermodynamically.

The fuel used during this experiment was unleaded gasoline having an octane number of 93, density of 0.7328 g/cm^3, lower heating value of 43.64 MJ/kg, and hydrogen/carbon ratio (H/C) of 1.92. The exhaust gas recirculation (EGR) flow into

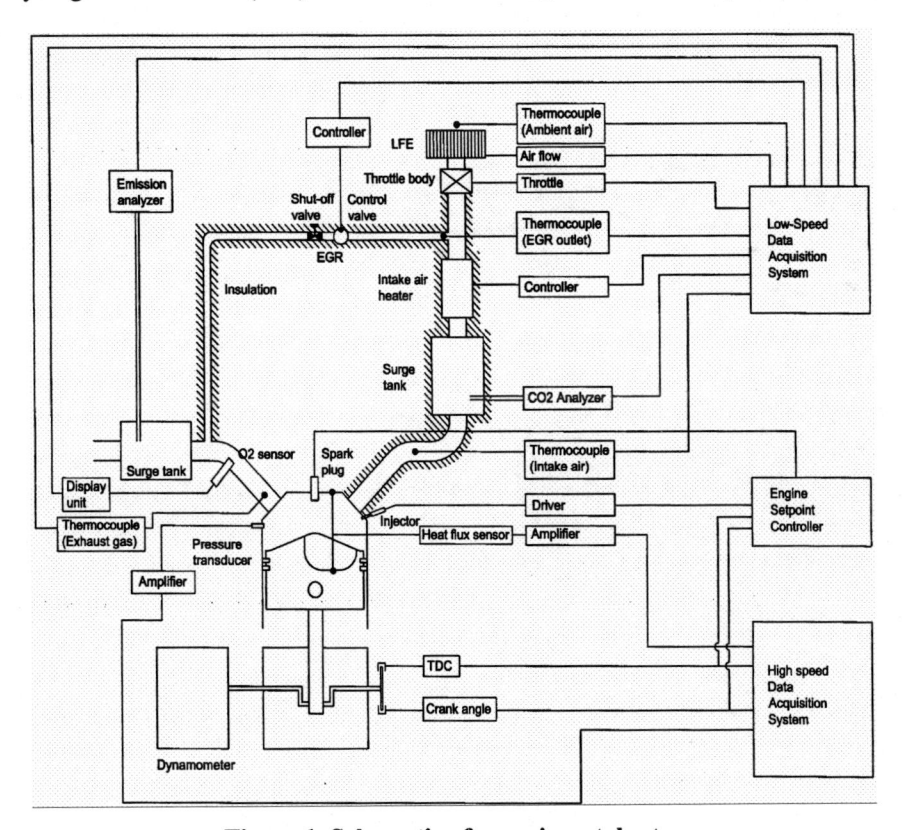

Figure 1. Schematic of experimental setup

the engine was electronically controlled using a stepping motor through feedback control system. The EGR line was insulated to reduce the possibility of water condensation. The schematic of experimental setup is shown in Figure 1.

2.2 Fast response thermocouple

The instantaneous combustion chamber surface temperatures were measured by J-type coaxial fast response thermocouples. The thermocouple consists of a thin wire of Constantan coated with a ceramic insulation of high dielectric strength, swaged securely in a tube made of Iron. A vacuum-deposited metallic plate forms a metallurgical bond with the two thermocouple elements, thus forming the TC junction with 1~2 micron thickness over the sensing end of the probe. The response time is on the order of a microsecond [24]. Figure 2 is an illustration of the construction of a coaxial thermocouple.

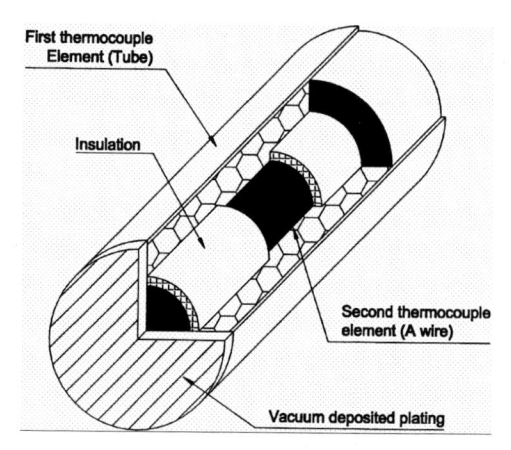

Figure 2. Construction of coaxial thermocouple. from [29]

2.3 Instrumented piston and cylinder head

Two fast response thermocouple probes were custom manufactured to fit into the instrumentation sleeves machined in the cylinder head. The probes are located close to the periphery of the combustion chamber and their respective positions are denoted as "H1" and "H2" - see Figures 3a. The sensing area is flush with the surface.

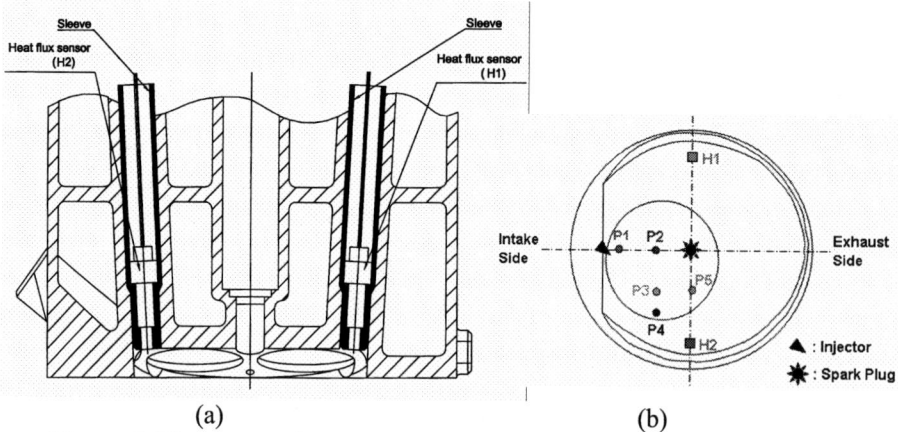

(a)   (b)
Figure 3. Thermocouple locations on: (a) cylinder head and (b) piston top

A DISI aluminum piston was instrumented with five fast response thermocouples press-fitted into the piston crown. The sensing area is flush with the piston surface. Figure 3b illustrates the locations of thermocouple junctions. The location of probe "P1" is on the intake valve side, near the fuel injector. Probe "P2" is located under the spark plug. Probes "P3", "P4", and "P5" are located in the bowl, along the fuel spray trajectory.

All probes include a second reference thermocouple (backside junction) located 4 mm from the tip ($\delta = 4$ mm) to facilitate determining of the steady-state component of heat flux.

2.4 Telemetry linkage system

Measurement of piston surface temperature in a firing engine relies on the ability to transmit thermocouple signals from the moving piston to the data acquisition system outside of the engine. The simultaneous crank-angle resolved measurements from multiple locations on the piston, with front and backside junctions at each location, require extremely high transmission rates. In addition, thermocouple signals are very weak, and thus susceptible to electric noise. The mechanical telemetry linkage system is much better suited to deal with these challenges, and was therefore adopted for this work.

Two-bar linkage system with fork joints was designed and made from Titanium due to its good strength-to-weight ratio. Major components are illustrated in Figure 4. The linkage system consists of an adaptor, an aluminum bracket, two links, and an anchor plate. The electric connectors for the signal wires are located at the aluminum bracket inserted into the adaptor. Cage guided needle roller bearings allow low-friction swivel motion at joints. The anchor plate attached to the side cover of the crankcase carries two electric connectors for transferring the signals to the data acquisition system.

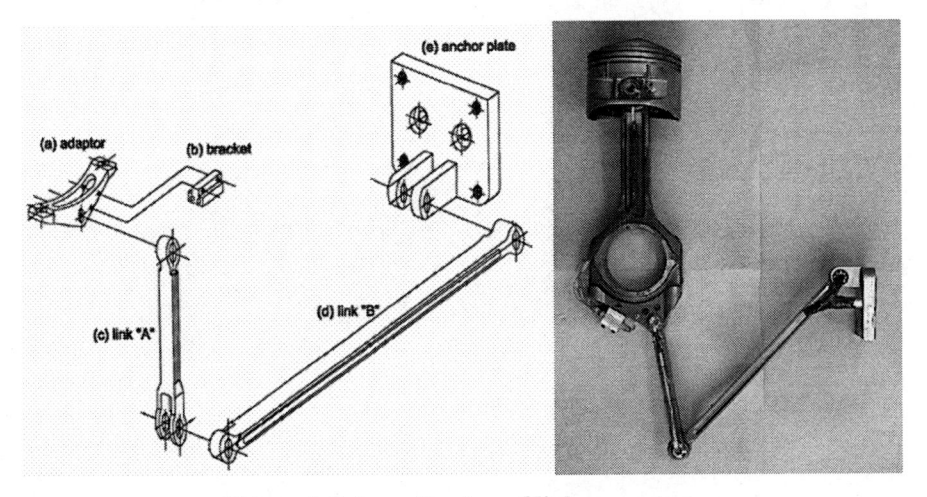

Figure 4. Schematic view of linkage system

Electric continuity of the signal wires is achieved by passing the signal wires directly through the dowel pins. This converts bending motion to torsion, therefore increasing the life of wires. The hook-shaped brass tubes are added at the pin joints of the links "A" and "B" to guide the wires in a way that releases the stress concentration of the signal wires. The brass tubes are attached to the links by steel wires that run through pre-machined holes at the links. Viton® tube encloses the signal wires at the pin joint between the link "B" and the anchor plate due to its superior properties at high temperature and oily environment.

2.5 Connecting the thermocouples to data acquisition - Law of intermediate metals

The thermocouple wire has low endurance when exposed to frequent flexing. Hence, it is not practical to run the thermocouple wire through the linkage and usage of proper wire is a crucial for developing a durable mechanical telemetry system. Using a wire made of a different material, e.g. stainless steel, introduces additional junctions in the circuit. The "Law of Intermediate Metals" tells that this is possible without distorting the signal, since "a third metal inserted between the two dissimilar metals of a thermocouple junction will have no effect upon the output voltage as long as the two junctions formed by the additional metal are at the same temperature". Past studies [25, 26] showed that the temperature of the underside of the piston can be maintained constant at given operating conditions; hence, establishing a reference thermocouple junction at that location and measuring the reference temperature with a thermistor allows determining the temperature of the hot junction. The special stainless steel wire with high resistance to flexing can then be used to transfer the signal from the reference junction, through the telemetry and out of the engine. Special isothermal aluminum plates were machined and installed on the inner side of skirts. The inner skirt surface was also machined in a way that ensures close contact and consistency of cold junction conditions - see Figure 5.

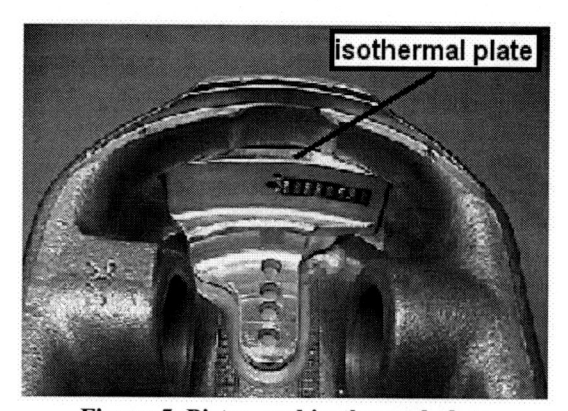

Figure 5. Piston and isothermal plate

The "Ultra Miniature Stainless Steel Medical Wire" was selected for transferring the signal from the isothermal plate onwards. This is a finely stranded stainless steel wire with nominal outside diameter of 0.356 mm, insulated with clear non-hygroscopic fluorocarbon and applicable in the range from -80°C to 200°C.

3 DATA PROCESSING

3.1 Heat flux calculation
The heat flux at the surface of the piston or the combustion chamber can be found by solving one-dimensional time dependent heat conduction equation with time varying temperature for surface boundary condition and a constant temperature at a depth beneath the surface [27, 28, 29, 30]. The experimental surface temperature variation can be expressed as:

$$T_w(t) = T_m + \sum_{n=1}^{N}[A_n \cos(n\omega t) + B_n \sin(n\omega t)] \quad (1)$$

where A_n and B_n are Fourier coefficients, n is a harmonic number, and ω is the angular frequency of the temperature cycle. T_m is the time-averaged experimental surface temperature. The heat flux can be expressed as:

$$q = \frac{k}{\delta}(T_m - T_\delta) + k\sum_{n=1}^{N}\varphi_n[(A_n + B_n)\cos(n\omega t) - (A_n - B_n)\sin(n\omega t)] \quad (2)$$

where $\varphi_n = \sqrt{(n\omega/2\alpha)}$, α is the thermal diffusivity of the wall material ($k/\rho c_p$), k, ρ, and c_p are thermal conductivity, density, and specific heat, respectively. In this study, the number of harmonic components (N) required for high accuracy was 40 [31].

3.2 Global heat loss calculation
The heat release calculations are critical for subsequent comparisons of the local surface measurements with global heat transfer parameters in the combustion chamber. After the net mass fraction burned is determined, the heat loss calculated using a selected heat transfer correlation is scaled to match the total energy released during the cycle. The latter is a determined from the fuel energy content multiplied by the combustion efficiency. The heat flux determined after such scaling of the heat transfer model represents accurately the global process affecting the bulk gas in the chamber. Comparing the local measurements with the global profile will indicate how well the classic correlations behave when applied to a DISI engine (see Section 4.2).

The analysis is based on a single-zone, ideal gas model of the combustion process. The algorithm is based on the work of Gatowski et al. [32] and the final form of the model given below:

$$\frac{dQ_{ch}}{dt} = \frac{\gamma}{\gamma-1}p\frac{dV}{dt} + \frac{1}{\gamma-1}V\frac{dp}{dt} - \frac{dQ_{loss}}{dt} \quad (3)$$

where $\frac{dQ_{ch}}{dt}$ is chemical heat release rate, γ is specific heat ratio, p is cylinder pressure, V is cylinder volume, and $\frac{dQ_{loss}}{dt}$ is heat loss rate. The global heat loss is:

$$\frac{dQ_{loss}}{dt} = hA_s(T - T_w) \qquad (4)$$

where h is the convective heat transfer coefficient, A_s is the cylinder surface area, T is the gas temperature, and T_w is the estimated cylinder wall temperature. The convective heat transfer coefficient is estimated using one of the well-known global heat transfer models for internal combustion engines. Correlations proposed by Woschni [22] and Hohenberg [23] were considered.

Once the convective heat transfer coefficient has been estimated with the global heat transfer models, the heat loss scaling factor (β) is determined such that the sum of the integrated net heat release energy and the heat loss is equal to the expected gross energy release from the fuel.

$$\beta = \frac{\eta_c \times E_{fuel} - \int Q_n}{\int Q_{loss}} \qquad (5)$$

where η_c is combustion efficiency, E_{fuel} is the energy content in the fuel (m_{fuel} x LHV). Q_n is the net heat release rate and Q_{loss} is the heat loss rate. The expected energy release from the mass of the fuel (η_c x E_{fuel}) sets the upper bound for the cumulative heat release and the calculated peak gross heat release ($\int Q_n + \int Q_{loss}$) should ideally reach this value. The crevice loss is neglected. The heat loss scaling factor is a good measure of the validity of the heat release calculation.

4 RESULTS AND DISCUSSION

4.1 Spatial variations of heat flux

To simulate steady-state fully warmed-up operation of the engine, the engine coolant and oil temperatures were maintained at ~90 °C and the intake charge temperature at ~95 °C for all firing tests. A 2000 rpm / part load (NMEP=~300 kPa) point was selected for testing under both homogeneous and stratified conditions. In case of stratified operation, MAP was ~95 kPa, and when running homogeneous mode, the throttle was adjusted to provide the same NMEP value. Engine idle at 600 rpm was added to the stratified DISI runs. Each engine operating condition was optimized based on the combustion stability, (COV)$_{NMEP}$ ≤ 3, and exhaust gas emission constraints, e.g. EINO <30 g/kg fuel and EIHC <60 g/kg fuel. A summary of operating conditions and performance data is given in Table 2. It clearly illustrates the efficiency advantage of lean and unthrottled operation as the DISI engine achieved a slightly higher NMEP at 2000 rpm using 18% less fuel when operating in a stratified mode vs. homogeneous. The aftertreatment challenge is visible too, as the NOx and HC emissions increased over the concentrations measured during homogeneous stoichiometric operating mode.

Table 2. Engine operating conditions and performance data

Speed	[rpm]	2000	2000	600
Mode		Homogeneous	Stratified	Stratified
Fuel Flow Rate	[mg/cycle]	12.3	10.4	6.6
A/F	[-]	14.6	29.8	35.9
Φ	[-]	1.00	0.50	0.41
MAP	[kPa]	45.4	93.1	92.3
EGR	[%]	14.2	20.9	30.4
T_{intake}	[°C]	94.6	97.1	96.2
$T_{coolant}$	[°C]	90.1	88.4	85.1
EOI	[°BTDC]	270.0	67.0	46.0
Spark Timing	[°BTDC]	22.0	30.0	31.0
Swirl index	[-]	3.35	1.55	1.55
NMEP	[kPa]	289.2	312.6	157.1
$(COV)_{NMEP}$	[-]	0.92	2.14	2.41
η_c	[%]	93.4	91.2	90.3
$T_{exhaust}$	[°C]	664.9	384.1	231.9
EI NO	[g/kg fuel]	19.3	27.6	28.9
EI HC	[g/kg fuel]	22.7	56.5	63.2
PCP	[bar]	20.2	33.0	27.5
LPP	[°ATDC]	14.2	4.9	1.5

4.1.1 Homogeneous operation

Figures 6a and 6b show average cylinder surface temperature and heat flux histories of 50 consecutive cycles at 2000 rpm / part load case. Overall temperature level of the location close to the center of the piston top always shows higher values than other locations. The temperature swing and its phasing appear to be similar at all three piston locations (P2, P4, and P5). Consequently, the spatial variations of heat fluxes on the piston are relatively small. The peak value at location P4 is higher most likely due to the swirl pushing the flame front in that direction. The heat flux profiles on the cylinder head (locations H1 and H2) are retarded and their magnitudes are smaller. These probes are closer to the periphery of the chamber; hence, it takes longer for the flame to arrive there.

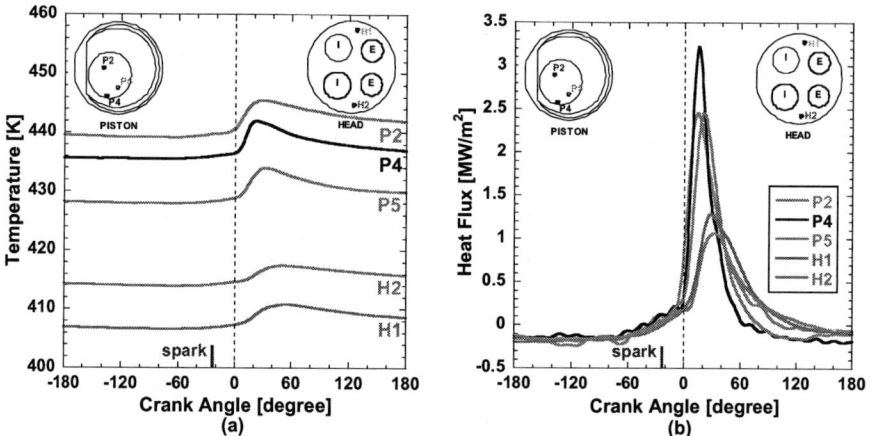

Figure 6. Histories of (a) temperature and (b) heat flux for
2000 rpm-homogeneous mode

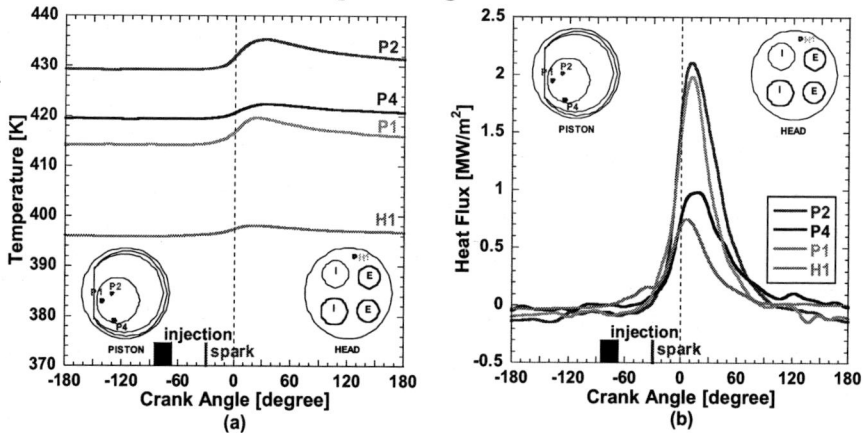

Figure 7. Histories of (a) temperature and (b) heat flux for
2000 rpm-stratified mode

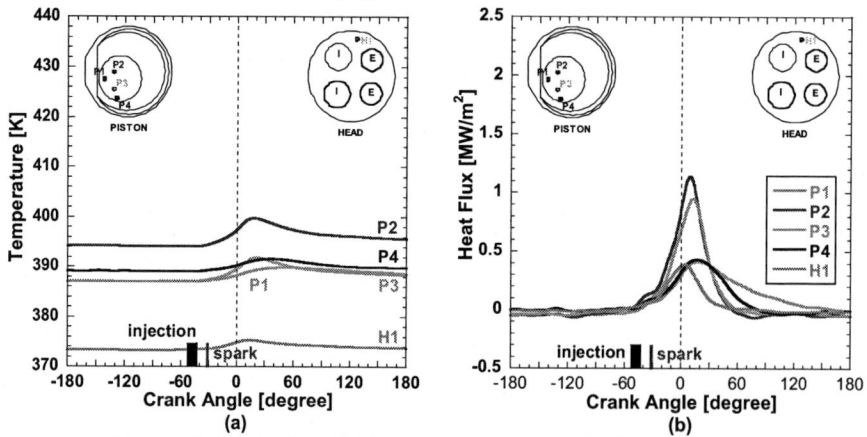

Figure 8. Histories of (a) temperature and (b) heat flux for
600 rpm-stratified mode

4.1.2 Stratified Operation

Temperature and heat flux histories at 2000 rpm / part load, stratified operating mode and averaged over 50 cycles are given in Figures 7a and 7b, respectively. Corresponding profiles obtained at 600 rpm / idle, stratified operating mode, are shown in Figure 8. Injection takes place late in the compression stroke as indicated on graphs in Figures 7 and 8.

General trends that apply to both cases are as follows: overall temperature level of the location near the spark plug (P2) is always higher than other locations, while the temperature levels at the cylinder head are significantly lower. The peak heat flux is also significantly higher at the location near the spark plug. There are significant differences in the magnitudes and phasing of the heat fluxes among the probes in the piston bowl and this will be discussed further. A complete summary of important combustion, surface temperature and heat flux parameters is shown in Table 3.

Table 3. Primary comparison between motoring and stratified operations

Speed	2000 rpm				
Location	Operation	(Tw)swing [°C]	LP Temp [°ATDC]	Peak HF [MW/m^2]	Loc. PHF [°ATDC]
H1	Motoring	1.4	11.5	0.46	0.5
	Stratified	2.2	21.5	0.70	6.0
P2	Motoring	2.1	25.5	0.63	2.0
	Stratified	6.1	34.0	2.05	10.5
Speed	600 rpm				
Location	Operation	(Tw)swing [°C]	LPT [°ATDC]	PHF [MW/m^2]	LPHF [°ATDC]
H1	Motoring	1.1	7.5	0.35	-3.0
	Stratified	2.0	14.0	0.39	3.5
P2	Motoring	1.5	13.0	0.26	-7.0
	Stratified	5.6	18.0	1.14	9.0

The spatial variations are quite different than what was observed under homogeneous conditions. As an example, while peak heat flux for homogeneous operation was the highest at P4 (see Figure 6b), it is much lower than the peak heat flux at other two piston locations when the engine runs stratified – (see Figure 7b). This could be explained mainly by the position of the ignitable fuel/air mixture during the rapid burning phase, as well as the cooling effect of the fuel spray impinging near location P4. The peak value of heat fluxes on the cylinder head in the stratified operating mode is lower than under homogenous conditions. Furthermore, the phasing advances and the peak is located practically at TDC. Therefore, it could be inferred that the temperature fluctuations on the cylinder head surface were not caused by the combustion at those locations, but could rather be attributed to the mixture compression due to the piston movement and the flame expansion. Overall, heat flux measurements provide good indirect indication of mixture stratification and local phenomena affected by charge motion.

Comparing stratified operation at 600 rpm and 2000 rpm reveals large differences in overall magnitudes of heat flux and similarities in trends related to spatial variations. However, when it comes to differences between locations on the piston top, a notable exception is location P1. The magnitude of heat flux at P1 at 600 rpm is relatively low and the heat flux profile is retarded, while at 2000 rpm the magnitude of P1 heat flux at is almost the same as P2, its profile is sharp and standard deviation high. This could be explained in two ways. First, the amount of fuel injected at 600 rpm was only 6.6 mg/cycle, and it was not enough to fill the piston bowl with rich mixture. Instead, the fuel-rich zone remained near the spark plug – see strong heat fluxes near P2 and P3. Secondly, the injection timing and the spray/charge-motion interaction is different at low speed, thus causing fuel impingement at location P1 and slower fuel film movement towards P4 due to swirl.

In summary, overall spatial variations of the heat flux are much higher in the stratified operating mode than in the homogeneous operating mode. Peak heat flux values are much higher under homogeneous operation. Flame arrival times are critical in homogenous operation, while the location of the fuel rich zone and the cooling effect of the impinging fuel spray injected late in compression become a factor in stratified operation. The level of stratification is such that heat flux swings at the periphery of the cylinder head seem to be mostly the result of the mixture compression rather than the flame effect. Therefore, the local heat flux profiles provide very valuable qualitative information about mixture stratification that can be used in conjunction with CFD or laser diagnostics to fully understand the complex mixing/combustion phenomena in a late-injection DISI engine.

4.2 Comparison of local heat transfer and predictions of global models
Classic heat transfer correlations were all developed for well mixed, homogeneous SI operation. However, no comparisons with global heat transfer correlations have so far been reported for stratified DISI engines. Observed differences in spatial variations of measured heat fluxes when the engine operation is switched from homogenous to stratified mode indicate that the overall heat transfer process must vary too. Therefore, this section focuses on evaluation of the classic heat transfer models applied to the DISI engines. We realize that the local and global, area-averaged heat fluxes should not be compared directly for the purpose of quantitative validation. However, examining the global variable in the light of multiple local measurements can provide a very valuable insight into the overall process and guidance for future application of models in DISI simulation studies.

Three data sets described before are analyzed for the purpose of evaluating the heat transfer correlations, namely the 2000rpm / part load, homogeneous and stratified mode, as well as the 600 rpm / idle, stratified operation. Details of operating conditions are given in Table 2.

Two global heat transfer correlations, which represent the mean value of heat flux from the gas to the entire combustion chamber surface, introduced by Woschni [22] and Hohenberg [23] were considered in this study. The heat flux obtained using global heat transfer correlations is scaled using factor (β), based on the heat release analysis (see section 3.2). In other words, the profile as a function of crank-angle shows how well the model behaves on a qualitative basis, while the value of β required for closing the energy balance provides the quantitative assessment. The scaling factors obtained from the heat release analysis are shown in Table 4.

Original Woschni's and Hohenberg's correlations overestimate in stratified operating mode (scaling factor < 1), while in homogeneous operating mode they underestimate the heat loss (scaling factor > 1). The reason for the former must be highly stratified compositional field, leading to a hot core zone and very lean, cool regions close to the wall. When the degree of stratification is higher, e.g. very late injection of a small amount of fuel at 600 rpm, this effect is more prominent and the scaling factor for Woschni drops from ~0.9 to ~0.6. Therefore, special considerations are required when applying classic global heat transfer correlations to stratified DISI engines. Interestingly, Woschni seems to behave better during the transition from homogenous to stratified operations at medium speed (2000 rpm) part load, as illustrated by a relatively smaller change of the scaling factor, while Hohenberg shows advantages during the stratified operation. In other words, very small variation of a scaling factor for Hohenberg's correlation between the two stratified points, namely 600 rpm and 2000 rpm, shows promise of a robust behavior for late-injection cases.

Table 4. Comparison of heat loss scaling factors

Speed [rpm]	2000	2000	600
Mode	Homogeneous	Stratified	Stratified
Woschni's correlation	1.235	0.904	0.603
Hohenberg's correlation	1.358	0.845	0.749

Figure 9 compares the 50-cycle averaged measured values of instantaneous heat flux with the predictions obtained by the global heat transfer correlations for (a) 2000 rpm, homogeneous operating mode, (b) 2000 rpm, stratified operating mode, and (c) 600 rpm, stratified operating mode, respectively. This provides qualitative insight into trends, and complements the previously discussed model scaling. All predictions are shown after proper scaling.

For 2000 rpm, homogeneous operating mode, the peak heat flux values calculated by the correlations of Woschni and Hohenberg are significantly smaller than the measured peak values in the piston bowl. The shape of profiles measured on the piston and predicted correlates well. However, the phasing of measured profiles on the cylinder head, closer to the periphery, is retarded. Therefore, we can hypothesize that the overall heat transfer is driven by the burn rates, while flame propagation dominates local measurements. The points on the bowl are obviously in touch with the flame very early and local peaks are exaggerated.

The picture for 2000 rpm, stratified operating mode, looks quite different – see Fig. 9b. The rapid increase of heat flux starts earlier in the cycle, due to the asymmetric shape of the burn rate in a stratified mode. The local measurements at locations P1 and P2 show peaks much higher then predictions, thus indicating significant stratification with a relatively small fuel-rich zone close to the spark-plug. The surface area exposed to cooler gases is obviously much larger and more dominant when it comes to the global value. When all other (cooler) zones are taken into account, the global heat loss turns out to be much smaller.

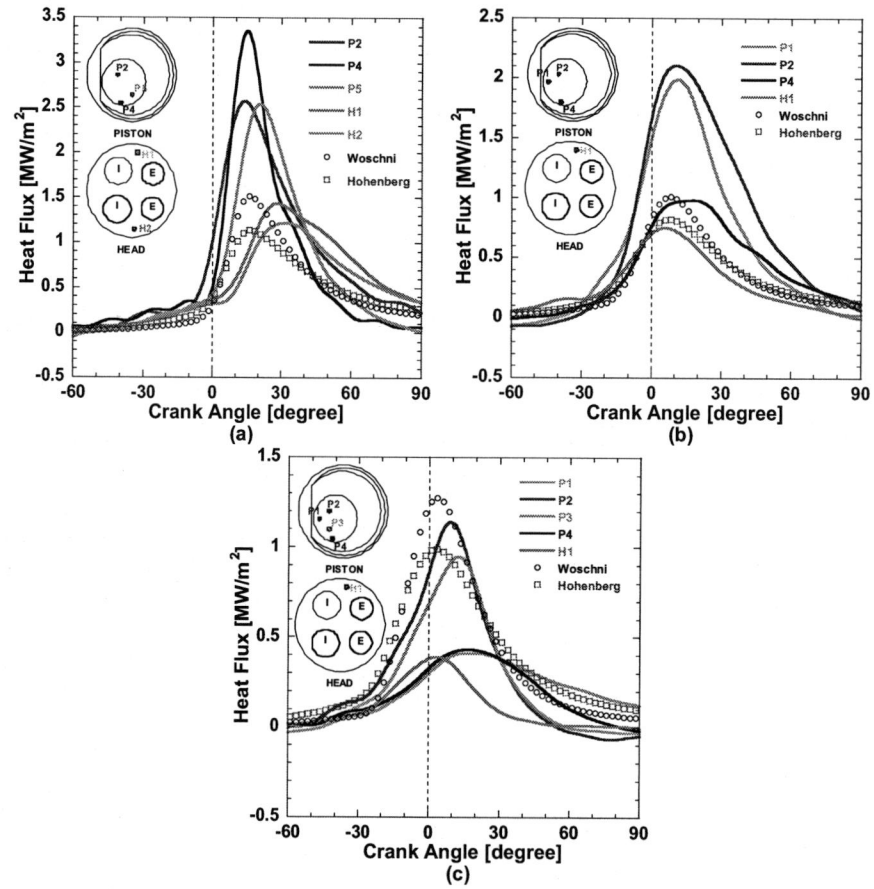

Figure 9. 50-cycle averaged heat flux with global heat transfer correlations for (a) 2000 rpm-homogeneous mode, (b) 2000 rpm-stratified mode, and (c) 600 rpm-stratified mode

Engine idling at 600 rpm produces dramatic differences in trends when compared to the 2000 rpm case. As seen in Fig. 9c, the predicted peak heat flux values are comparable with measured peaks at the locations P2 and P3 on the piston bowl surface. The heat flux fluctuations at the locations P1 and P4 show smaller peaks and retarded phasing mainly due to the combined effect of fuel spray and apparent movement of flame away from the location P4. This indicates that the amount of injected fuel (6.6 mg/cycle) is not enough to fill the piston bowl with a rich mixture, and the hot burned zone appears to linger closer to the cylinder head and in the vicinity of the spark plug. It was not possible to install probes in that area, but we can infer that the heat fluxes must be high on the head surface close to the spark that zone in order to offset the lower values elsewhere. The measured peak heat flux values at the periphery of the cylinder head (H1) are very low and phased with the compression pressure, an indication that the charge is extremely lean and cold in that region. In closing, great care should be taken when classic models are applied to the DISI stratified operation. Overall heat flux values are expected to be lower by roughly thirty to fifty percent when compared to homogenously operation of the same engine, since large zones on the periphery remain lean and cold, thus reducing heat loss to walls. The effect of stratification on heat transfer becomes more exaggerated at very low loads. The Woschni correlation behaves better under homogenous conditions, while the Hohenberg correlation seems to better handle the variations of load during stratified operation.

5 CONCLUSIONS

An experimental study has been conducted to investigate the in-cylinder heat transfer and combustion characteristics resulting from different injection strategies for a Direct Injection SI engine. Investigations included both homogeneous and stratified operation. Fast response thermocouples were used on the piston and the cylinder head to measure instantaneous surface temperatures at different locations of the combustion chamber. A mechanical telemetry system, involving a two-bar linkage mechanism, was used to provide safe routing of the thermocouple signal wires from the hot junctions on the piston top surface to the data acquisition system outside the engine. Each engine operating condition was optimized based on the combustion stability, $(COV)_{NMEP} \leq 3$, and NO_x and THC emissions constraints.

The stratified operating mode yields much higher spatial variations than the homogeneous operating mode. This behavior is directly caused by the main features of stratified combustion, i.e. vigorous burning of a close-to-stoichiometric mixture near the spark plug, and cool, extremely lean mixture at the periphery. Significant spatial variation of heat flux was observed even among the probes located in the piston bowl during the stratified operating mode. Higher fueling caused relatively high heat fluxes in the bowl, with the exception of a point exposed to impinging spray during late injection. Idling with stratified mixture reduces the size of the hot, fuel-rich zone and changes the impingement pattern. This is illustrated by very low peak heat fluxes and retarded instantaneous temperature signals at two locations in the bowl. Conditions at the periphery of the chamber are so lean in the stratified mode that mixture compression, rather than flame itself, determine the phasing of the surface temperature and the heat flux profiles at locations far away from the spark.

Comparison between the calculated global heat fluxes and measured local heat fluxes were performed in order to access the behavior of the two classic heat transfer models, namely Woschni and Hohenberg. Two important factors were taken into account when comparing the global and the local heat fluxes, i.e. (i) the global heat transfer correlation represents "area-averaged" heat fluxes and heat transfer coefficients, where as the measurements represent the instantaneous local values, and (ii) flame propagation in SI engines separate the hot burned zone from the unburned zone, whereas determination of the global heat transfer coefficient from the heat release analysis utilizes the bulk gas temperature. Nevertheless, comparisons between the global and local heat fluxes provide additional insight into spatial variations, and indications about the suitability of classic models for investigations of the heat transfer in a DISI engines.

Predictions of global models were scaled in all cases to satisfy the energy balance for the cycle. The scaling factors required for individual operating points indicate that both models in their original form underpredict the heat loss for homogeneous part load operation, and overpredict under stratified operating mode. Overall heat flux values are lower by more than thirty percent when compared to homogenously operation of the same engine at the same load, since large zones on the periphery remain lean and cool, thus reducing the heat loss to the walls. This effect is more pronounced at very low loads, such as idle. The Hohenberg correlation seems to better handle the variations of load during stratified operation, while Woshni behaves better under homogenous conditions.

Closer examination of local heat flux profiles indicates that extremely diluted conditions at 600 rpm lead to very low heat fluxes even in the piston bowl. Thus, the fuel-rich zone seems to be limited to the zone close to the cylinder head surface and in vicinity of the spark plug. For 2000 rpm / part load conditions, for both the homogeneous and stratified operation, the calculated global heat flux profile falls between the extremely high heat fluxes measured on the top of the piston, and much lower heat fluxes detected at the periphery of the cylinder head, but closer to the latter. In other words, the locally high peaks of heat fluxes for points close to the spark are limited to a relatively small area exposed to the flame in the early phase of burning, and do not necessarily dominate the bulk gas phenomena.

6 ACKNOWLEDGEMENTS

This work is supported through General Motors Corporation Collaborative Research Laboratory at the University of Michigan. The authors also would like to express their gratitude to Dr. Alexandros C. Alkidas for his valuable suggestions and comments throughout this project.

REFERENCES

1. **Eichelberg, G.** Some New Investigations on Old Combustion-Engine Problems. *Engineering*, 148, 1939, 463-466, 547-560
2. **Enomoto, Y., Ohya, T., Ishii, M., Enomoto, K., Kitahara, N.** Study on Analysis of Instantaneous Heat Flux Flowing into the Combustion Chamber Wall of an Internal Combustion Engine (Examination in the Case of Consideration of Heat Storage Term and the Temperature Dependency of the

Thermocouple's Thermophysical Properties). *JSME International Journal Series II,* Vol. 35, No. 4, 1992

3. **Assanis, D. N., Badillo, E.** Evaluation of Alternative Thermocouple Designs for Transient Heat Transfer Measurements in Metal and Ceramic Engines. SAE paper 890571, 1989
4. **Tree, D. R.** *Development of a Heat Flux Gauge for a Partially Insulated Internal Combustion Engine.* M. S. Thesis, Purdue University, 1988
5. **Alkidas, A. C., Myers, J. P.** Transient Heat-Flux Measurements in the Combustion Chamber of a Spark-Ignition Engine. *Journal of Heat Transfer,* Vol. 104, 1982, 62-67
6. **Alkidas A. C., Puzinauskas, P. V., Peterson, R. C.** Combustion and Heat transfer studies in a Spark-ignited Optical Engine. SAE paper 900353, 1990
7. **Enomoto, Y., Furuhama, S., Minakami, K.** Heat Loss to Combustion Chamber Wall of 4-Stroke Gasoline Engine (1st Report, Heat Loss to Piston and Cylinder). Bulletin of JSME, Vol. 28, No. 238, 1985
8. **Enomoto, Y., Furuhama, S.** A Study of the Local Heat Transfer Coefficient on the Combustion Chamber Walls of a Four Stroke Gasoline Engine. *JSME International Journal Series II*, Vol. 32, No. 1, 1989
9. **Enomoto, Y., Kitahara, N., Takenaka, T.** Thermal Load of Four Cylinder Gasoline Engine with Carburetor. JSAE Review, Vol. 11, No. 1, 1990
10. **Hayes, T. K., White, R. A., Peters, J. E.** Combustion Chamber Temperature and Instantaneous Local Heat Flux Measurements in a Spark Ignition Engine. SAE paper 930217, 1993
11. **Harigaya, Y., Toda, F., Suzuki, M.** Local Heat Transfer on a Combustion Chamber Wall of a Spark-Ignition Engine. SAE paper 931130, 1993
12. **Myers, J. P., Alkidas, A. C.** Effects of Combustion-Chamber Surface Temperature on the Exhaust Emissions of a Single-Cylinder Spark-ignition Engine. SAE paper 780642, 1978
13. **Syrimis, M.** *Characterization of Knocking Condition and Heat Transfer in a Spark Ignition Engine.* Ph. D. Thesis, University of Illinois, 1996
14. **Toda, F., Harigaya, Y., Ohyagi, S., Iwamoto, S., Yoshida, M.** Instantaneous Heat Flux in a Side Valve Type Spark-Ignition Engine. JASE Review, Vol. 7, No. 1, 1986, 16-25,
15. **Wimmer, A., Pivec, R., Sams, T.** Heat Transfer to the Combustion Chamber and Port Walls of IC Engines – Measurement and Prediction. SAE paper 2000-01-0568, 2000
16. **Assanis, D. N., Friedmann, F.** A Telemetry Linkage System for Piston Temperature Measurements in a Diesel Engine. SAE paper 910299, 1991
17. **Ogawa, H., Kimura, S., Enomoto, Y.** A Study of Heat Rejection and Combustion Characterisitcs of a Low-Temperature and Pre-mixed Combustion Concept Based on Measurement of Instantaneous Heat Flux in a Direct-Injection Diesel Engine. SAE paper 2000-01-2792, 2000
18. **Uchimi, T., Taya, K., Hagihara, Y., Kimura, S., Enomoto, Y.** Heat Loss to the Combustion Chamber Wall in a D. I. Diesel Engine – First Report: Tendency of Heat Loss to Piston Surface. JSAE Review 21, 2000, 133-135
19. **Sihling, K., Woschni, G.** Experimental Investigation of the Instantaneous Heat Transfer in the Cylinder of a High Speed Diesel Engine SAE paper 790633, 1979

20. **Steeper, R. R., Stevens, E. J.** Characterization of Combustion, Piston Temperatures, Fuel Sprays, and Fuel-Air Mixing in a DISI Optical Engine. SAE paper 2000-01-2900, 2000
21. **Kato, N., Moritsugu, M., Shimura, T., Matsui, J.** Piston Temperature Measuring Technology Using Electromagnetic induction. SAE paper 2001-01-2027, 2001
22. **Woschni, G.** A Universally Applicable Equation for the Instantaneous Heat Transfer Coefficient in the Internal Combustion Engine SAE paper 670931, 1967
23. **Hohenberg, G. F.** Advanced Approaches for Heat Transfer Calculations. SAE paper 790825, 1979
24. **Medtherm Corporation** Bulletin 500, Huntsville, Alabama, 2000
25. **Furuhama, S., Enomoto, Y.** Heat Transfer into Ceramic Combustion Wallof Internal Combustion Engine. SAE paper 870153, 1987
26. **Li, C. H.** Piston Thermal Deformation and Friction Considerations. SAE paper 820086, 1982
27. **Overbye, V. D., Bennethum, J. E., Uyehara, O. A., Myers, P. S.** Unsteady Heat Transfer in Engines. SAE Transaction 69, 1961, 461-494
28. **Alkidas, A. C.** Heat Transfer Characteristics of a Spark-Ignition Engine. Journal of Heat Transfer, Vol. 102, 1980, 189–193
29. **Borman, G., Nishiwaki, K.** Internal-Combustion Engine Heat Transfer. *Prog Energy Combust Sci*, Vol. 13, 1987, 1-46
30. **Heywood, J. B.** *Internal Combustion Engine Fundamentals.* 1988 (McGraw-Hill, New York)
31. **Cho, K.** *Characterization of Combustion and Heat Transfer in a Direct Injection Spark Ignition Engine through Measurements of Instantaneous Combustion Chamber Surface Temperature.* Ph. D. Thesis, University of Michigan, 2003
32. **Gatowski, J. A., Balles, E. N., Chun, K. M., Nelson, F. E., Ekchian, J. A., Heywood, J. B.** Heat Release Analysis of Engine Pressure Data. SAE paper 841359, 1984

DIESEL

The potential of downsizing diesel engines considering performance and emissions challenges

Thomas Körfer, Matthias Lamping, Andreas Kolbeck, Torsten Genz
FEV Motorentechnik GmbH, Germany

Stefan Pischinger, Hartwig Busch, Dirk Adolph
Institute for Internal Combustion Engines (VKA), Germany

ABSTRACT

The steadily increased specific power and torque output leads to improved driving performance of Diesel powered vehicles. In the context of the CO2 discussion, the downsized displacement is an option, which needs to be considered for future applications. In this paper, the different aspects of a downsizing concept will be discussed. Based on gas-exchange calculations with respect to compatibility to the targeted full load performance of 80 kW/l, a 2-stage boosting concept was chosen and the EGR-path and the combustion system were enhanced to meet this significant challenge. Finally the potential of the downsizing concept during the NEDC is assessed.

1. INTRODUCTION

More precise predictions about climate change in the future and increased costs for individual mobility through a continuous raise of the fuel price drive engine manufacturers to achieve an additional reduction of real world fuel consumption. Future SI-Engines show a clear trend towards downsized, highly boosted engines with direct injection. These concepts show potential to reduce fuel consumption by approximately 10% during the NEDC, in some vehicles even a reduction up to 20% can be realized. The main driver for the improved fuel economy with such an engine concept is the reduced engine displacement. In this context two questions arise for the HSDI-Diesel engine: Is a comparable approach feasible? Which benefit can be realized for a diesel engine?

The HSDI-Diesel engine as power source for powerful and fuel efficient passenger cars has achieved a market share of approximately 50% by the year 2005. To continue this success and to conquer new markets meeting future emission standards does not suffice. Besides emissions, an excellent fuel economy compared to competing powertrains in combination with an attractive performance characteristic is absolutely mandatory. The implementation of emission control systems, necessary

© *FEV Motorentechnik GmbH 2007*

to complying with future and current emission standards, cannot be realized without fuel consumption penalties (3). Additional technological effort is therefore necessary to achieve better fuel economy compared to today's engines in order to keep the diesel engine attractive compared to future SI-engines and hybrid powertrains. When developing new diesel engine concepts, also the costs of diesel engines have to be considered, as diesel engines are significantly more expensive in development and production than typical SI-engines.

In the last decade, the power density of passenger car diesel engines was greatly improved. A state-of-the-art 2.0l-Diesel engine achieves nowadays a driving performance in compact and middle class cars, which was previously attributed to sports cars. This increase in power density allows a reduction of engine size while keeping the power output constant. Additionally the trend towards the installation of the same base engine in many different vehicle platforms requires small, easy to integrate and powerful propulsion units. This challenge is made even more difficult, as modern diesel engines also face the requirement of lowering production costs. Within this paper, different challenges and influencing parameters concerning the topic "downsizing" are discussed, analyzed and finally assessed.

2. STATE OF TECHNOLOGY

The modern HSDI-Diesel engine has been greatly improved in the past 25 years. Considering the increasing costs for individual mobility, the diesel engine represents a highly attractive power unit for passenger cars today, as also driving and NVH performance have been vastly improved.

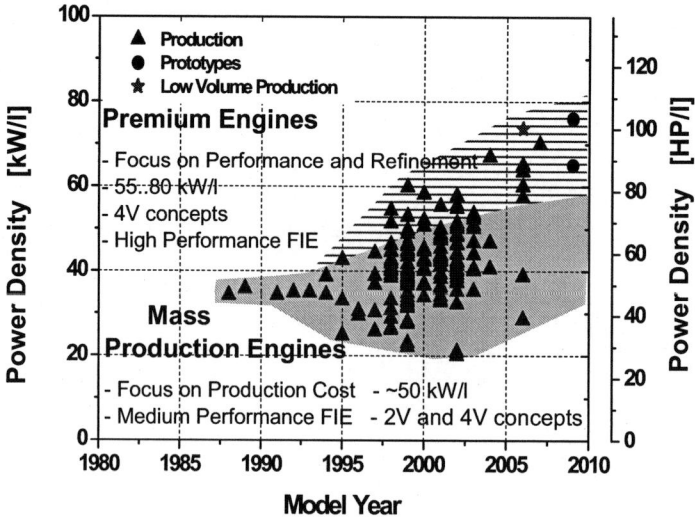

Fig. 1: Development of specific power output

Driven by improvements in power density of premium engines, also high volume, mass-market engines achieve increased specific performance figures. In Fig.1 the trend to higher power density for selected engines is shown for the last decade. It is noteworthy, that even older engine designs could be upgraded with new technologies and with some optimization achieve improved performance as well as the latest emission standards. The trend towards higher power outputs was accompanied by a silent downsizing trend. While ten to fiveteen years ago a 2.0l diesel engine was usually utilized in compact and mid-class sedans and station wagons, today the application range for 2.0l engine concepts covers almost the complete vehicle portfolio from small cars to large MPV's.

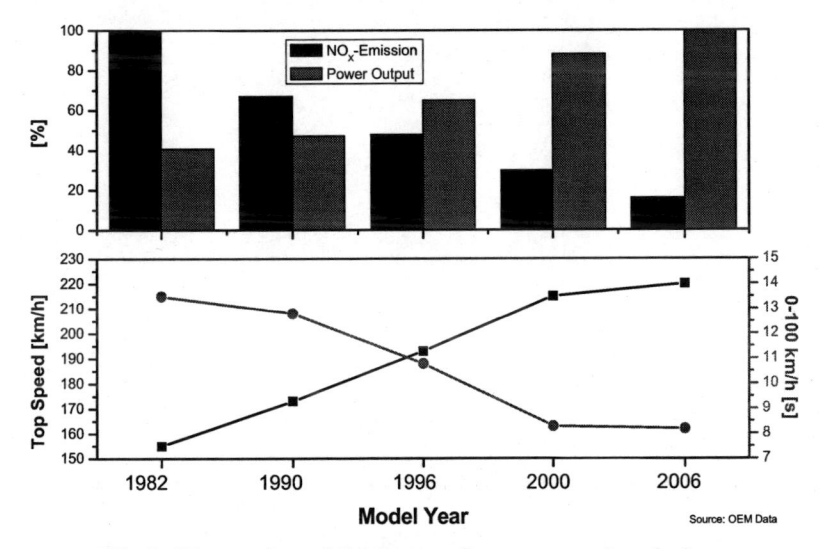

Fig.2: Illustration of driving performance and emissions

In spite of the continuously raised standards the introduction of modern charging and fuel injection concepts allowed continous improvement of driving performance. Fig.2 also illustrates, that the emission performance has been improved in the same time partly even without complex emission control systems.

The modern diesel engine has a fuel consumption advantage of ~25% compared to a conventional gasoline engine with PFI and of about 17% compared to modern DI concepts. The advantage of the HSDI-Diesel engines over SI-engines is reduced to values down to 8% and less compared to new, technological ambitious approaches for highly charged SI-engines with reduced displacement. Furthermore, there is still a gap of 30 g/km to the aspired fleet-average in Western Europe of 130 g/km CO_2 for 2012. Therefore a significant reduction of fuel consumption also for diesel engines is necessary in order to meet future fleet-average fuel consumption targets as well as to keep the diesel engine attractive to customers. In order to find the most efficient way to reduce fuel consumption, simulation results for a 20% reduction of drag, weight, road resistance and engine displacement are exemplarily shown in Fig.3. While reducing drag and road resistance results only in a relatively small fuel consumption gain of approximately 3%, reducing vehicle weight shows with 7% fuel consumption benefit a much higher potential. However, it is questionable, whether such a significant reduction of vehicle weight is feasible for production, as a 20% reduction of vehicle weight means in absolute values a reduction of approximately 320kg for a compact car which would have to be achieved without enormous drawbacks regarding comfort, passive safety and production costs.

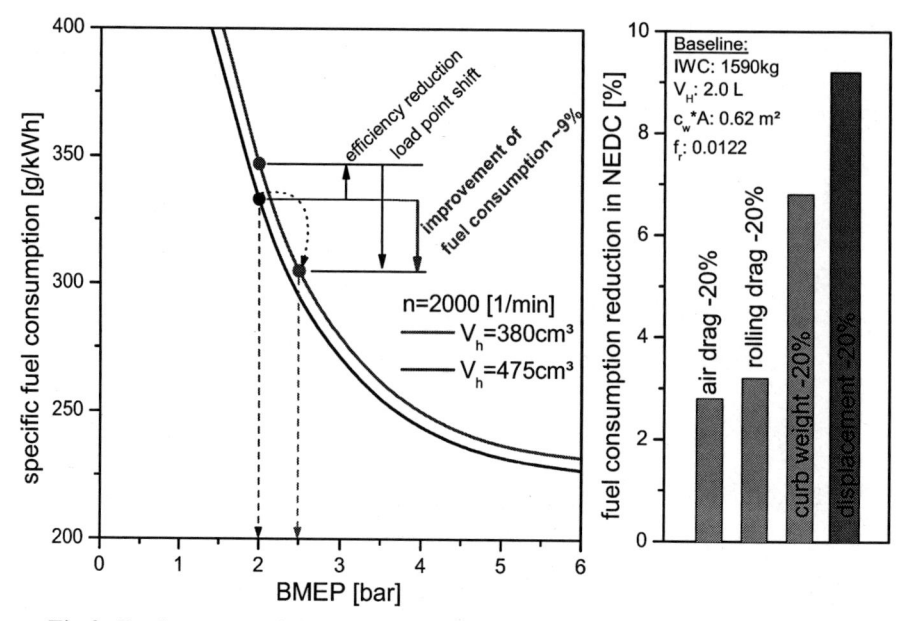

Fig.3: Fuel consumption reduction potential with decreased displacement

In contrast a reduction of engine displacement offers an even larger fuel consumption potential of 9%, despite a lowered thermal efficiency with reduced cylinder displacement. Therefore the downsizing of the combustion engine seems to be a very efficient and cost attractive solution for meeting future fuel consumption targets.

3. BOUNDARY CONDITIONS FOR A DISPLACEMENT REDUCTION

The reduction of the cylinder displacement and simultaneously increased specific power output results in the following challenges for concept definition and engine architecture:

- Engine architecture and cylinder head design: Increasing peak pressures and rising heat flow densities resulting from higher specific power outputs intensify thermo-mechanical stress, especially in the cylinder head. Driven by increased cylinder peak pressures also crankcase and crank train face higher loads.
- Injection system: Higher maximum injected fuel quantities and lower minimum injected fuel quantities demand a larger fuel quantity spread and increased injection pressure maxima, while demanding more precise metering.
- Combustion system: An adaptation for higher specific power output is necessary. Shifting the relevant engine operating points to higher loads requires an

appropriate raw emission level, which can only be achieved by a thorough optimization of the combustion system, including all subsystems.
- Charging system: To generate the necessary boost pressure advanced charging concepts are needed.
- Low-end torque: As most charging systems cannot supply a sufficient boost pressure level at low engine speeds for an adequate torque output, this drawback has to be compensated by additional measures.
- Engine configuration: Reducing engine displacement can be realized through smaller cylinder units or less cylinders. Reducing the number of cylinders offers advantages with respect to surface to volume ratio resulting in improved efficiency and lower engine raw emissions. However, drawbacks with respect to NVH performance and possibly customer perception also have to be taken into account.

This summary shows that downsizing requires a high engineering effort in all fields of engine development. The following chapter focuses especially on the layout of the boosting system required for downsizing a 2.0l engine with a specific power output of 65 kW/l with a displacement of 1.6l.

4. DEFINITION OF BOOSTING CONCEPT

In order to allow a significant downsizing level even for high performance engines with a power density of 65 kW/l, a significant increase of specific torque and power output is necessary. Consequently, a 20% reduction of engine volume requires a specific power output of ~80 kW/l in order to achieve the same absolute performance figures. This requires the upgrade of the boosting system to generate the demanded boost pressures in the range of 3.0 bar (abs.). In order to find the best solution for a high performance downsizing concept, three different boosting systems have been evaluated with GT-Power. All calculations were based under the boundary condition, that a modern 2.0l HSDI-Diesel engine should be replaced with a 1.6l engine capable of the same performance and emission levels.

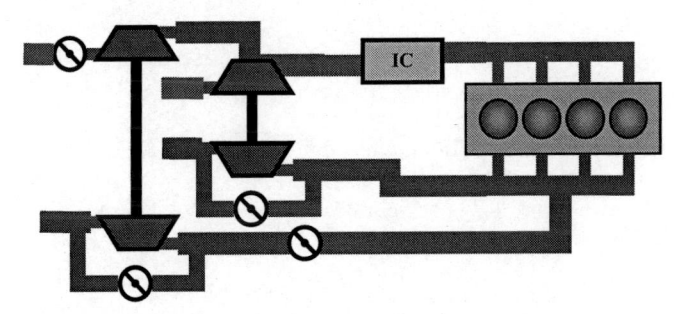

4.1. Potential of parallel sequential boosting system

Fig.4: Parallel-sequential boosting system

Recent developments (1) of boosting concepts showed a promising performance of parallel sequential boosting systems as shown in Fig.4. An analysis concerning the feasibility of such a layout for a downsized engine concept showed, that the desired rated performance cannot be realized, as the needed boost pressure for a specific power output of >80 kW/l can hardly be delivered by one compressor stage.

4.2. Advanced VNT

Recent developments for low volume production engines (2) show that specific power output of 74 kW/l can be realized by using an improved VNT-turbocharger. Fig.7 highlights that such a concept can achieve excellent specific power and torque outputs. The low end torque performance however is relatively poor compared to other engines equipped with a different TC-matching. Although such a performance characteristics is well accepted for sports cars, a broad acceptance of poor low-end-torque characteristics in view of various vehicle applications in the mass market is questionable. Therefore, this concept is not feasible for a wide range of downsizing applications.

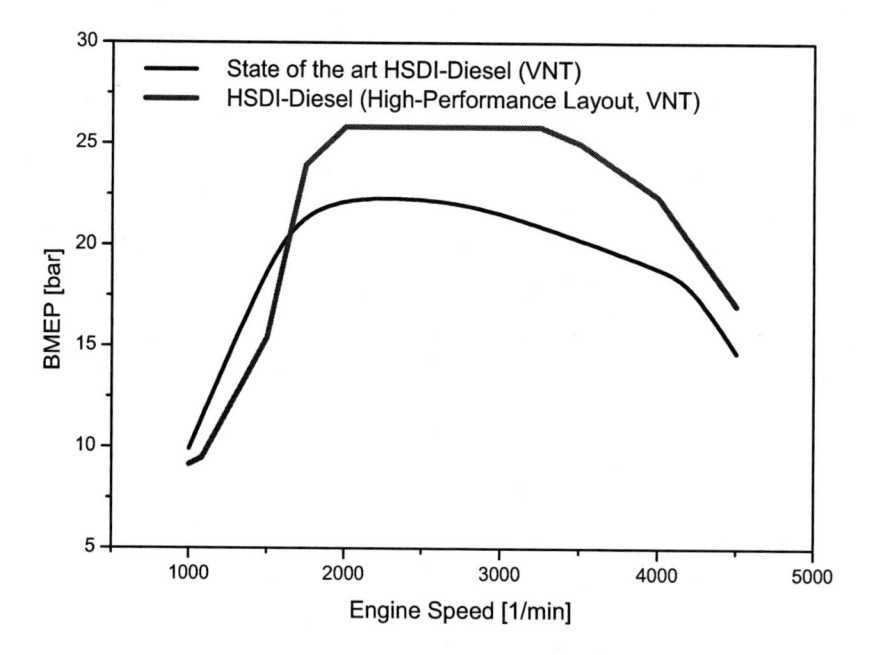

Fig.5: Comparison of torque output for different VNT-TC´s

4.3. 2-stage Boosting

Another solution for high specific outputs is a 2-stage boosting system featuring two turbochargers. The layout of a 2-stage boosting system allows many different arrangements concerning TC-configuration (FG, WG, VNT) and intercooling downstream LP-compressor. All in all the simulations included seven different concepts.

The best overall performance could be achieved with a concept, which uses a small HP-VNT turbine with turbine bypass and FG HP-turbine. Additionally a second intercooler is used between both compressor stages. So it is possible to use 2-stage boosting over the entire engine map. A comparison of the new 1.6l engine version, equipped with this boosting concept and a modern, state-of-the-art 2.0l engine is shown in Fig.10.

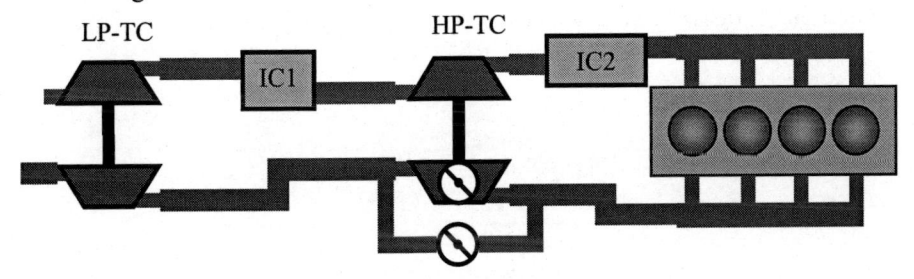

Fig.6: FEV 2-Stage boosting system

With the chosen charging concept, the realization of a torque output comparable to larger displacement engine is possible. At low engine speeds up to 1400 1/min even an improvement is possible, enabling an even better driving characteristic than the baseline engine. In the medium speed range between 1400 and 3000 1/min minor drawbacks of ~15 Nm in the torque profile can be observed. In the rated power area, both concepts achieve the same figures. Therefore the described 2-stage boosting concept has been chosen for the realization of a high performance downsizing concept.

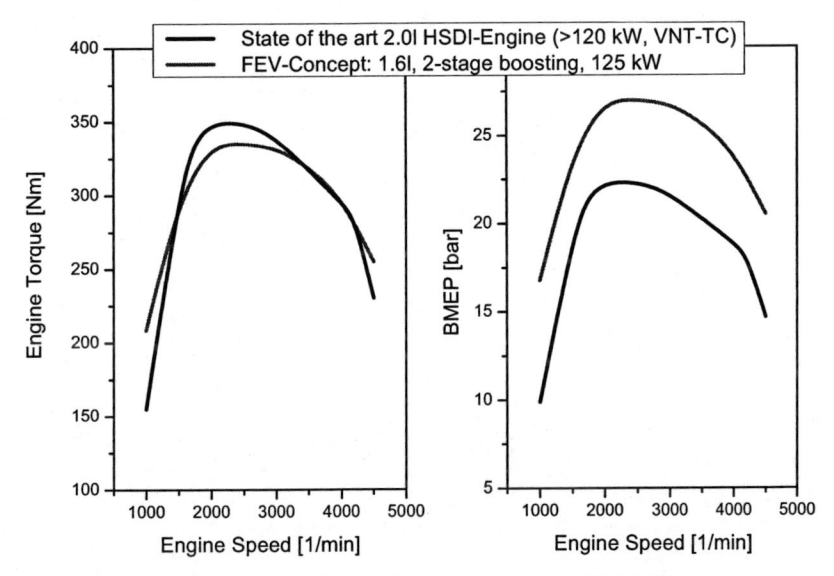

Fig.7: Full load behaviour of a small diesel engine with FEV combustion process

5. PART LOAD EMISSION POTENTIAL WITH DOWNSIZED ENGINE

By downsizing the engine, the operating points are shifted towards higher loads in typical test cycles such as the NEDC. The operation at higher loads typically leads to lower CO- and HC-emissions and an improved oxidation catalyst light-off. However, also higher NO_x- and PM-emissions are normally a result of engine operation at higher load. In Fig.8, these characteristics are exemplary shown for a typical driving condition over the EUDC the acceleration from 100 to 120 km/h in top gear.

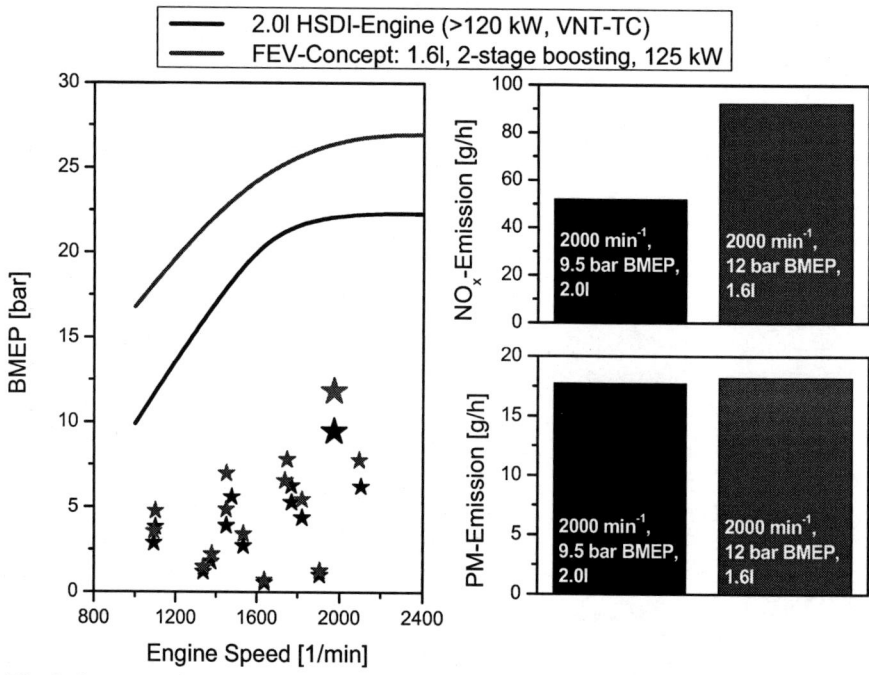

Fig.8: Impact of downsizing on engine operating area and high load emissions

The downsizing level of the engine results in an equal increase of BMEP, if the gear ratios remain unchanged. This higher engine load leads to increased NO_x and PM emissions. To define appropriate countermeasures with the goal to achieve the same raw emission performance as well as to protect the engine concept for future, more stringent NO_x limits, a fundamental understanding of the NO_x-formation is necessary. In Fig.9 eight characteristic operating points, representing the emission relevant operation area for typical passenger car applications are shown with respect to NO_x-emission and smoke. In the upper row, the NO_x-concentration is shown as a function of smoke number (left) and the O_2-concentration in the intake manifold (right). In the lower row the relationship between smoke and relative A/F ratio and smoke vs. intake manifold O_2-concentration is plotted. This layout clearly shows two main influencing parameters for pollutant formation. The NO_x-concentration shows for all operating points a clear dependency on the O_2-concentration of the intake charge, the smoke level shows a clear correlation to the relative A/F ratio. For the smoke level, it is especially noteworthy, that smoke emissions rapidly increase with relative A/F ratios lower than 1.3. With detailed knowledge of these dependencies, two targets for the development of future HSDI-Diesel combustion systems can be defined. At first, a massive reduction of the intake O_2-concentration is necessary to enable low NO_x-emissions. Additionally, it must be ensured, that the relative A/F ratio remains above $\lambda=1.3$ in order to avoid excessive PM formation.

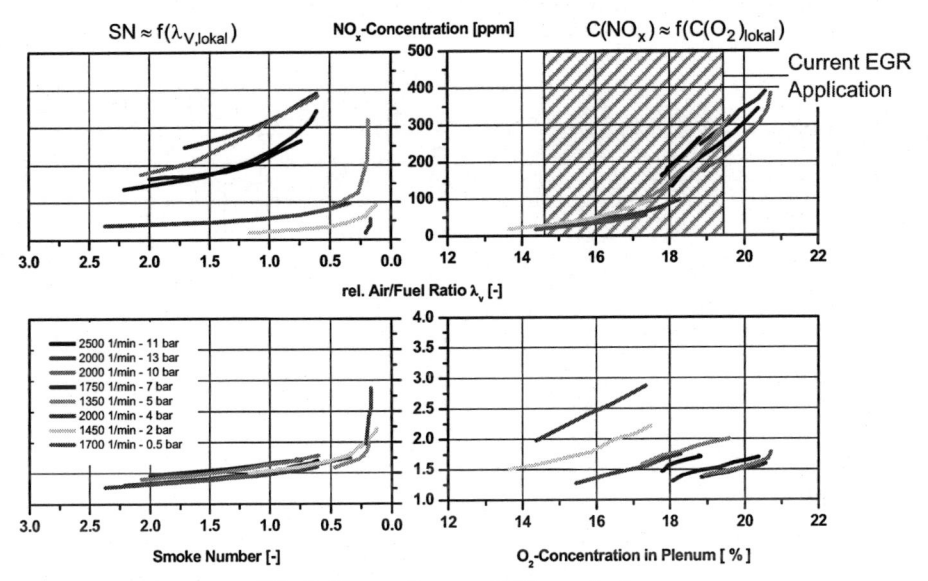

Fig.9: Dependence of NOx emissions

These requirements can be achieved with increased EGR-Rates to bring down the intake manifold O_2-concentration and increased boost pressure and cooling to avoid operation under low air/fuel ratios. By operating the engine with different inlet charge compositions and densities, also the other aspects of the combustion system such as combustion chamber geometry, fuel injection equipment etc. are affected. Based on the requirement to operate the engine at higher EGR-rates to reduce the NO_x-emissions, FEV conducted thorough part load testing of the small engine. The results at high load operating condition (i.e. acceleration from 100 to 120 km/h in top gear) are shown in Fig.10. In a first step, the combustion system was upgraded with respect to FIE definition and a lowered compression ratio was implemented. In order to be able to run higher EGR-rates, a more efficient EGR-cooler was implemented.

Even with this simple layout regarding the air system approach the emission performance could be improved significantly compared to the 2.0l baseline engine. However, with this technology package an improvement of fuel consumption in these operating conditions is not possible, as the operation at high EGR-rates in combination with high boost pressures results in severe gas exchange losses. In a second step, the potential of the 2-stage boosting concept was investigated. Compared to the initial tests with VNT-TC the boost pressure could be increased by 350 mbar and the gas exchange losses could be significantly reduced. The reduced gas exchange losses in combination with a leaner combustion due to the increased boosting level results in an additional improvement of emission performance and a 5% benefit in fuel consumption even at high load operation.

These results indicate that a downsizing concept is fully compatible with future emission legislation. Additional improvements compared to the currently achieved level can be expected by adapting a LP-EGR system, which enables an even better charge cooling and helps avoiding gas exchange losses.

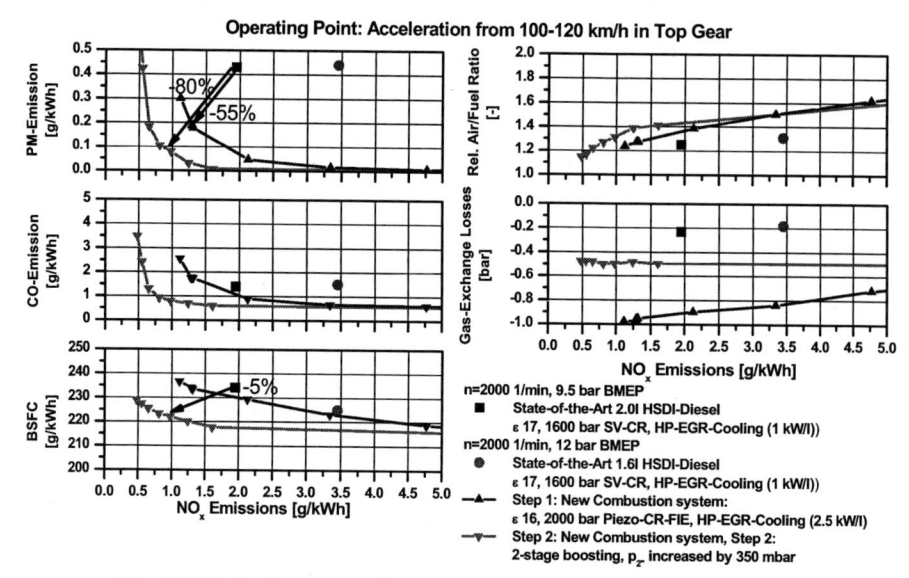

Fig.10: Emission-potential of optimized combustion processes

6. CONCLUSION

This report summarizes the main development steps, necessary for the realization of a small, highly boosted HSDI-Diesel engine, which offers compared to a current state-of-the-art engine, at least a comparable performance characteristic but in combination with significantly reduced fuel consumption. Fig.11 shows the future EU5 emission legislation and the realized CO2-reduction compared to current powertrains. Downsizing the displacement by 20% and utilizing the described technology package, a fuel consumption benefit of 17% is achieved while upgrading the emissions level to EU5 legislation.

For the realization of these goals, the air handling system (including boost and EGR) is a key component with respect to realization of ambitious full load performance targets as well as substantial emission reduction. The chosen boosting concept enables an excellent low-end-torque characteristic as well as an outstanding rated power of ~80 kW/l. This enables at least the same driving performance compared to the baseline engine with larger capacity. Additionally, the boosting system shows a high potential for minimized NO_x raw emissions by realization of high EGR-rates in combination with reasonable A/F ratios. The highly optimized combustion system represents a promising basis for next-generation specific power output in combination with lowest NO_x-, PM-, HC- and CO raw emissions.

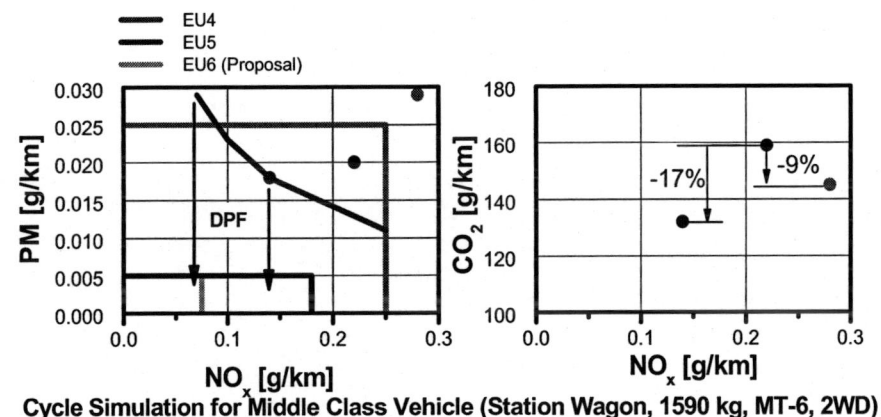

Fig.11: Emission- and CO2-potential of a small high charged diesel engine in mid class vehicles.

Cycle Simulation for Middle Class Vehicle (Station Wagon, 1590 kg, MT-6, 2WD)
- Baseline Engine (2.0l)
- Downsized Engine (1.6l)
- Downsized Engine (1.6l) w/ Technology Upgrade:
  2-stage boosting, 2000 bar Piezo FIE, Intensified EGR-Cooling,
  Optimized Combustion System w/ lowered compression ratio ($\varepsilon=16$)

7. ACKNOWLEDGEMENT

The described work has been performed within the project "CO_2-optimierter rußfreier Dieselmotor", supported by the Bundesministerium für Wirtschaft und Technologie (BMWi). The authors like to thank Mr. P.-J. Heinzelmann, BMWi, and Dr. B. Koonen, TÜV Rheinland Group, for supporting the project.

8. BIBLIOGRAPHY

(1) J. Portalier, J.C. Blanc, F. Garnier, N. Schorn, H. Kindl, J. Galindo, D. Jeckel, P. Uhl, J-J. Laissus, Twin Turbo Boosting system design for the new generation of PSA 2,2 liter HDI Diesel engines
(2) http//www.alpina-automobiles.com
(3) Lamping, M.; Körfer, T.; Pischinger, S.: Correlation between emissions reduction and fuel consumption in passenger car DI diesel engines, MTZ 2007/01

Delphi's 2000 bar common rail development for the Multec™ diesel common rail system

Rainer W. Jorach, D. Schoeppe, R.T. Nevard,
I.R. Thornthwaite, N.D. Wilson
Delphi Diesel Systems Ltd, UK

ABSTRACT

The global market for diesel engines is expected to continue to grow as Common Rail Diesel Systems are the most effective means to achieve more stringent emissions' legislation and improved engine performance.

Delphi developed the second generation of high pressure fuel pumps, the DFP3 & DFP4 pump family and further components, to meet this need; offering:
- 2000bar capability
- Robustness against extended endurance and poor fuels
- Flexible packaging with high component commonality

The generic development process used aggravated testing and reliability forecasting to shorten the development cycle.

This publication will show the potential of this Delphi fuel injection system.

1. INTRODUCTION

Delphi is a key supplier for fuel systems and many components for car and truck applications. This paper will present the latest diesel technology introduced to the light duty and medium duty markets.

2. DIESEL MARKET

The diesel engine market is growing globally as consumers are discovering the benefits of diesel vehicles and especially Common Rail. In Europe, approximately every second new car sold is powered by a diesel engine. There are many reasons for this surge in popularity. Consumers are discovering that diesel engines offer:

© *Delphi Diesel Systems Ltd 2007*

- Better fuel efficiency: Light-duty diesel engines use ~30% less fuel than gasoline engines of similar power under similar circumstances
- More torque: Diesels produce more drive force at low engine speeds than gasoline engines under similar circumstances, making diesels more fun to drive
- Lower greenhouse gas emissions: Less fuel consumed translates to ~15% less carbon dioxide emissions.
- Less noise: Multiple injection technology combined with lower engine rpm make in particular common rail diesels quieter than equivalent petrol vehicles.

Increasing system pressure is beneficial as it enables at the same time increased specific power of diesel engines and decreased emissions and particulate matter in light load condition. Increased power density also allows for smaller engine sizes – and enhanced fuel economy – without sacrificing power. Developments to meet the new demands of increased pressure include low leak injector design, high fatigue resistance rail and introducing a new common-rail pump family capable of delivering fuel pressures up to 2000 bar. This pump family is called Delphi DFP3 and DFP4.

The diesel market is segmented by engine size and so are the fuel systems. DFP3 and DFP4 pumps fit onto light- and medium-duty applications respectively. Light duty passenger cars and light commercial vehicles are typically within the range of 50-350hp which means cylinder capacities of less than one litre per cylinder and overall engine capacities of less than 5 litres. Medium duty on- and off-highway applications are usually within the power range of 100-350hp, which means cylinder capacities of less than 1½ litres per cylinder and overall engine capacity of less than 9 litres. These applications are light and medium sized trucks as well as agricultural machinery and construction vehicles. Above that range there is a heavy duty diesel market with the power range of 215-600 hp.

3. EMISSIONS LEGISLATION

Over several decades developing emissions legislation has become the main hurdle for the car makers to be able to enter the markets. Fuel injection, exhaust turbo-charging, and exhaust after treatment are the key engine technologies tuned to achieve the different emission limits [fig. 1].

Amongst these key technologies fuel injection is the most significant because of its major impact on the internal combustion process itself and on formation of air pollutants. Looking at the European emissions legislation for passenger vehicles, even from the Eu 4 to the Eu 5 legislation, there is a requirement to decrease Nitrogen Oxides (NOx) by 28% from 0.25 grams/km to 0.18 grams/km. This comes with the decrease of Particulate Matter (PM) by 88% from 0.025 g/km to 0.003 g/km. Since Eu 5 obviously favours a reduction of PM the next step change, which will be Eu 6, will reduce NOx further. This will be a further reduction of 55% down to 0.080 g/km. High pressures of 2000 bar offer more freedom to precisely shape the fuel injection and the combustion process to meet these requirements.

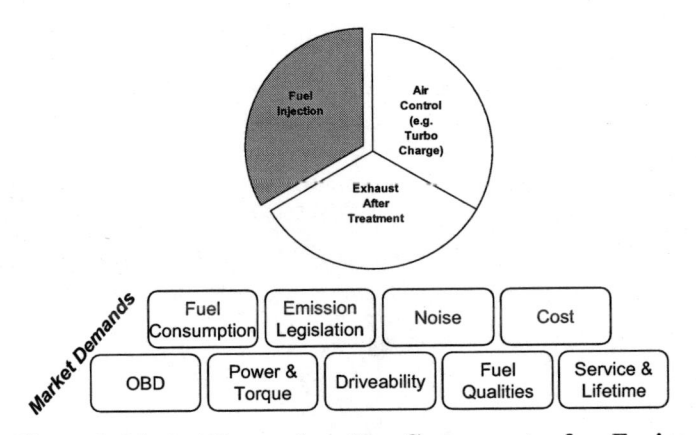

Figure 1: Market Demands vs Key Components of an Engine

4. COMMON RAIL DIESEL FUEL SYSTEM

The Delphi Multec™ diesel fuel system consists of the fuel pump, injectors, rail and piping, as well as the electronic control unit (ECU) and software [fig.2].

Common Rail Fuel Systems consist of:
- **Fuel Pump**
- **Injectors**
- **Rail & Piping**
- **Electronic Control Unit (ECU) + Software**

Flexible modular pump
- Designed for up to 2000 bar capability
- Key focus on robustness and durability
- Complete pump family supporting various packaging and drive mechanism

Figure 2: Delphi Multec™ Diesel Common Rail - Features

There have been three key drivers for the development of the flexible and modular pump for the second generation system: (a) the design rail pressure has been increased from 1600 bar on the first generation to 2000 bar capability for the second generation. (b) The key focus has been on the pump's robustness and durability. (c) To satisfy the engine market a complete pump family with the capacity between 0.5 and 1.5cc/rev was needed. Each pump type within the pump family then had to support various drive mechanisms and packaging needed for the different customers' engine applications.

5. PUMP PACKAGING

The pump selection criteria are drive ratio, capacity, positioning on the engine, and the engine drive type. The DFP3 and DFP4 family offers two different plunger sizes of 6.5mm and 7.5mm diameter. By having 2-plunger and 3-plunger variants for each of the plunger sizes this gives four different pump types [fig.3].

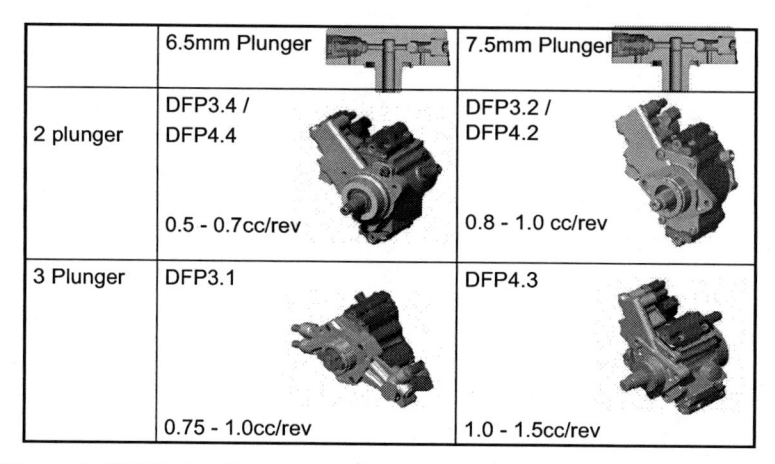

	6.5mm Plunger	7.5mm Plunger
2 plunger	DFP3.4 / DFP4.4 0.5 - 0.7cc/rev	DFP3.2 / DFP4.2 0.8 - 1.0 cc/rev
3 Plunger	DFP3.1 0.75 - 1.0cc/rev	DFP4.3 1.0 - 1.5cc/rev

Figure 3: DFP3 Family - Overview of DFP3 Components and Functions

The DFP3.4 is a 2-plunger pump of 0.5-0.7cc/rev capacity, while the DFP3.2 and DFP4.2 offer up to 1.0cc/rev capacity with two plungers of Ø7.5mm. The small 3-plunger pump, DFP3.1, offers 1.0cc/rev while the bigger DFP4.3 offers up to 1.5cc/rev capacity. All DFP3s are typically passenger car pumps while the DFP4's offer extra devices for the medium duty market, e.g. extended endurance and extra low pressure connections to use transfer pump pressure for filtration.

The pump drive ratio on typical passenger cars is between 1:2 and 3:2 vs the engine speed. The majority of car applications today ask for half engine speed. This leads to a 2-plunger, single lobe pump design to synchronise the pump events with injection events on four cylinder engines. Depending on the capability of injector and software this may support low emissions strategies, especially when it comes down to many injection pulses per cycle. For cost reasons there is a visible trend to increase the drive ratio to 1:1 drives, which means that a smaller and cheaper pump with fewer cylinders is able to feed a highly rated engine. Delphi has prepared for these types of application with maximum nominal pump speeds of 4000prpm today and the prospect of further increases.

Although there is no existing standard for the fuel pump packaging envelope on an engine there is an economic need to apply small pumps with high commonality to various engines. While this results in a more compact pump, it required more development with respect to drive torque, fluid flow and bearing load.

The various car makers follow different technical directions for their drive types, which is why Oldham coupling, spur gear, helical gear, chain and belt drive all exist in parallel. So the drive train components need to cope with the forces and torques resulting from these.

6. DFP3.x PUMP FAMILY/DESIGN FEATURES

A technology roadmap exists for all Delphi product lines. DFP1 was the first generation Delphi Common Rail Pump, launched in the year 2000 and limited to 1600 bar rail pressure at that time. The DFP3 family was launched in 2005 at a design pressure of 2000 bar. While the first DFP3 applications were Eu 4 engines the later applications cover Eu 5 and Eu 6 and Tier 2 bin 5 emissions legislation.

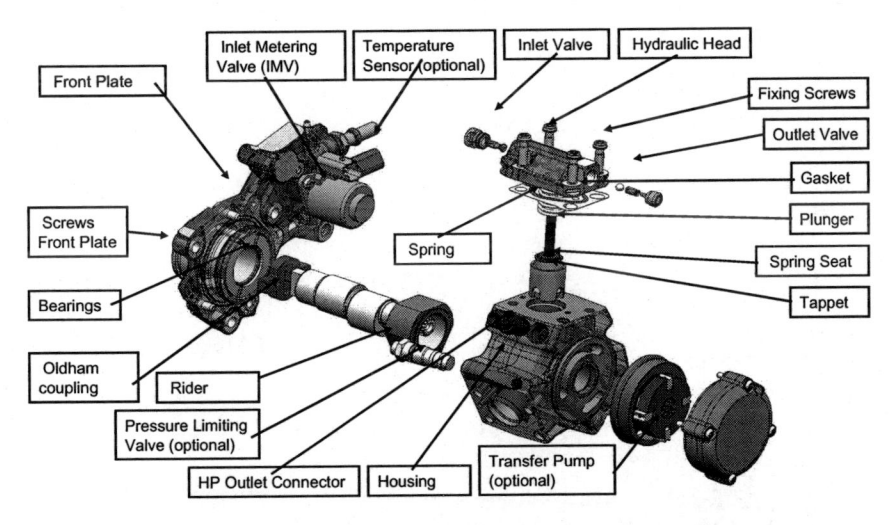

Figure 4: Principal Common Rail High Pressure Pump Components

Figure 4 shows the principal DFP3 pump components. The delivery control of any Delphi Common Rail Pump uses an inlet metering valve (IMV) so that only the amount of fuel actually needed by the engine is compressed. This guarantees a high efficiency of pump and engine. The pump internal drive train is a slipper tappet design which offers nominal pump speed of up to 4000 prpm and a pump over speed of 5000 prpm. Driven by the engine the drive shaft runs inside bearings pressed into the housing and the front plate. A drive polygon called a "rider" is operated by the movement of an eccentric part of the driveshaft. This rider has two or three flats depending on the number of plungers. Turning the shaft causes the rider flats to translate, driving the slipper tappets, which slide on the flats, up and down. The pump plungers sit inside these tappets so that any outwards movement will lead to compression of the fuel volume above the plunger, while on the return stroke they will fill.

The plunger cylinders are formed in the hydraulic heads. These are made from forged and hardened steel material and bolted to the pump housing. They each carry an inlet valve and outlet valve. The inlet valve is designed as a cone valve in order to guarantee precise timing and to be durable as well with poor fuels and over long endurance. The outlet valve is a spring actuated ball valve. The head to housing gasket needs to have both high and low pressure sealing areas: A Viton layer printed onto the metal gasket forms the low pressure seal towards the environment. Since the high pressure from each hydraulic head goes through drillings in the housing to the high pressure outlet on the housing each head needs to have a high pressure seal as well. Three concentric grooves formed in the same metal gasket around the high pressure drilling provide this functionality.

The engine interface is provided by the front plate. It is specific to each engine application and carries the IMV, low pressure connectors and any optional temperature sensor or venturi.

7. DEVELOPMENT AND VALIDATION

The validation of DFP3 & DFP4 was based on a rigorous process and on a standardised suite of tests. The Design FMEA and Process FMEA were very important tools to ensure all principal failure modes were tested by the set of validation tests. The DFMEA and PFMEA are also used to predict reliability.

Generic validation of early 1600bar systems started at an aggravated rail pressure of 1800 bar and 3500 rpm based on both car and medium duty boundary conditions. Individual applications orientated tests have been added to the generic ones

The Delphi principal tests are those which provide maximum confidence to demonstrate the product's reliability by exciting most failure modes. These are:
- 2600 hrs Speed and Load Cycle
- Highly Aggravated Hot Fuels Test
- High Pressure Fatigue Tests
- Worst Case Fuels Tests

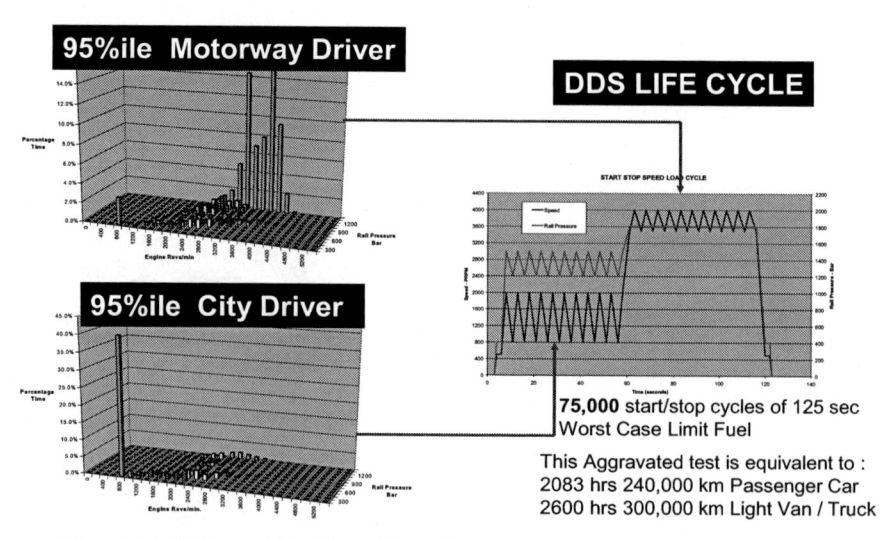

75,000 start/stop cycles of 125 sec Worst Case Limit Fuel

This Aggravated test is equivalent to :
2083 hrs 240,000 km Passenger Car
2600 hrs 300,000 km Light Van / Truck

Figure 5: Delivery Van Duty Data Used to Create DDS Principal Test

Further additional tests have been conducted, e.g. vibration and thermal cycling.

This set of tests is applied to all pump applications during design validation and repeated during process validation.

The aggravated speed and load cycle test is the primary durability measure for the pump and is based on real vehicle usage data [fig. 5]. It consists of a lower speed part covering the 95th percentile city driver and a high speed part covering the 95th percentile motorway driver. 2,083 hours of testing are equivalent to 240,000 kilometres on a passenger car, while 2,600 hours are equivalent to 300,000 kilometres of a light van and truck application, depending on speed and load aggravation factors.

Component Condition on Completion of 2600 hrs Test

Rider   Tappet   Housing Bearing

Component Condition on Completion of 4300 hrs Test

All Parts In Excellent Condition

Figure 6: Component Condition on Completion of Speed & Load (start/stop) Test

Based on investigations following the tests a Service Life Catalogue has been put in place describing Service Life Ratings for each of the key components.

All components have been analysed after test completion and any findings investigated using an 8D quality system to drive improvement. Pump robustness is illustrated by figure 6 which shows minimal wear on the highly loaded drive train components even after increasing the test duration from the standard 2600hours to 4300 hours. After completion of the process validation individual components have been through the Production Part Approval Process (PPAP) as well as the whole pump.–However early designs revealed a number of engineering concerns which were investigated using Finite Element Analysis, Shainin based development and aggravated testing.

8. STATISTICAL METHODS

From a statistical point of view, the confidence in product reliability grows with the number of validation hours run. At the beginning of a validation programme the reliability increases very quickly so at the 50% point in the start stop cycle 90% reliability confidence has been built. This means that the second half of the testing builds the confidence level by 5%, to reach 95% reliability. 95% reliability confidence at target life is typical of current customers' requirements for fuel systems.

Six Sigma engineering methodology (Define, Measure, Analyse, Improve and Control) has been applied to assess pump reliability. Using the, Reliasoft, Weibull++ software to perform the "Measure and Analyse" phases of the investigation to and predict the reliability. Failures in the start stop test, regardless of failure mode, were plotted into a log-log plot [fig 7] at the running hours. Plotting a best fit line, with confidence bounds, through all these occurrences allowed a forecast of reliability to be extrapolated to higher running hours. This allowed early identification of potential concerns. Hence more design iterations and a faster rate of improvement was achieved. The chart shows the effect of the design changes that were possible between early development and the final development stages with the steeper line showing much improved reliability which meets the target life at 2000 hours aggravated testing.

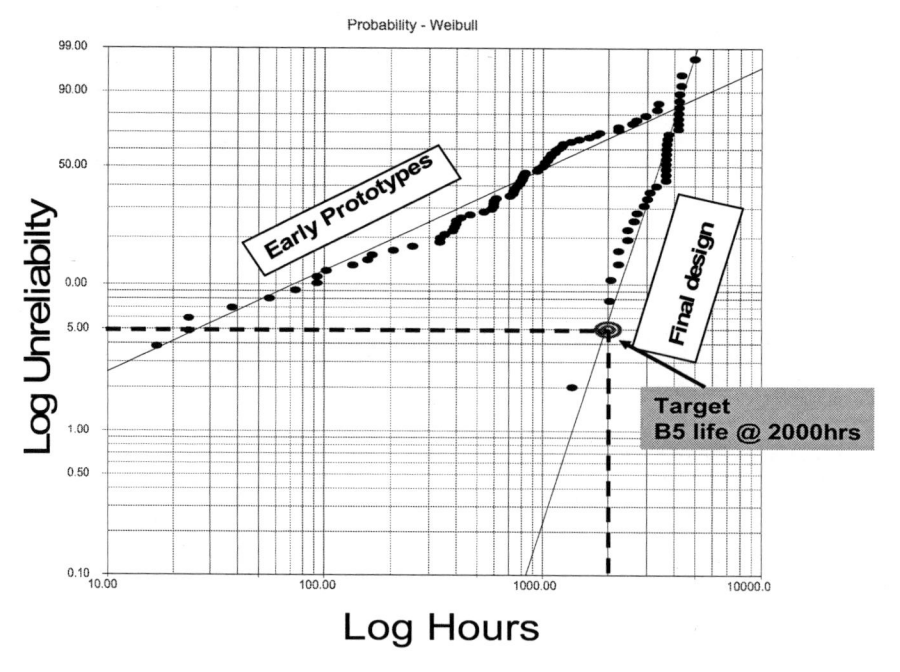

Figure 7: Six Sigma Benefit to Pump Launch on Failure Results

To enable objective comparison of component wear assessments, a Service Life Catalogue has been put in place. All important pump components and functions have been described with the ratings between 0 and 5. Where 0-3 means that the part is acceptable and 5 means a significant failure and the part no longer functions.

On completion of each test the pumps are stripped, investigated and a service life rating made for the components. This ratings tally offers a quantifiable indicator of the comparative wear or different pump components and excellent indication for potential failure modes needing further attention [fig.8]. This has been used to further extend pump robustness and endurance.

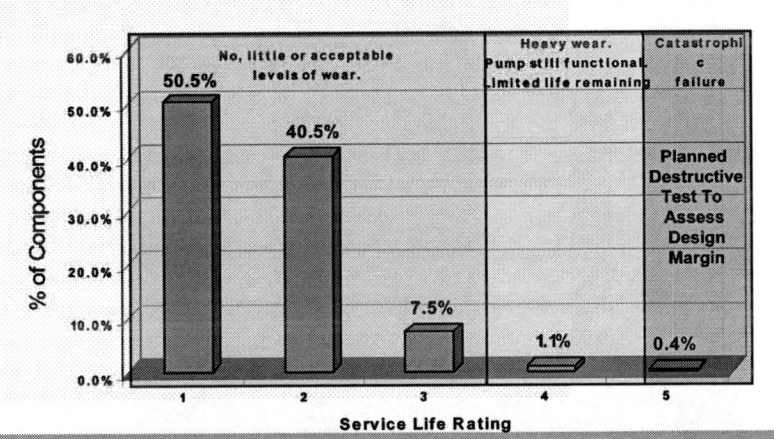

DFP3 has demonstrated a high level of service life robustness

Figure 8: End of Test Component Service Life Assessment

At launch 98.5% of all ratings during the entire validation programme have been between 1 and 3. Just 1.1% had a Service Life Rating of 4 with limited life remaining while the pump was still functional. The validation programme included a number of tests to failure (e.g. fatigue tests) these resulted in the 0.4% of parts rated 5.

9. CONCLUSIONS

The Diesel Market is growing significantly and Delphi is a major contributor to this trend.

- About 400,000 new generation Delphi DFP3 pumps are in the field since launch in 2005.
- The DFP3 design is robust up to 2000bar and uses generic components plus application specific optimisation.
- Robustness and durability have been enhanced by the use of six sigma and statistical tools during development and validation.
- DFP3 Launch has been flawless with excellent product quality.
- A future roadmap is in place for continued product enhancement.

The Delphi DFP3 and DFP4 Common Rail pumps fulfil today's and future demands with one common design for various car, light duty and medium duty applications.

10. ACKNOWLEDGEMENT

The authors would like to thank all those who supported us and helped reach the high targets, i.e. our customers, motorists, as well as our colleagues in Delphi. The authors would especially like to thank the DFP3 team in Product and Manufacturing Engineering, Purchasing, Quality, Logistics, Sales, as well as in Gillingham Technical Centre and Sant Cugat plant.

11. REFERENCES

Schoeppe, D:
Common Rail Technology for the Future Low Emission Medium Duty Engine Market
3rd AVL International Commercial Powertrain Conference
Graz, Austria, 2005

Schoeppe, D; Spadafora, P; Guerrassi, N; Greeves, G; Geurts, D:
Einspritzsysteme für Motoren mit hoher Leistungsdichte
Dresdner Motorenkolloquium 2005

Zuelch, S; Schoeppe, D; Jorach, R W; Judge, R:
Das neue Delphi Common Rail Einspritzsystem für den Einsatz in Medium Duty Dieselmotoren
Aachener Kolloquium Fahrzeug-und Motorentechnik 2006 9-11.10.2006
Editor: Stefan Pischinger, Hennig Wallentowitz.
Aachen: VKA/IKA 2006 pp 577-596 (Band 1)

Winterbourn, M; Balin. M; Jorach, R W; Soteriou, C; Tang, W; Zuelch, S:
The Virtual Pump: Integrated Simulation of the High Pressure Diesel Common Rail Pump
In: Diesel- und Benzindirekteinspritzung IV – Anwendungen – Zukunftsentwicklungen - Messtechnik – Simulation.
Editor: Ulrich Brill, Helmut E Tschoeke, Renningen—Malmsheim: Expert Verlag 2007 pp 156-165 (Fachbuchreihe/Haus der Technik: Bd 77)

UK Petrol and Diesel Demand, Energy & Emissions: Effects of a Switch to Diesel – March 19 2007

PDT OConnor:
Pratical Reliability Engineering 3rd Edition revised 1995
J Wiley & Sons

Application of JCB Dieselmax high performance technology to future off-highway diesel engines

Andrew Banks, Matt Beasley, Andy Skipton Carter
Ricardo Consulting Engineers Ltd, UK

Alan Tolley, Tim Leverton
JCB Power Systems Ltd, UK

ABSTRACT

The JCB Dieselmax LSR engine that powered the successful land speed record vehicle includes a range of technologies that are also being developed to meet the requirements of the future off-highway (Tier 4) emissions legislation. These include a two-stage boosting system, high performance common rail fuel injection system and a low temperature combustion system.

This paper describes how this technology is being developed to enable Tier 4 emissions levels to be achieved. Results from Ricardo and JCB advanced engineering programmes are presented and a technology roadmap is defined. The paper focuses on achievement of the Tier 4 interim emissions levels and also outlines a strategy for achieving the challenging Tier 4 final emissions levels.

ABBREVIATIONS

ECU	Engine Control Unit	CFD	Computational fluid dynamics
NOx	Nitrogen Oxides		
DPF	Diesel Particulate Filter	EGR	Exhaust gas recycle
LNT	Lean NOx Trap	BMEP	Brake mean effective pressure
HCCI	Homogeneous charge compression ignition	FGT	Fixed geometry turbine
HPCC	Highly premixed cool combustion	VGT	Variable geometry turbine
		IEGR	Internal exhaust gas recycle
VVA	Variable valve actuation	OBD	On board diagnostics
SCR	Selective catalyst reduction	NTE	Not to exceed
NRTC	Non road transient cycle	AECC	Association of European catalysts
FIE	Fuel injection equipment		
		LION	Low emission combustion

© Ricardo UK Ltd 2007

1 INTRODUCTION

The 23rd Aug 2006, concluded 19 months of intensive development by JCB and Ricardo which achieved the new diesel land speed record. The car "Dieselmax" (1) achieved a speed of 350 MPH at the Bonneville salt flats USA. The Dieselmax car was powered by two diesel engines designated JCB LSR, one driving the front wheels in front of the driver, and the other driving the rear wheels at the rear of the driver. The engines used in the land speed record were based on the current JCB 444 TCA engine, introduced in 2004 and currently used in a range of products. The engines originally designed in conjunction with Ricardo, had been developed by the team over the 19-month period, and resulted in the world's highest specific output automotive diesel engine. The table below compares the then current JCB 444 engine rated at 93kW with the final LSR engine rated at 559 kW.

Table 1 JCB 444-TCA and JCB – LSR comparison.

Specification	JCB 444 – TCA	JCB 444 – LSR
Power	93 kW	559 kW
Peak Torque	525 Nm	1500 Nm
BMEP	11.7 bar	35.7 bar
Rated Speed	2200 rpm	3800 rpm
Bore	103 mm	109 mm
Stroke	132 mm	134 mm
Swept Volume	4.4 litres	5.0 litres
Boost	2.2 bar abs	6.0 bar abs
Fuelling	79 mg/injection	356 mg/injection
Engine Mass	470 kg	395 kg

To achieve the power levels from the JCB 444 – LSR engine significant development of the 2 stage intercooled boosting system was carried out, along with high-pressure common rail technology and high performance and low emission combustion development. The LSR engine also used Diesel Particulate Filters (DPF) to control particulate emissions.

The technologies successfully utilised during the Dieselmax land speed record are currently being developed to meet future Tier 4 legislation.

1.1 Tier 4 legislation – USA EPA 40 CFR Part 1039

Tier 4 is the designation for the US non-road legislation, being similar to Stage 3B and Stage 4 in Europe. Tier 4 legislation is a phased introduction of emission limits for power ranges from <8kW upto 560 kW. Appendix 1. The power range is split into six power bands. We focus here on the 56-129kW and the 130kW to 560kW bands that cover the mid range off highway sector. Tier 4 is broadly split into two phases Tier 4 interim, primarily focussed at reduction in Pm, and Tier 4 final which in addition to Pm reduction forces a significant reduction in NOx emissions.

What is Tier 4 Legislation?

Emissions Legislation for Tier 3, Tier4 interim and Tier 4

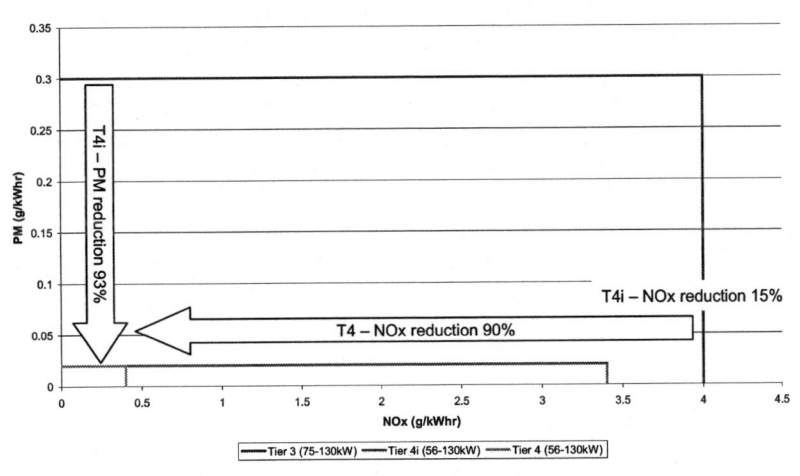

Fig 1 Tier 4 interim and final legislation challenge

The challenge facing the off highway industry over the next seven years is approximately a 93% reduction in Pm emissions and a 90% reduction in NOx emissions for engines of power greater than 130 kW. The introduction of Tier 4 final will see the off highway industry technically aligned with on highway legislation.

Unlike Tier 3 legislation, which Tier 4 replaces, Tier 4 introduces the non-road transient cycle (NRTC) ensuring that transient emissions are regulated in addition to the ISO – 8178 C1 8 mode steady state cycle. The NRTC includes a cold start cycle (ambient temperature) with a cycle weighting of 5% cold and 95% hot. EU stage 3B/4 has a 10% cold cycle element.

NRTC v 8 mode

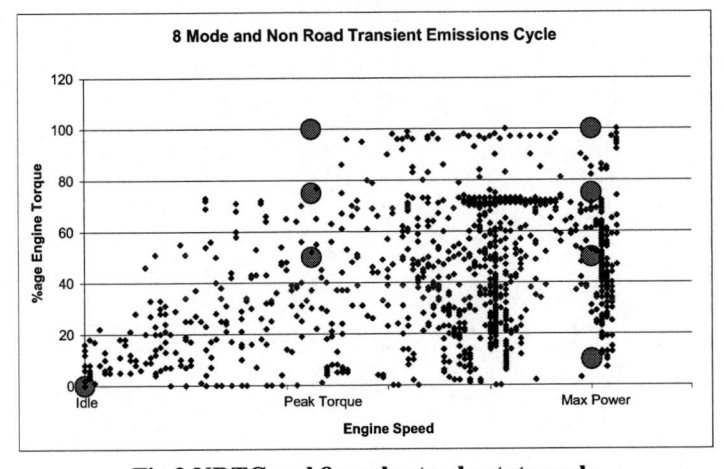

Fig 2 NRTC and 8 mode steady state cycle

1.1.1 Tier 4 legislation will also introduce the following criteria:

- On board diagnostics (OBD) – the EPA have just started to investigate the adoption of OBD to the non road market, at this stage it is anticipated that OBD legislation will be introduced into the non road sector, and as such OBD should be considered as part of the Tier 4 strategy. (2)
- Not to exceed limits (NTE) – Tier 4 specifies emission limits outside the emission testing points, not to exceed between 1.25 to 1.5 times the cycle limit, this also includes altitude upto 1676 m.
- Deterioration factors (DF) for both the engine and aftertreatment equipment must be demonstrated.
- Closed crankcase ventilation (CCV), the Tier 4 legislation will allow open breathers for turbocharged engines (closed for naturally aspirated), however open breather systems will have to measure the emissions and add to the exhaust emissions.
- US sulphur levels in fuel will reduce from 500 ppm today to 15ppm around 2011 timeframe.

2 MACHINE INTEGRATION

Tier 4 legislation will result in significant vehicle engineering and integration of the technologies adopted. The best-case scenario is that existing vehicle platforms will be able to accept Tier 4 engines, however in some applications it will require a major redesign of the vehicle to accept the Tier 4 package. The following areas will require careful consideration and review:

- Extra heat rejection from the engine, due to the adoption of EGR, analysis carried out on the JCB engine in WAVE predicts an EGR cooler heat rejection of 35kW at rated power of 130kW with 25% EGR rate. The vehicle cooling system will need to be sized to accommodate this extra heat to coolant.
- Packaging of engine and aftertreatment systems, the challenge to package the Tier 4 envelope including aftertreatment will require early consideration.
- Duty cycles – varying from high load factor, e.g. Excavators and tractors to low load factor e.g. telehandler or backhoe loader. The large variety of infield cycles will need to be reviewed with respect to aftertreatment effectiveness e.g. DPF regeneration.

Optimisation of machine performance will require extensive in machine calibration and testing effort.

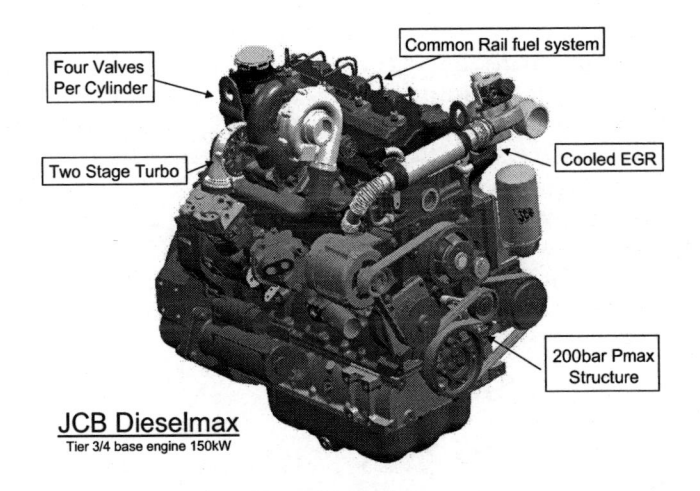

Fig 3 JCB Dieselmax Tier 3/4 Base engine and aftertreatment

3 OFF HIGHWAY ROADMAP – TIER 4 INTERIM AND TIER 4 FINAL

The roadmap shows the likely technologies adopted for a given timeline, it is anticipated that post Tier 4 final timeline, that emission legislation will focus on CO_2 reduction.

The non-road technology road map is split into 4 areas:
- Combustion and Performance Technology
  Ricardo are developing specific non road combustion systems utilising high EGR rates, as well as low NOx combustion systems targeting engine out Tier 4 emission levels

- Boosting Technology
  The JCB dieselmax utilised 2-stage boosting, this technology is currently being developed by JCB for Tier 3 ratings, and will be utilised for Tier 4 at power levels greater than 130 kW. Single stage boosting will also be part of the boosting technology, in either fixed or variable geometry turbine configuration.
- Fuel Injection Equipment (FIE)
  Mechanical systems used today will be replaced by electronically controlled systems, and this will be complemented with higher injection pressures with increased functionality. Emission control and aftertreatment NOx engine control strategy control strategy and aftertreatment technology for both Pm and NOx

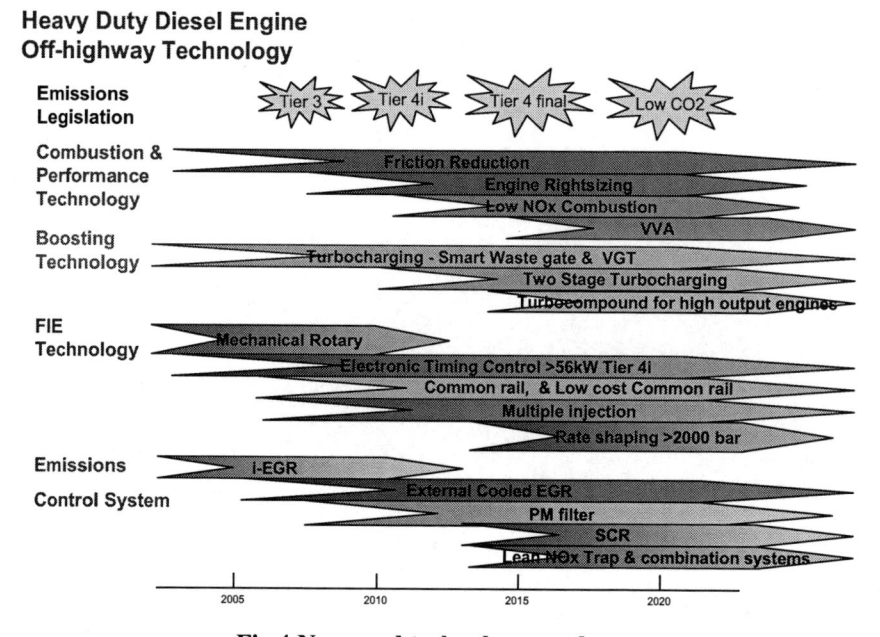

Fig 4 Non road technology roadmap

4 TIER 4 TECHNOLOGIES UNDER DEVELOPMENT

JCB and Ricardo are active on a number of Tier 4 activities with engine ratings being maintained at Tier 3 ratings.

4.1 Combustion system

The combustion system evaluated was a new design targeting maximum EGR rates upto 25%, based on guideline compression ratio and inlet swirl. During the assessment and bowl decision process, 3D CFD analysis was used to assess the combustion system performance and support chamber profile selection. Figure 5 below shows the outline chamber profiles assessed (blue and green with recessed chamber, and a square lip design in green).

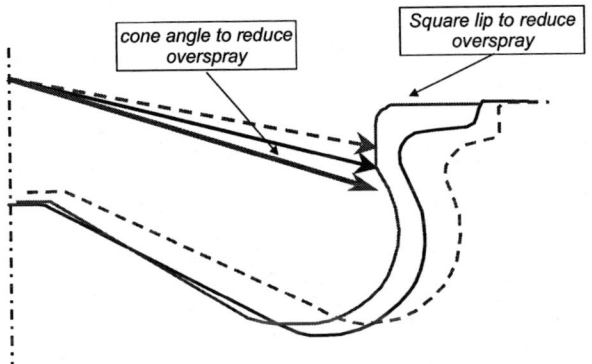

Fig 5 Combustion bowl designs

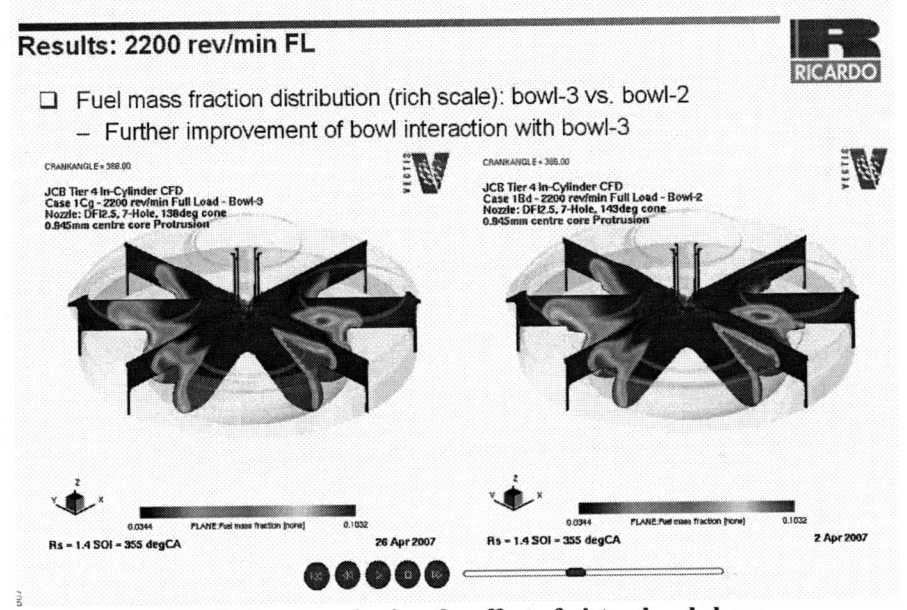

Fig 6 CFD run reviewing the effect of piston bowl shape

The designs were evaluated using VECTIS, and assessment focused on mixture preparation and transport within the chamber, and the impact on combustion behavior. The recessed chamber design showed improved mixture retention and the square lip profile decreased the in bowl mixing resulting in increased regions of rich mixture leading to increased soot. This analysis process selected the optimum bowl design to carry forward to engine development.

4.1.1. Low NOx combustion
Although not the main subject of this paper, Low NOx combustion is currently research based, and is primarily focused at Tier 4 final and beyond. Using single cylinder research engines, JCB and Ricardo are reviewing the possibility of achieving engine out NOx levels <0.4 g/kWh. The work carried out under action ref

(3) is currently focussing both on highway and off highway. This activity includes Low NOx combustion (HPCC, HCCI), and VVA in single and multi cylinder configurations. Results will be published in 2008.

4.2 Boosting system

Alternative boosting technology has been evaluated using WAVE 1d simulation code. This has both steady state and transient capability, and is a valuable tool for technology assessment prior to procuring and testing expensive hardware. As well as turbocharging evaluation of FGT, VGT and 2 stage systems, the intercooling performance of 2 stage systems has been reviewed, including split cooling systems of engine and EGR cooler.

4.3 EGR circuit

Unlike the Dieselmax engine, NOx control for Tier 4 interim will be controlled via external cooled EGR, which has resulted in a new EGR circuit being designed. The configuration adopted is manifold to manifold to avoid compressor fouling. Extensive use of 3D CFD has resulted in improved EGR distribution, optimum EGR valve location and even mixing cylinder to cylinder. The guideline is that EGR distribution cylinder to cylinder should be +/-5%. This has been achieved under the following EGR rates.

Table 2 Target tier 4 interim and final EGR rates > 130 kW

Emission level >130 kW	NOx target g/kWh (engine out)	EGR rate %
Tier 4i	1.8	15-22
Tier 4F	1.1	25-30

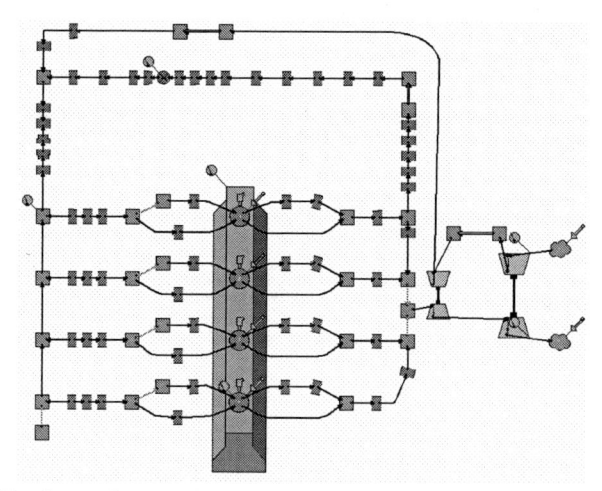

Fig 7 WAVE schematic showing 2 stage boosting technology and EGR circuit.

A further analysis has been carried out to review the impact of inlet throttling, this can increase the efficiency of the system and also aid in the DPF regeneration strategy. In the configuration analysed, an increase of 3% EGR rate over non-throttled design was achieved at peak torque speed.

4.4 FIE

The current Tier 3 JCB TCA engine uses common rail technology; this will be maintained and updated for Tier 4. Pressures >1800 bar capability and multiple injection will be adopted, since this has been shown to reduce Pm levels. Up to 5 injections per cycle will be adopted, including pilot, main and post injection.

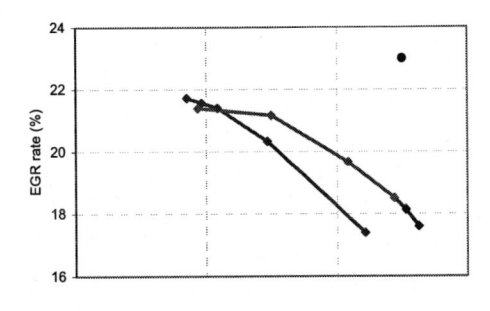

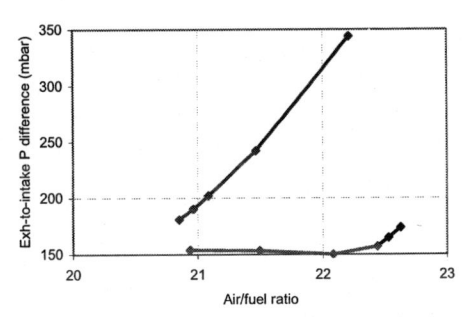

Fig 8 Intake throttling on engine EGR

Research carried out as part of the European funded Framework 6 Green Project on a 2L/cylinder single cylinder research engine (Proteus) with high pressure fuel system > 2250bar ref (4) shows the benefit on Pm emissions with the adoption of a close coupled post injection.

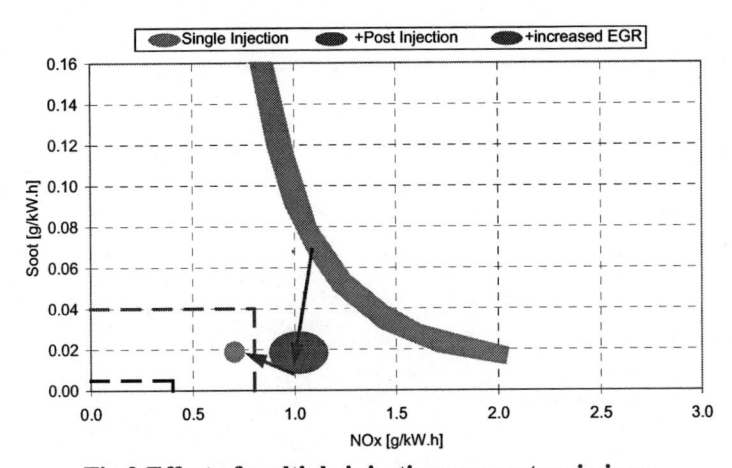

Fig 9 Effect of multiple injections on soot emissions.

Figure 10 shows the NOx Pm trade off curves for Tier 3 and tier 4 with advanced FIE.

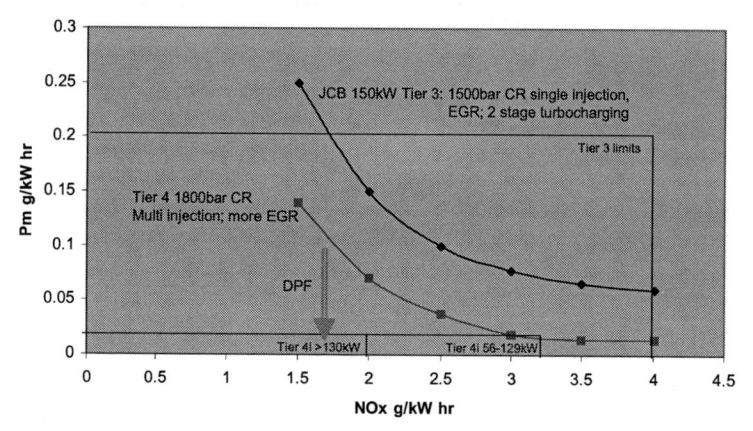

Fig 10 NOx Pm trade off curves Tier 3 and Tier 4.

4.5 Aftertreatment

4.5.1 DPFs are currently available in certain non-road markets e.g. Switzerland and as retrofit solutions. Tier 4 interim legislation will introduce DPFs for control of Pm emissions. The regeneration of the DPF will move from passive today to active regeneration strategies, and align with similar technologies being adopted on highway for US07 and in Europe on passenger car applications.

Currently 3 methods of regeneration of DPFs are under evaluation:
- Late post injection from the common rail fuel system
- Separate HC doser in the exhaust system, which is a current heavy duty diesel solution
- Auxiliary fuel burner, which again is a current heavy duty diesel solution

Research carried out in 2004 ref (5) concluded that the HC doser strategy prevents the risk of oil dilution compared to a late post injection strategy. This strategy will be adopted for Tier 4.

There is opportunity to achieve <130kW requirements without DPF. Further development is required to confirm production feasibility.

4.5.2 NOx aftertreatment – Tier 4 final
The most likely strategy for NOx aftertreatment will be selective catalytic reduction (SCR). This technology is already in production in Europe on highway, however the technology relies on urea as areductant to allow NOx conversions. This is injected onto the catalyst via a separate tank on the vehicle. The infrastructure to support this aqueous urea (Ad Blue) has been established on highway. For off highway the urea infrastructure is expected to be available in the 2014 timeframe. Liebherr have declared that they will utilise SCR for NOx aftertreatment for Tier 4 final. Ref (6), as yet no manufacturer has declared a SCR only solution for Tier 4 interim.

Tier 4 final emissions have been demonstrated with DPF and SCR systems, using a US07 engine as part of the Association of European Catalyst Consortium (AECC) research programme. Ref (7)

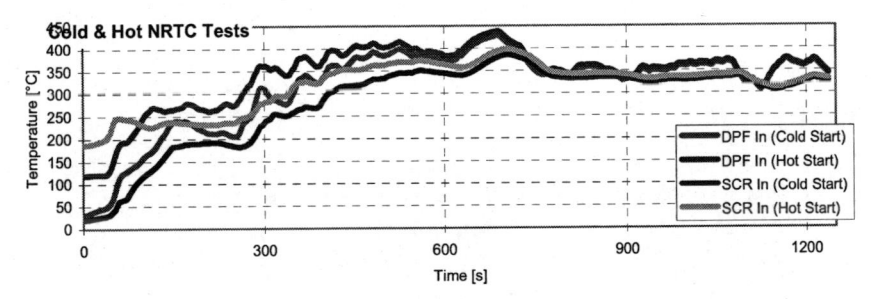

Fig 11 Aftertreatment temperatures across hot and cold NRTC

During the NRTC NOx conversion does not occurs when the SCR temperature is below ~200degC.

Since the NRTC test is conducted cold (ambient temperature soak) followed by hot, the SCR system on the NRTC cold cycle reaches 200degC in under 300seconds. This compares with a temperature of only 150 degC over the US FTP cycle in similar time. This time delay for NOx conversion to occur may result in the need for thermal enhancement of the system when running NRTC cold.

4.5.3 Lean NOx Traps (LNT)
This technology is still advantageous over SCR at lower engine ratings due to system cost. LNT do not rely on any urea infrastructure and operate independent of any driver input. LNT technology has been demonstrated as a solution for on highway 2010 emissions ref (8). The regeneration strategy for LNT can have a negative effect on the engine fuel consumption compared to SCR systems. Hence LNT's are likely to be adopted on low load cycle applications, where NOx conversion efficiency is less demanding and also fuel consumption is less of a concern to the operator.

5 SUMMARY

This paper highlights the required technologies required for Tier 4 interim and Tier 4 final legislation.

It is likely that Tier 4 interim technology solutions will be EGR for NOx control and DPF aftertreatment for Pm control. This is aligned with on highway US 07 solutions. There is opportunity to achieve <130kW requirements without DPF. Further development is required to confirm production feasibility. Tier 4 final will see the adoption of NOx aftertreatment (SCR and LNT), which will be aligned with on highway 2010 technical solutions.

JCB are actively pursuing a number of Tier 4 technologies, some of which were clearly demonstrated during the Dieselmax success, and will allow JCB to maintain their technical leadership within their market place.

APPENDICES

Appendix 1 Tier 4 legislation figures

Power Range	Compliance Dates		Emissions Limits [g/kW.h]			
[kW]	Level	Date	NOx	HC	CO	Pm
Tier 3						
37<P<75	Tier 3	2008	4.7		5.0	0.40
75<P<130	Tier 3	2007	4.0		5.0	0.30
130<P<560	Tier 3	2006	4.0		3.5	0.20
Tier 4						
P<8	Tier 4	2008	7.5		8.0	0.40
8<P<19	Tier 4	2008	7.5		6.6	0.40
19<P<37	Tier 4 int.	2008	7.5		5.5	0.30
	Tier 4 final	2013	4.7		5.5	0.03
37<P<56	Tier 4 int.	2008	4.7		5.0	0.30
	Tier 4 final	2013	4.7		5.5	0.03
56<P<130	50% of sales	2012-2013	as Tier 3		5.0	0.02
	50% of sales	2012-2013	0.4	0.19	5.0	0.02
	Alternative 1	2012-2013	2.3	0.19	5.0	0.02
	Alternative 2	2012-2013	3.4	0.19	5.0	0.02
	Tier 4B	2014	0.4	0.19	5.0	0.02
130<P<560	50% of sales	2011-2013	as Tier 3		3.5	0.02
	50% of sales	2011-2013	0.4	0.19	3.5	0.02
	Alternative	2011-2013	2.0	0.19	3.5	0.02
	Tier 4 final	2014	0.4	0.19	3.5	0.02

ACKNOWLEDGEMENTS

The author would like to thank JCB and Ricardo directors for being able to publish this paper.

REFERENCES

1. Powertrain Development for an High Speed Off-Road Vehicle - The JCB DIESELMAX Land Speed Record Car Alan Tolley J. C. Bamford Excavators David Hoyle JCB Transmissions Matt Beasley Ricardo UK Ltd ICPC 2007 - 4.3

2. George Lin, OBDII Technical Lead, Medium Core Engine Systems, Caterpillar Electronics, Caterpillar OBD "TopTec" Symposium 2006

3. Advanced Diesel Technology to Achieve Tier 2 Bin 5 Emissions Compliance in US Light-Duty Diesel Applications Brian Cooper, Ian Penny, Matt Beasley, Adrian Greaney and Jackie Crump Ricardo Consulting Engineers SAE 2006-01-1145

4. Investigation of Fuel Injection Strategies on a Low Emission Heavy-Duty Diesel Engine with High EGR Rates Andrew J Nicol and Chris Such, Ricardo UK Ltd Ulla Sarnbratt, Volvo Technology Corp, Sweden ImechE to be published December 2007

5. Potential of Exhaust Mounted Injector to Generate Conditions for Active Regeneration of DPF Systems In Heavy Duty Applications Imeche combustion conference 2004 M H Niven, C H Such, Ricardo Consulting Engineers Ltd, Shoreham-by-Sea, UK

6. Liebherr BAUMA 2007 publication PG 6

7. The Application of Emissions Control Technologies to a Low-Emissions Engine to Evaluate the Capabilities of Future Systems for European and World-Harmonised Regulations. Aachener Kolloquium Fahrzeug- und Motorentechnik 2007 Mr John May, Mr Dirk Bosteels AECC, Brussels, Belgium Mr Andrew Nicol, Mr Jon Andersson, Mr Chris Such Ricardo UK Ltd, Shoreham, UK

8. S Whitacre, B J Adelman, M P May, J G McManus, "Systems Approach to Meeting EPA 2010 Heavy-Duty Emission Standards Using a NOx Adsorber Catalyst and Diesel Particle Filter on a 15LEngine"SAE2004-01-0587

GASOLINE

Controlled Auto Ignition based on GDI – strategies derived from experiment and simulation

André Kulzer, Jean-Pierre Hathout, Christina Sauer, Ansgar Christ
Robert Bosch GmbH, CR/AEE, Germany

ABSTRACT

The controlled auto-ignition (CAI), also known as gasoline homogeneous charge compression ignition (HCCI), improves dramatically the efficiency of a gasoline engine without penalties in emissions. Although CAI produces negligible NO_x, and a simple three-way catalyst suffices, it depends strongly on judiciously operating the engine dynamically. Gasoline direct injection (GDI), variable valve actuation (VVA), exhaust gas recirculation (EGR), turbo-charging (TC) and advanced controls based on combustion state sensing are key factors to enable the dynamic operation of the CAI combustion process. Test bench investigation on a single-cylinder research engine were undertaken to investigate these actuation strategies, using thermodynamic analysis and thermo-kinetic simulation, in defined operating points to show the effects that influence the self-ignition and combustion process of CAI.

1 CONTROLLED AUTO-IGNITION (GASOLINE HCCI)

CAI allows for significant efficiency gains in combination with extremely low NO_x-emissions when compared to conventional spark ignited combustion concepts {1, 2}. Low carbon monoxide and hydrocarbon emissions are also expected from a lean CAI concept as shown in this last paper. The option of lean operation direct injection engine with a simple exhaust gas after-treatment system (three-way catalyst) makes CAI appealing.

Several challenges arise for stable and effective operation of an engine with CAI. First, the lack of spark ignition to phase CAI combustion events puts a heavy burden on cleverly controlling the thermodynamic and species concentrations state of the cylinder charge before compression. Second, the essential combustion mode switch to conventional spark ignition (SI) to cover high load demands should be performed without diluting the fuel efficiency gains of CAI. This could only be achieved by defining operating and mode switching strategies that provide optimal combustion rendition which cover the full operating engine load-speed range, rather than optimising an individual combustion mode, CAI, limited to a portion of the engine operation map.

The combustion phasing remains as the most important parameter to control to overcome the first caveat mentioned above. Phasing is strongly affected by the initial mixture state (composition, pressure, temperature), the compression rate and heat transfer rate. Several methods have been proposed to influence the combustion phasing for CAI including pre-heating the intake {3}, affecting the fuel composition or dilution {4}, varying the compression ratio {5}, and recycling of exhaust gas {6, 7}. The latter achieved with variable valve actuation, thus minimizing the pumping losses, has shown to be the most attractive to sustain CAI {8}. Using variable valve actuation and external exhaust gas recirculation {9} in combination with suitable direct injection events could influence both the initial fresh charge and temperature for each CAI cycle, and hence fix the combustion phasing. Additionally, supercharging can significantly increase the engine operation map region where CAI is manageable {10}. A strategy to preserve the benefits in fuel consumption gained in the limited CAI operating mode is pivotal in making this an attractive concept in multi-mode engine operation. Thus, our stance in optimising combustion for the full operating engine load-speed map.

In this paper, we analyse several control strategies for CAI using different actuators: flexible variable valve-train, direct fuel injection, external exhaust gas recirculation and supercharging. Test bench investigation using a single-cylinder research engine with gasoline direct injection (DI), variable valve actuation (VVA), exhaust gas recirculation (EGR) and supercharging were undertaken. The goal is to investigate these actuation strategies and their effects influencing self-ignition and combustion process of CAI, using thermodynamic analysis and thermo-kinetic simulation.

2 MODELLING CONTROLLED AUTO-IGNITION

In contrast to conventional spark ignited (SI) gasoline engines, the start of the ignition with CAI combustion cannot be induced by a spark plug. To trigger CAI combustion, compression of an air, fuel and residuals mixture has to reach a determined temperature and pressure that will drive the mixture into self ignition.

To understand thermodynamic and reaction kinetics phenomena ongoing during a CAI operating point, a thermo-kinetic model was used {11} and further developed {12}. Briefly, the model has following properties: the in-house coded thermodynamic model (TECSim) is coupled with Chemkin to account for reaction kinetics; the model accounts for gas exchange (0D), residuals recompression, compression stroke, combustion and expansion stroke, i.e., the complete 720°CA cycle calculation; the reaction kinetics modelling is active during all phases; single or split injection is modelled and evaporation enthalpy is considered; as a surrogate fuel iso-octane was used, which was modelled with a reduced reaction mechanism comprising 37 reactions and 26 species {11}.

The analysis presented in the subsequent chapters will be supported by the thermo-kinetic simulation to explain the effects influencing thermodynamic values and species concentrations, thus affecting self-ignition and combustion phasing.

3 ANALYSIS OF EVC EFFECT USING THE RESIDUALS TRAPPING STRATEGY

The experimental single-cylinder engine specifications are described in Table 1, where centrally mounted direct injection (DI) is used with a multi-hole injector and VVA with an electro-hydraulic valve-train system.

Table 1 - Single-cylinder research engine data

Stroke [mm]	85
Bore [mm]	82
Compression Ratio [-]	11.5
Displacement [cm^3]	449
Connecting Rod [mm]	145
Fuel	EuroSuper 95 RON
Injector	Bosch multi-hole DI injector
Valve-train	Bosch electro-hydraulic valve-train

When using the residuals trapping strategy, realized with a negative valve overlap, the exhaust valve closing timing (EVC) determines essentially the residual mass fraction. An early EVC induces a higher residual mass fraction, which leads to a lower air mass fraction, i.e., to a richer air/fuel mixture. At the end of the compression stroke, the elevated temperature from high compression ratio and high hot residuals mass fraction leads to an earlier auto-ignition and shorter burn duration, due to favourable conditions for the reaction kinetics.

To demonstrate the influence of such parameter, Figure 1 shows an experimental variation of exhaust valve closing timing between 73 to 92 CAD before gas exchange top center at 2000 rpm, constant injection and intake conditions. The influence on combustion phasing, i.e., 50% mass fraction burned or heat release (MFB$_{50\%}$), excess air ratio (λ), temperature at IVC (T$_{IVC}$) and net indicated mean effective pressure (NMEP) is shown. The load accounts for around 3 bar NMEP. The analysis of fuel efficiency and emissions with such variation was already analysed in {7} and the focus here goes into the understanding of the mechanisms behind the EVC control strategy. A comparison of experimental and simulation data was performed: using the boundary conditions (for example EVC, start of injection (SOI), fuel mass, thermodynamic conditions) used in the experiment, simulation based on the thermo-kinetic model described in the previous chapter was performed. Load NMEP and excess air ratio λ show a valid trend and good agreement regarding absolute values. The combustion phasing MFB50% also shows the right trend, but with a small offset. The deviations observed in the model occur among other reasons due to the simple gas exchange modelling (0D) and high temperature sensitivity of reaction kinetics. Nevertheless, the trends can be successfully predicted, and also the gas mixture temperature at IVC shows the right trend is in good agreement with the experimental data.

Summarizing, the EVC strategy shows a strong temperature influence on the thermodynamic process, thus giving a combustion phasing control knob: the earlier the closing time of EVC, the more hot residuals are kept in the combustion chamber, leading to a higher mixture temperature at the beginning of the compression stroke, thus resulting in a reduced ignition delay time and earlier combustion phasing, and vice-versa. More detailed information about the combustion process thermodynamic split of losses analysis can be found at {2}.

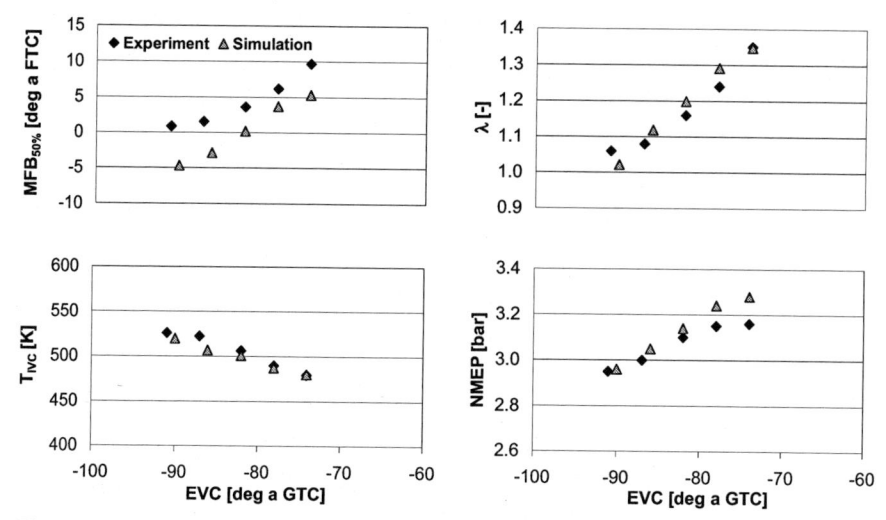

Figure 1 – Comparison of EVC variation between experiment and simulation at 2000 rpm and 3 bar NMEP

4 ANALYSIS OF SOI EFFECT USING THE RESIDUALS TRAPPING STRATEGY

A variation in injection timing has influence on auto-ignition and combustion duration. This is tightly coupled to the temperature profile which in turn is affected by evaporation enthalpy, change in thermodynamic gas mixture properties, reaction kinetics and slight heat release especially at very early injection timings (before gas exchange top center). An early injection timing in the compressed residual gas results in earlier auto-ignition and faster combustion. Injection timing is one of the easiest ways to control CAI. This is due to the fast combustion response to injection; resulting in combustion in-cycle control. Also, injection allows a superior cylinder individual control. Hence, the various control strategies enabled by GDI allow for the realization of the CAI combustion concept even with camshaft based variable valve-trains which are limited in flexibility compared to fully variable concepts (e.g., EHVS {13}).

An investigation regarding start of injection (SOI) variation was undertaken at a speed of 2000 rpm and a load of approximately 3 bar NMEP, using constant injected fuel mass and constant valve-train timings (Figure 2, where m_a represents air mass and $MFB_{5\%}$ 5% heat released). The data can be confirmed by the thermo-kinetic simulation results. An early injection during the residuals recompression phase lead to an earlier combustion. A variation of injection timing during the air intake phase has nearly no effect on combustion phasing, because no significant differences in temperature occur; there will be only a slight mixture preparation difference. Early injection during the residuals recompression phase lead to higher temperatures at intake valve closing timing (IVC), i.e., lower air mass and richer mixture. This results in reduced auto-ignition delay.

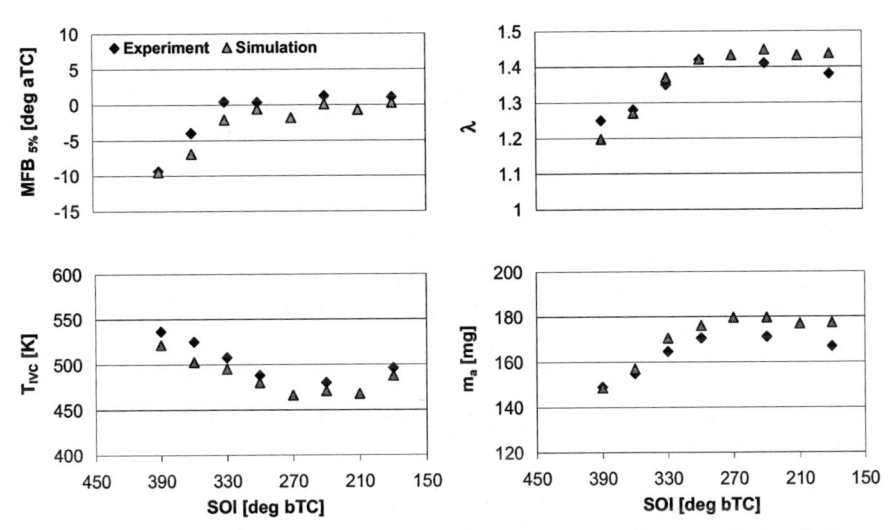

Figure 2 – Comparison of SOI variation between experiment and simulation at 2000 rpm and 3 bar NMEP

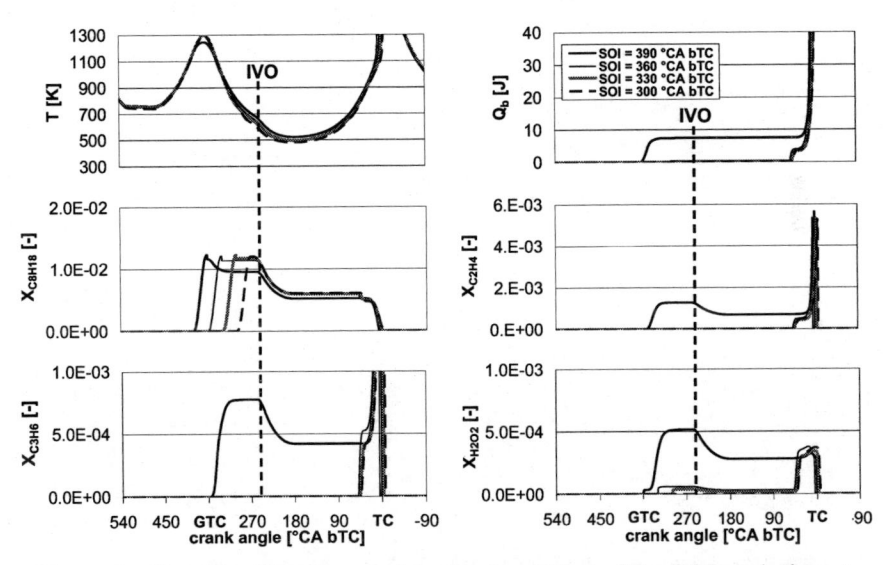

Figure 3 – Results of the thermo-kinetic simulation of the SOI variation at 2000 rpm and 3 bar NMEP

Figure 3 shows the simulation of temperature (T), heat release (Q_b), iso-octane as surrogate fuel, propene, ethylene and hydrogen peroxyde mole fractions profiles for the SOI variation between 390°CA before ignition top centre (ITC) already shown in

Figure 2. The latter shows the injection timing effect on the temperature profile. One of the reasons for the temperature increase lies on the reduction of wall heat capacities ratio, i.e., polytropic exponent. The second is the heat release observed at very early injection event before gas exchange top centre (GTC) coupled with available oxygen in the trapped residuals mass, due to the lean operation. For example, when injection is placed after the GTC. Temperature reduction occurs due to evaporation enthalpy. The fuel also reduces the polytropic exponent leading to a lower expansion level. The earlier the SOI after GTC occurs, the longer a lower polytropic exponent dominates the residual trapping expansion phase, leading to a higher temperature of the residuals-fuel mixture at the intake valve opening timing (IVO). When injection occurs before GTC, the polytropic exponent is reduced already during the compression phase, leading to a lower IVO temperature than for a GTC injection. On the other hand, the injected fuel before GTC starts to react with the available oxygen in the residuals, due to dwell time at temperatures significantly over 1000 K. This leads to the production of smaller hydrocarbons and other species and to a small but sensible heat release, which leads after the expansion phase to higher IVO temperatures than the GTC injection. The earlier the injection takes place, the more time there is for reaction kinetics, assuming the temperature is high enough (> 1000 K) and assuming there is enough oxygen in the residuals.

The effect of available oxygen in the trapped residuals considering injection timing is presented in Figure 4, where combustion phasing (MFB50%), maximum pressure gradient (dp/dφ), burn duration (BD) and excess air ratio (λ) are shown. An experimental variation of SOI at three different residual fractions at 2000 rpm and a load of 2 bar NMEP is shown. A higher residuals fraction means at the same time a lower air mass fraction, thus, lower oxygen fraction in exhaust residuals for early reaction kinetics in the residuals recompression phase.

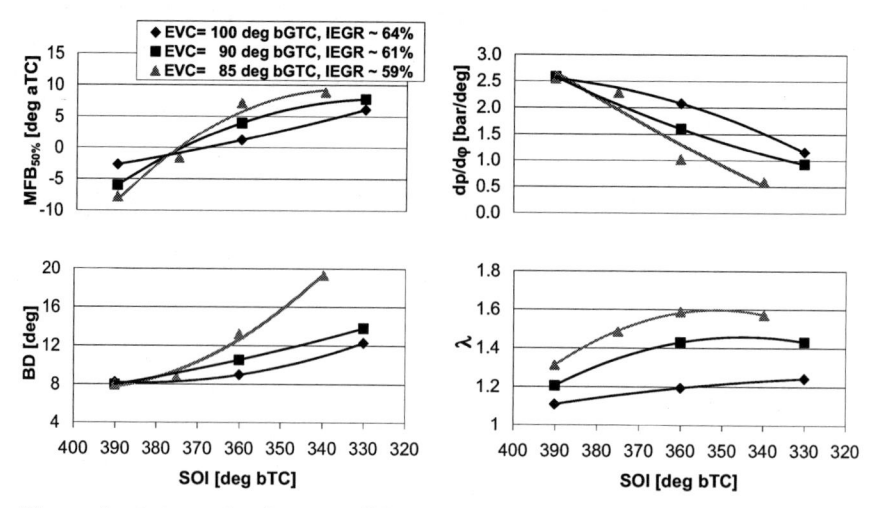

Figure 4 – Interaction between SOI and EVC at 2000 rpm and 2 bar NMEP

If the injection occurs after GTC, the main combustion control effect is coupled with the residuals fraction. The higher the residuals amount through earlier EVC, the earlier the combustion phasing. Now, if the injection occurs before the GTC, the combustion phasing control effect is reversed. For example for SOI at 390°CA before ITC, the earliest combustion phasing is achieved with the lowest residuals fraction. This means that the main knob here is the injection timing combined with the higher oxygen concentration in the residuals, leading to reaction kinetics driven heat release in the GTC. Using early direct injection enables you not only to have a control knob for the combustion phasing, but it allows you additionally to widen the CAI operation map towards lower loads. At these loads the low exhaust temperature would limit auto-ignition. Using this early direct injection or even split injection strategies – one very early pilot injection in the residuals recompression phase and a later second injection during expansion or intake phase – higher temperatures at start of compression are induced, mainly due to heat release in the residuals trapping phase, enabling driving CAI even for very low loads and speeds {12}.

5 ANALYSIS OF EXTERNAL EGR AND CHARGING EFFECT

Adding to the fact that the main knobs for CAI control are SOI and EVC, EGR and supercharging or turbocharging can support for additional NO_x free load range. To analyse the external EGR effect on the controlled auto-ignition, an operating point at 2000 rpm and a load of about 4.6 bar NMEP in CAI mode without external EGR was used. Holding constant fuel injection mass of 12.1 mg, a variation in external EGR was undertaken. Combustion phasing was also held fairly constant mainly using start of injection timing (SOI). Additionally, when SOI stopped having any effect on combustion phasing, exhaust valve closing time (EVC) had to be adapted towards higher internal EGR rates (Figure 5, where m_f represents the fuel mass).

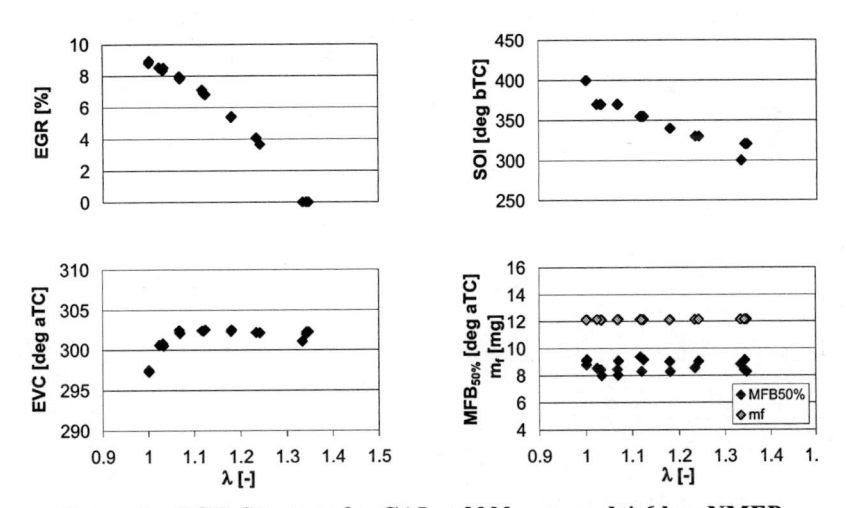

Figure 5 – EGR-Strategy for CAI at 2000 rpm and 4.6 bar NMEP

The trend of having to compensate the increase of external EGR with earlier SOI and earlier EVC can be explained through chemical kinetics. Adding external EGR reduces the air mass, i.e., the inert gas in the combustion chamber increases and the oxygen decreases, resulting also in a lower specific heat ratio for the compression phase. These facts (higher inert gas concentration, lower oxygen concentration and lower specific heat ratios) lead to slower reaction kinetics in the combustion chamber despite of slightly higher mixture temperature at IVC due to exchange of air with external EGR. It can be observed that driving CAI with external EGR causes a reduction in fuel efficiency ISFC of about 6%. I.e., NMEP shows a similar reduction (Figure 7) due to constant injected fuel mass. During the external EGR variation, HC-emissions stay nearly constant. Otherwise, CO-emissions have a significant increase when running $\lambda = 1$ surely due to slight inhomogeneities in mixture quality. NO_x-emissions show a reduction at $\lambda = 1$ probably due to the lack of excess oxygen for NO_x-formation. When examining maximum pressure gradient one can find no relevant changes. Yet, the combustion stability, i.e., the standard deviation of NMEP (σ_{NMEP}), shows a significant improvement, from 0.10 to 0.06 bar, when adding external EGR, driving $\lambda = 1$.

To analyse the supercharging effect on CAI, an operating point at 2000 rpm and a load of about 5.1 bar NMEP in CAI mode without external EGR and no supercharging was used. Holding constant fuel injection mass of 13.4 mg, supercharged air was gradually added. EVC was changed to keep enough internal EGR for the auto-ignition in the combustion chamber. Additionally, SOI was corrected to keep constant combustion phasing (Figure 6, where p_2/p_1 represents the charging level).

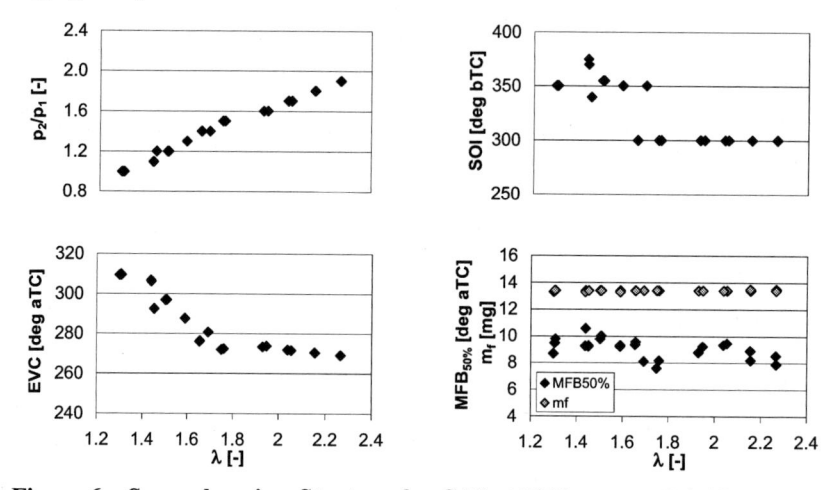

Figure 6 – Supercharging-Strategy for CAI at 2000 rpm and 4.6 bar NMEP

It can be observed that driving CAI with supercharging causes an increase of fuel efficiency of about 9% (Figure 7, where ISFC represents the indicated specific fuel consumption). Note that, in this case, the energy needed for charging is not taken into account. The higher charging shows a remarkable increase of CO-emissions which derive from colder combustion temperatures, as can be observed in the NO_x-emissions, which are reduced to a near zero level, even at this high CAI load operation. At a certain level, the increasing CO-emissions limit the fuel efficiency improvement. The maximum pressure gradient also shows significant reduction, from 4.5 bar/°CA to 2 bar/°CA, with supercharging the CAI combustion concept.

The combustion stability shows a trend towards slightly worse values, from 0.09 to 0.17 bar, when supercharging CAI extremely. This correlates with the CO-emissions and lower combustion temperatures.

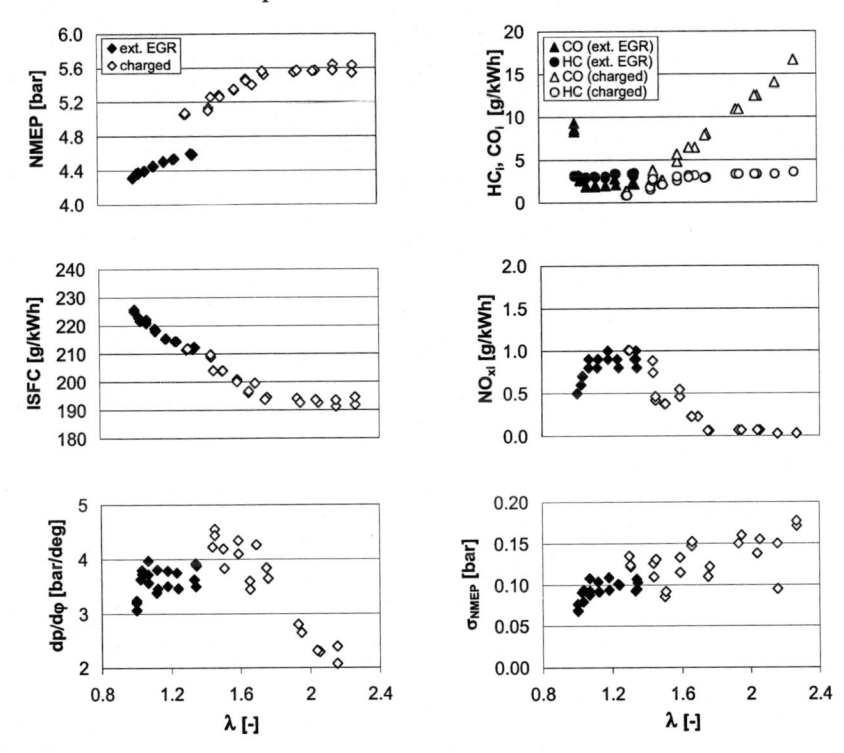

Figure 7 – Results using EGR- or Supercharging-Strategy for CAI at 2000 rpm and 4.6 bar NMEP over λ

Analysing both external EGR and supercharging over the air/fuel-ratio λ, a jump in load at λ = 1.3 is recognizable – this comes from the different fuel mass used in each case. Nevertheless a comparison is possible. The increase in external EGR, as already mentioned, reduces efficiency with reducing the mixture quality λ. On the other side, increasing air charging increases λ and efficiency. In both cases and in a certain range, the efficiency increase correlates with λ. The advantages of supercharging in comparison to the external EGR lie on: higher λ and therefore higher efficiency; higher mass in the process, reducing even more the combustion temperature, which can be read from the decrease in NO_x; decrease in maximum pressure gradient. The disadvantages of supercharging are the higher CO-emissions and the slight lower combustion stability.

In Figure 8, a comparison of CAI without external EGR (λ > 1), with external EGR (λ = 1) and supercharging (λ > 1) over a certain load range of 3.2 to 5.5 bar NMEP is shown. One recognizes again the fuel efficiency penalty of using external EGR – that is the price to pay for NO_x conversion in the three way catalyst. Combustion phasing was held fairly constant and combustion stability was of the same order.

Interesting is to see that the pressure gradient almost linearly increases with load and only the charged CAI operation shows slight benefit in comparison to the non-charged operation. On the emission side, external EGR and charging show worse results regarding CO at lower loads.

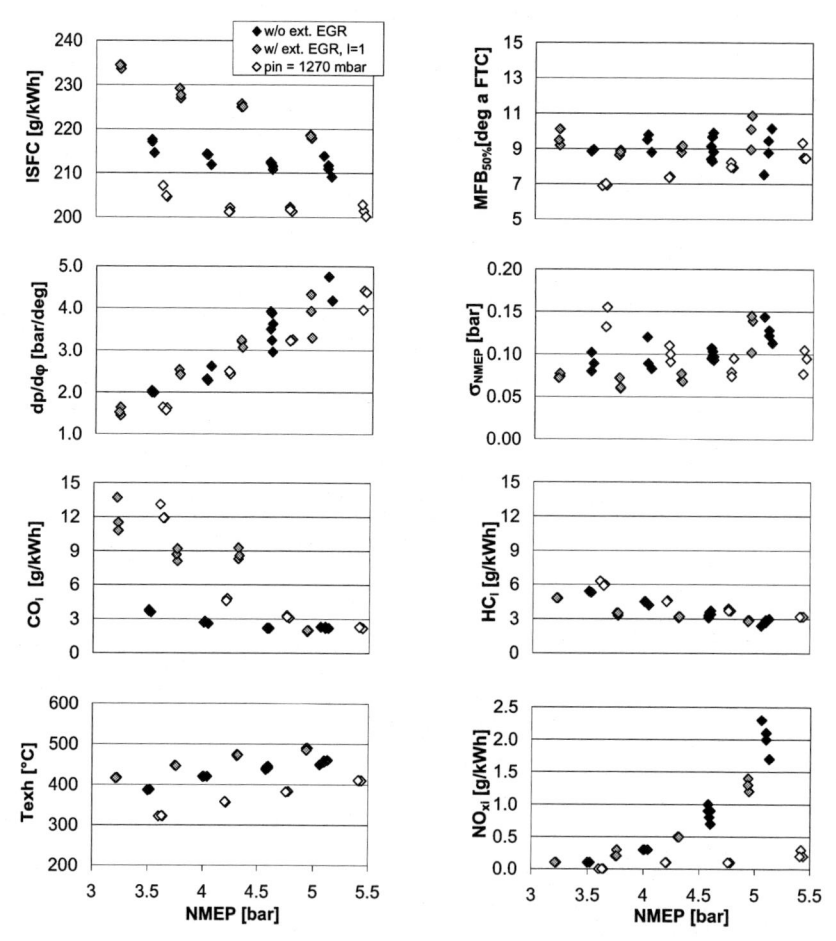

Figure 8 – Results using EGR- or Supercharging-Strategy over NMEP for CAI at 2000 rpm over NMEP

At loads over 4.8 bar NMEP results are similar to naturally aspirated without external EGR (operation with $\lambda > 1$). This effect could be explained by the lower combustion temperature at the lower loads. HC emissions are rather similar. In the exhaust gas temperature (T_{exh}), the charged operation is clearly recognized by the lower temperature induced by the higher air mass taking part of the thermodynamic process. The NO_x values between CAI with and without external EGR are in the same way exponentially dependent of the load. Only through the higher mass in the thermodynamic process using supercharging, the NO_x value is maintained at a very low level. Higher charging levels make NO_x negligible. If this strategy of higher boosting costs too much fuel efficiency, then a λ = 1 strategy with lower charging level has to be used for these high CAI loads.

In Figure 9, a comparison of three CAI operating points at 2000 rpm and 5.1 bar NMEP is shown. The pressure profile puts supercharging in evidence. Combustion heat release of CAI operation with EGR shows a slightly faster profile and higher temperature than the one with supercharging. The temperature profile of the supercharged CAI is lower due to higher process mass. The thermodynamic split of losses {2} emphasize these effects. The higher temperature profile and the EGR, when running CAI with EGR and λ = 1, are responsible for the worse thermodynamic properties (real charge, real gas properties) and the higher wall heat loss effects. The corrected injection timing for the constant desired combustion phasing shows some disadvantages in the pumping losses. When looking at the supercharged CAI operating point, advantages in thermodynamic properties of the mixture are recognized due to the lower temperatures (higher mass in process) and the higher λ value (excess air). Wall heat losses are not lower because of the higher pressure profile, which is favourable regarding convection. Injection timing correction for the constant desired combustion phasing shows also here a slight disadvantage in the pumping losses.

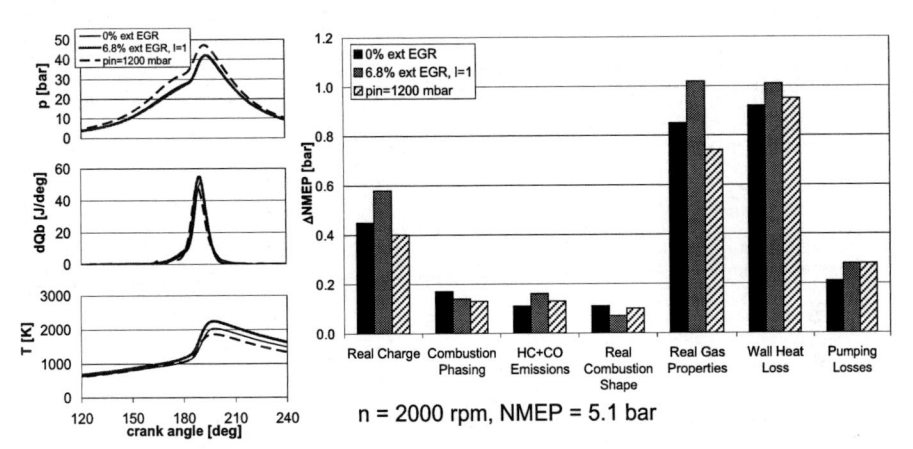

Figure 9 – Results using EGR- or Supercharging-Strategy for CAI at 2000 rpm and 5.1 bar NMEP

CONCLUSIONS

Controlled Auto Ignition is a very promising combustion concept combining at the same time, high fuel efficiency with low raw emissions, especially NO_x. Thus, allowing for a state of the art three-way catalysis. Single-cylinder research engine investigations were undertaken using thermo-dynamic analysis and thermo-kinetic simulation to clarify the effects behind self-ignition and combustion process of CAI. Gasoline direct injection (GDI) and variable valve actuation (VVA) will be the key players regarding ignition control of CAI, due to its direct influence on mixture temperature, which determines auto ignition timing. Additionally, exhaust gas recirculation (EGR) and turbo-charging (TC) allow for widening CAI operating range whilst keeping after-catalyst NO_x low, either by running EGR with $\lambda = 1$ or by charging with significantly lean λ for high load CAI with near zero raw NO_x. Next steps are the combination of these operation strategies with combustion state sensing based controls to enable dynamic operation of the CAI combustion concept.

REFERENCES

{1} Thring, R.: "Homogeneous-charge Compression Ignition (HCCI) Engines", SAE Paper 892068, 1989
{2} Kulzer, A.; Christ, A.; Rauscher, M.; Sauer, C.; Würfel, G; Blank, T.: "Thermodynamic Analysis and Benchmark of Various Gasoline Combustion Concepts", SAE Paper 2006-01-0231
{3} Yang, J.; Culp, T.; Kenney, T.: "Development of a Gasoline Engine System Using HCCI Technology - The Concept and the Test Results", SAE Paper 2002-01-2832
{4} Olsson, J.-O.; Tunestål, P.; Ulfvik, J.; Johansson, B.: "The Effect of Cooled EGR on Emissions and Performance of a Turbocharged HCCI Engine", SAE Paper 2003-01-0743, 2003
{5} Haraldsson, G.; Tunestål, P.; Johansson B., Hyvönen, J.: "HCCI Combustion Phasing in a Multi-Cylinder Engine using Variable Compression Ratio", SAE Paper 2002-01-2858
{6} Wolters, P.; Salber, W.; Geiger, J.; Duesmann, M.; Dilthey, J.: "Controlled Autoignition Combustion Process With An Electromechanical Valve Train", SAE Paper 2003-01-0032
{7} Kulzer, A.; Rauscher, M.; Sauer, C.; Orlandini, I.; Weberbauer, F.: "Methods for Analysis of SI-HCCI Combustion with Variable Valve Actuation", 6th International Stuttgart Symposium, 2005
{8} Kulzer, A.; Hathout, J.-P.; Sauer, C.; Karrelmeyer, R.; Fischer, W.; Christ, A.: "Multi-Mode Combustion Strategies with CAI for a GDI Engine", SAE Paper 2007-01-0214
{9} Cairns, A.; Blaxill, H.: "The Effects of Two-Stage Cam Profile Switching and External EGR on SI-CAI Combustion Transitions", SAE Paper 2007-01-0187
{10} Christensen, M.; Johansson, B.; Amnéus, P.; Mauss, F.: "Supercharged Homogeneous Charge Compression Ignition", SAE Paper 980787, 1998
{11} Orlandini, I.; Kulzer, A.; Weberbauer, F.; Rauscher, M.: "Simulation of self ignition in HCCI and partial HCCI engines using a reduced order model", SAE Paper 2005-01-0159
{12} Sauer, C., Kulzer, A.; Rauscher, M.; Hathout, J.-P., Bargende, M.: "Strategies for a CAI gasoline engine derived from experiments and simulation", 7th International Stuttgart Symposium, 2007
{13} Mischker, K.; Denger, D.: "Anforderungen an einem vollvariablen Ventiltrieb und Realisierung durch die elektrohydraulische Ventilsteuerung EHVS", 24th International Vienna Engine Symposium 2003

A comparison study of different NVO strategies in a diesel HCCI engine using 3D CFD

Ming Jia, Zhijun Peng
Department of Engineering and Design, University of Sussex, UK

ABSTRACT

A three-dimensional modeling was carried out to investigate effects of negative valve overlap (NVO) on gas exchange and fuel-air mixing processes in a diesel homogeneous charge compression ignition (HCCI) engine with early fuel injection. Four cases with different EVO and IVC were calculated to compare the effects of the begin time and the end time of the gas exchange process on HCCI engines with NVO firstly. And then three NVO strategies with different maximum valve lift and valve timing were compared to make it clear that the effects of valve lift and valve timing on retaining residual and control of the in-cylinder temperature, and their further influences on air motion and mixture preparation process.

1   INTRODUCTION

Because of the potential to lower NOx and particulate emissions while improving fuel economy, homogeneous charge compression ignition engine (HCCI) is being widely investigated around the world. Unlike combustion in spark-ignition or diesel engines, the ignition and combustion of HCCI engines are mainly controlled by chemical kinetics, so controlling the combustion phase and rate becomes more difficult.

Various control strategies, such as variable compression ratio (1), variable exhaust gas percentage (2, 3) and variable valve timing/actuation (VVT/VVA), have been investigated by many researchers. Adjusting the residual gas fraction by VVT/VVA is considered to be the most promising and feasible way for achieving the HCCI combustion control (4).

For diesel HCCI engines, the required compression ratio is typically too low for satisfactory performance in cold start and low load situations, where the temperature at the end of compression is critical and the preparation of homogeneous charge is very hard. One possible way is changing the compression ratio by using IVC. Shi et

© *University of Sussex 2007*

al. (5) studied the diesel HCCI combustion by directly injecting the diesel fuel into the in-cylinder residual gas during NVO interval, and found that a large amount of residual with a high temperature benefited fuel vaporization and mixing. It was concluded that increasing the NVO benefited the combustion stability and fuel vaporization at low loads.

However, the influence of VVT/VVA on the in-cylinder air motion and the interaction of air motion with the fuel spray are not fully understood yet. It is useful to study VVT/VVA and its effects on the behavior of gases during the induction and exhaust strokes, which will help to understand gas dynamics effects on the fuel-air mixing, the engine volumetric efficiency, emissions and performance (6). In this study, a three-dimensional modeling was employed to simulate the gas exchange process and compression process up to top dead centre (TDC) in a diesel HCCI engine with NVO and early in-cylinder injection. Several NVO strategies were compared and discussed in more detail.

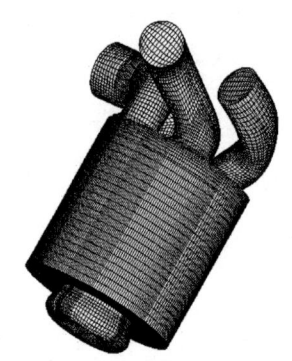

Fig. 1 Computational grid at BDC

2 NUMERICAL MODEL

The computational fluid dynamics (CFD) code is KIVA-3V (7). The turbulent flows within the combustion chamber are modeled using the RNG turbulence model, modified for variable-density engine flows (8). A moving-grid method is used to trace piston and valves movement. The computational domain includes intake ports and exhaust ports, the cylinder and the piston bowl, as shown in Fig. 1.

Table 1. Engine specifications

Displacement (single-cylinder)	0.5 L
Bore	86.0 mm
Stroke	86.0 mm
Connection Rod Length	160.0 mm
Squish Height	1.81 mm
Geometry Compression Ratio	14.3:1
Swirl Ratio	1.4
Speed	2000 rpm
Combustion Chamber	In-piston Mexican Hat
Fuel	Diesel
Wall Temperature	400 K
Firedeck Temperature	400 K
Piston Temperature	400 K

The computations were conducted from BDC at the end of expansion stroke (-180 °CA relative to intake TDC) to TDC at the end of compression stroke (360 °CA). Initial thermodynamic and turbulence quantities were assumed to be uniform in the ports and the cylinder. In order to have a reference for comparisons, the same initial and boundary conditions were used for all the runs. Constant pressure boundary conditions were assigned at inlets and outlets, so the dynamic effects were neglected.

3 TEST ENGINE

The test engine simulated in this paper is derived from a high-speed direct-injection (HSDI) diesel engine with a displacement of 0.5 liters/cylinder. A major modification to the original engine is decreasing the compression ratio from 18.2 to 14.3 by increasing the squish height in order to suit for HCCI combustion. The engine specifications are listed in Table 1 The engine is designed with two direct intake ports whose inlet is tangential to the wall of the cylinder.

The injector was located in the centre of the cylinder head. According to the research results from Lechner et al. (9) and Kim et al. (10), a narrow cone angle nozzle with spray cone angles of 60° in conjunction with early injection timing of 300 °CA and injection duration of 30 °CA was used in this study to reduce fuel deposition on the cylinder wall.

4 RESULTS AND DISCUSSIONS

4.1 The influence of EVO and IVC

In this section, four cases with different EVO and IVC were calculated to compare the effects of the begin time and the end time of the gas exchange process on HCCI engines with NVO. NVO was realized by varying early EVC coupled to a symmetrically varying late IVO, and reducing the maximum valve lift. Four cases were examined, for which the EVO were -240, -210, -180 and -150 °CA while the maximum valve lift and valve duration were kept constant. At the same time, IVO was delayed respectively in order to prevent excessive back-flow of trapped residuals into the intake manifold for the early EVC. The valve profiles can be found in Fig. 2. These cases are referred as EVO=-240°, EVO=-210°, EVO=-180° and EVO=-150° in the following sections.

The graphs depicted in Fig. 3-5 show the in-cylinder mass, temperature and pressure with these four EVOs, respectively, and provide an understanding of the processes occurring during the gas exchange and compression strokes. In order to fully understand those results, the figures are discussed jointly.

Because the exhaust valve durations are same for all cases, more early EVO results in less exhaust gas being expelled from the cylinder, which can be seen from Fig. 3. As EVO is delayed to -150 °CA, much gas is pushed out of the cylinder with high velocity due to the large pressure difference between in-cylinder and exhaust port. It can be seen from Fig. 3 that more hot residual gas is trapped in the cylinder for earlier EVO cases after EVC. Because the piston still moves up, the residual is compressed until TDC, which results in significant increments of temperature and pressure for all cases as shown in Fig. 4 and Fig. 5. As the piston moves down and intake valve opened, the combustion product in the chamber is pushed into intake port because the pressure in the chamber is still higher than that in intake ports for EVO= -210°, EVO= -180° and EVO=-150° as shown in Fig. 4. As a result, the in-cylinder mass decreases for these three cases, shown in Fig. 3. After this period, it comes into the true air induction process, and the fresh air is drawn into the combustion chamber. It is interesting to note that some of the in-cylinder gas flows back into the intake port in case of EVO=-240° at the end of intake stroke. It is mainly due to the low inflow velocity and unnoticeable inertia of the gas when much residual with high temperature are retained in the cylinder for EVO=-240°. From Fig. 5, it can be seen that the temperatures at 180 °CA (at IVC) are 366 K, 375 K, 370 K and 351 K for EVO=-240°, EVO=-210°, EVO=-180° and EVO=-150° respectively. Although there is much hot residual exists in the cylinder for EVO=-240°, the temperature at IVC is still very low due to late IVC and high heat loss from the wall during the time between EVC and IVO.

Fig. 2 Valve profiles used for simulation

Fig. 3 In-cylinder mass as function of crank angle

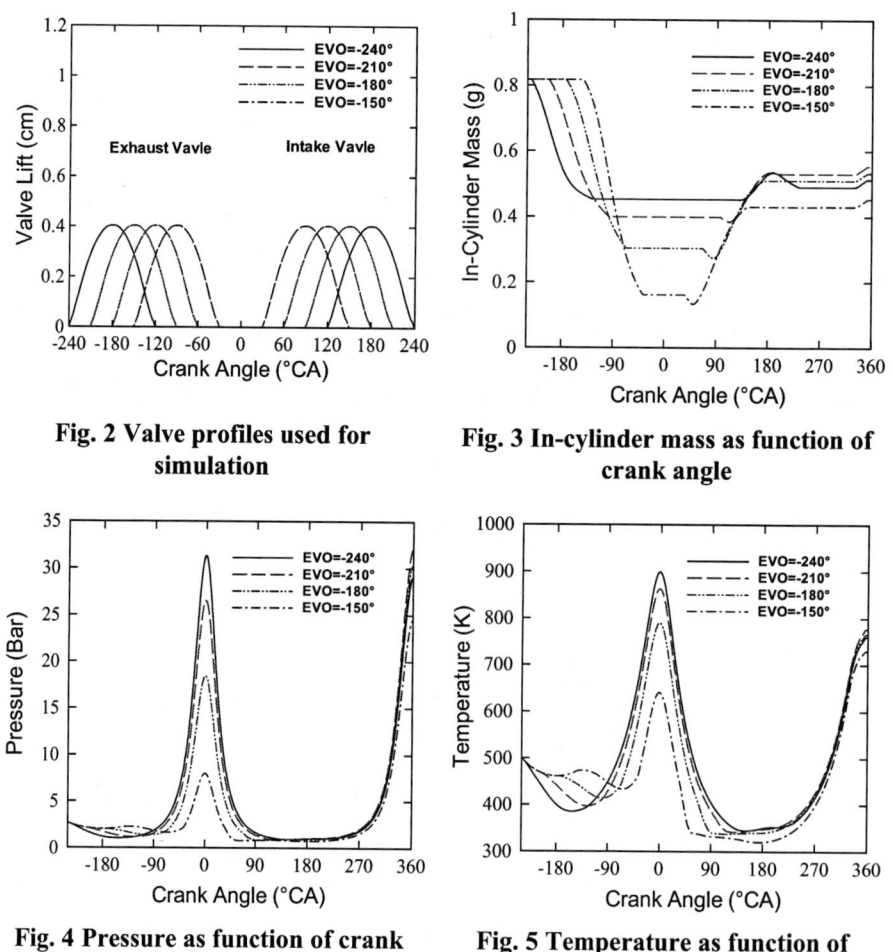

Fig. 4 Pressure as function of crank angle

Fig. 5 Temperature as function of crank angle

The computed swirl ratios for these four cases are shown in Fig. 6. The definition of swirl ratio, tumble ratio x and tumble ratio y can be found in (11). From Fig. 6 it can be seen that the swirl ratio decreases significantly during the exhaust stroke, and increases rapidly when the exhaust valve close until TDC due to the effect of compression. During the intake stoke, the swirl ratio drops for late IVC, and it is enhanced by the induction of inflow as soon as the intake valve opens. When it reaches the end of the intake stroke, the discharge velocity from the intake port to the cylinder decreases and swirl drops slowly for all cases. The reducing trend continues in the first part of compression stroke due to friction at the wall. However, approaching TDC, swirl is enhanced as the flow characters should preserve its angular momentum. Therefore, the last swirl ratio at the end of compression stroke is decided by both the effect of compression during the late exhaust stroke and the induction during the intake stroke for different NVOs. It can be found from Fig. 6 that, at the time of SOI, the case with earlier IVO leads to higher swirl ratio at SOI due to higher intake flow velocity. Hence, the induction process plays a more dominant role in the last swirl ratio.

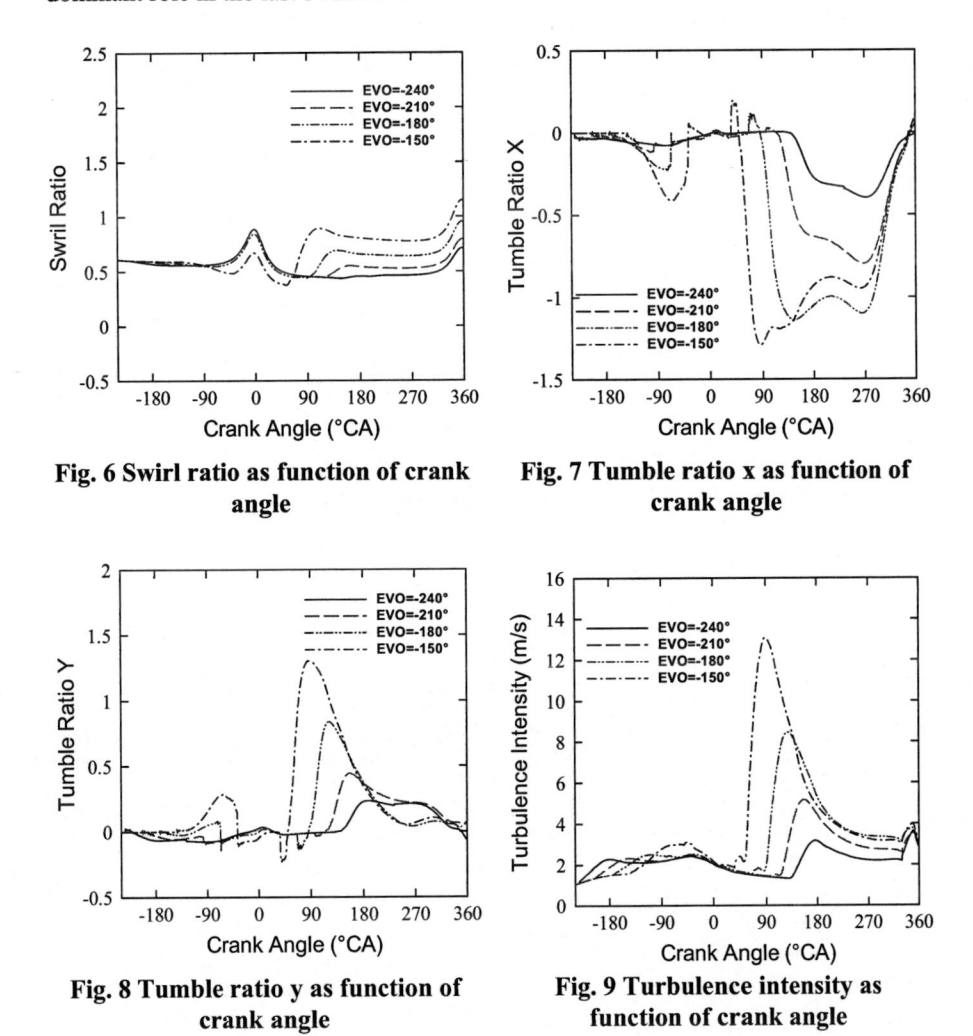

Fig. 6 Swirl ratio as function of crank angle

Fig. 7 Tumble ratio x as function of crank angle

Fig. 8 Tumble ratio y as function of crank angle

Fig. 9 Turbulence intensity as function of crank angle

The comparisons of the tumble ratio x and tumble ratio y are shown in Fig. 7 and Fig. 8. X and y are referred to the directions of x and y axes shown in Fig 2. During the exhaust stroke, the tumble ratio is generated by the exhaust flow. Then it decreases monotonically during the end exhaust stroke. At 0 °CA, the largest tumble ratio is established for the latest EVO case. Then the tumble ratio is well established in the whole cylinder, even in the combustion chamber during the intake stroke for all of EVOs. Especially for EVO=-150° and EVO=-180°, strong tumble motion is generated by the air jet flow through the valve curtain areas at the early induction stage. During the following compression stroke, as the piston moves from BDC to TDC, the well-developed tumble is broken down for all the four cases.

Fig. 9 illustrates the variations of turbulence intensity for four EVOs. It can be observed that the developments of turbulence intensity during the exhaust are very similar for all cases. In contrast to EVO=-240°, the latter IVO generates stronger flow fluctuation during the intake stroke due to the higher inflow velocity. During the compression stroke, the in-cylinder flow shows very little structure and was largely piston driven upward flow without strong active generation mechanism for turbulence kinetics energy, therefore, turbulence intensity decreases rapidly. The fuel injection into the cylinder at 330 °CA causes an increase in turbulence intensity at last. It should be noted that the flow fluctuations predicted by EVO=-180° and EVO=-150° increase 42% and 37% over which predicted by NVO=0° during the compression stroke. The enhanced fluctuations are considered to assist the mixing process, helping to form the homogeneous charge that is especially desirable for very early direct injection engines.

The fuel distributions in the cylinder after injection are shown in Fig. 10. The volume fractions within five different equivalence ratio ranges are used to evaluate the homogeneity of the fuel-air mixture. Because of high in-cylinder temperature and long injection duration, the fuel is almost completely vaporized at the end of injection timing for all four cases. It should be noted that the global equivalence ratio are not same due to the different amounts of residual gas being retained in the cylinder for the cases, although the same fuel is injected. Fig. 10 indicates that less rich fuel regions (equivalence rations is greater than 1.5) exist in cases of EVO=-180° and EVO=-150°, and it is mainly due to the higher turbulence intensity and higher in-cylinder velocity which benefit the mixing processes. However, the lean mixture regions with the equivalence ratio less than 0.0001 decrease for EVO=-240° and EVO=-150° because of the lower in-cylinder pressure which leads to higher spray penetration. It is also seen that the EVO=-180° and EVO=-150° have the higher volume fraction of cells in the equivalence ratio range of 0.5 to 1.5 which is more suitable for burn.

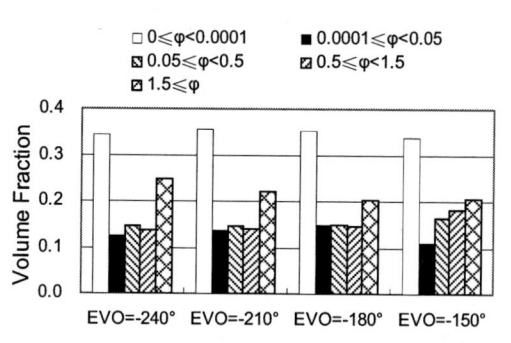

Fig. 10 In-cylinder volume distributions within four different equivalence ratio ranges at 350°CA

As far as power output is concerned, pumping loss are increased for EVO=-240° and EVO=-210° because high heat transfer from the in-cylinder mixture to the wall in the recompression-expansion process during the NVO period. And the EVO before BDC also leads to decrease of power output for these two cases. For EVO=-150°, the late EVO after BDC and early IVC before BDC result in increase of pumping work.

Therefore, from the considerations of reduction of pumping losses, formation of homogeneous mixture, increase of the turbulence intensity, and efficient control of residual fraction and in-cylinder temperature after IVC, the EVO and IVC are kept at -180°CA and 180°CA respectively in the following study.

4.2 The comparison of three different NVO strategies

In order to trap various quantities of residual gas and control the ignition point of HCCI combustion, several NVO systems have well documented in the literatures. Three different NVO strategies were compared in this section, and the details of the variable valve profiles can be found in Fig. 11. The base case is chosen with maximum valve lift of 0.6 mm, 0 °CA EVC and 0 °CA IVO. The first strategy is realized by reducing the maximum valve lift from 0.6 mm to 0.3 mm while keeping EVC and IVO, and the corresponding cases are referred as VMAX=0.5, VMAX=0.4, and VMAX=0.3 respectively. In the second strategy, the maximum valve lift is fixed at 0.6 mm, EVC timing is varied to retain moderate residual in the cylinder. According to EVC, the three case in Fig. 11(b) are named as EVC=-30°, EVC=-60°, and EVC=-90°. Three cases of NVO=60°, NVO=120° and NVO=180° are designed by varying maximum valve lift, EVC and IVO simultaneously in the last strategy (see Fig. 11(c)). In order to prevent the outflow of the residual into the intake port during the intake process, IVO is symmetrically varied with EVC in all cases. The aim of comparison of these ten cases here is to make it clear that the effects of

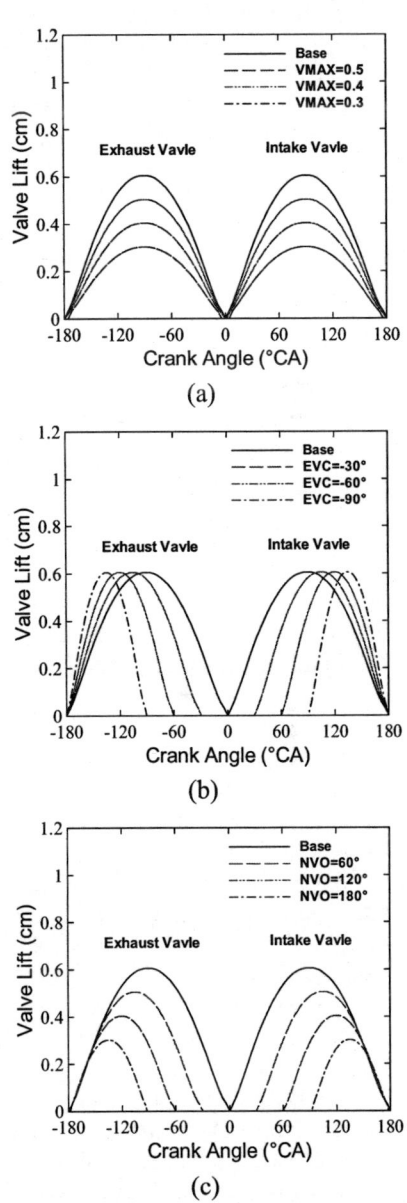

(a)

(b)

(c)

Fig. 11 Valve profiles used for three different NVO strategies

valve lift and valve timing on retaining residual and control of the in-cylinder temperature, and their further influences on air motion and mixture preparation process.

The mass of in-cylinder fresh air after IVC versus residual mass is shown in Fig. 12 for all the cases. It can be found that the VMAX strategy is not an efficient way to retain the residual gas in the cylinder only by decreasing the maximum exhaust valve lift for the cases with EVC close to BDC. However, in the VMAX strategy, the fraction of air in the cylinder after IVC is more sensitive to the valve lift.

When EVC is advanced, decrease of the maximum valve lift will play more important role in controlling the percent of residual gas than the timing of EVC, which can be found through the comparison of NVO=180° and EVC=-90°. In these two cases, the valve lift is changed without modifying the valve timing. By the combination of reducing the maximum valve lift and advancing EVC, the residual is efficiently kept in the cylinder for NVO=180°.

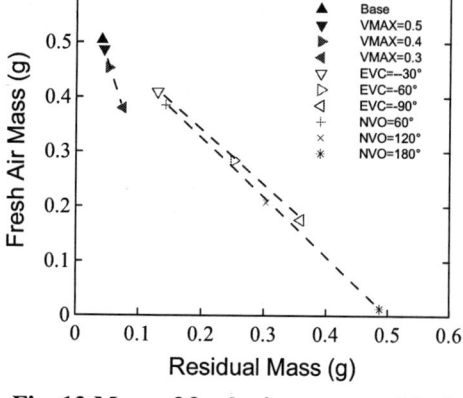

Fig. 12 Mass of fresh air versus residual mass at IVC

The relationship curves between residual mass and fresh air mass are very close for the last two strategies, therefore, it can be concluded that the intake valve profiles doesn't significantly influent mass of inflow air from intake ports when a certain mount residual must be retained in the cylinder for NVO conditions. Little fresh is induced into the cylinder when much residual is retained. It is mainly due to the fact that IVO should be delayed in order to void the outflow of internal residual into intake port. Although the valve lift can be increased, it is not an efficient way to solve this problem because the short duration of intake stroke and high in-cylinder pressure during the gas exchange process.

Fig. 13 illustrates the In-cylinder temperature versus residual mass at IVC. It can be seen that, for the VMAX strategy, the in-cylinder temperature increases when much residual is retained in the cylinder for VMAX=0.5 and VMAX=0.4 But the temperature decreases significantly for MAX=0.3 because of the high

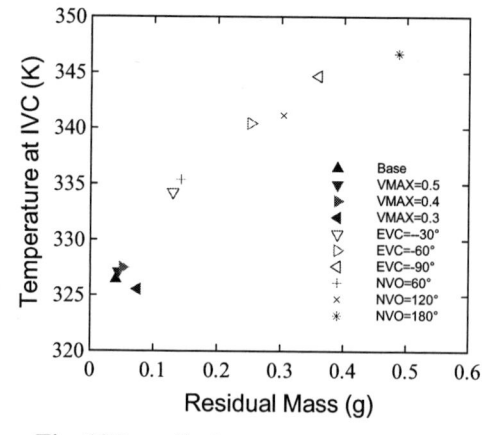

Fig. 13 In-cylinder temperature versus residual mass at IVC

resistance during the intake process which result in less air being induced. Although there are different mount of residual and fresh air in the cylinder for NVO strategy and EVC strategy, the temperatures are very close for NVO=180° and EVC=-90°, even for NVO=120° and EVC=-60°. Since the valve timing are the same for NVO=180° and EVC=-90°, as well as NVO=120° and EVC=-60°, and the lower valve lift could lead to more residual and less air exist in the cylinder at IVC, it can be said that the maximum valve lift of intake valve affects the temperature at IVC remarkably. The small intake valve lift increases the inflow resistance and decrease the in-cylinder temperature.

It is also found that the increase rate of temperature slows down when the residual mass is increased from 0.0 4g to 0.5 g for NVO strategy and EVC strategy. Although much residual is retained in the cylinder, the effect of valve profile on the temperature at IVC becomes week. It can be explained by the reasons of the increase of inflow resistance and high heat transfer from in-cylinder gas to the wall for the cases with large amount of residual.

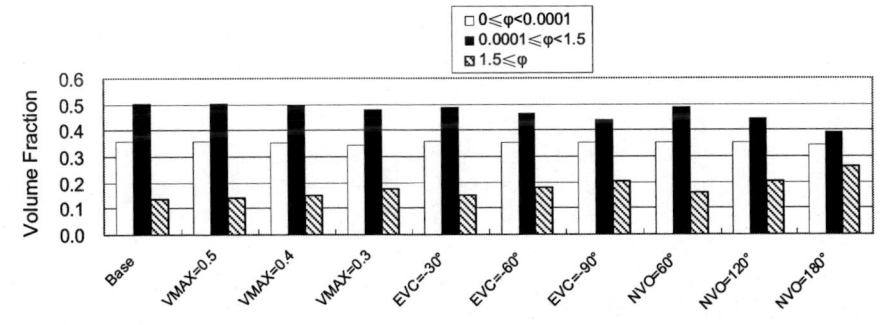

Fig. 14 In-cylinder volume distributions within four different equivalence ratio ranges at 350 °CA

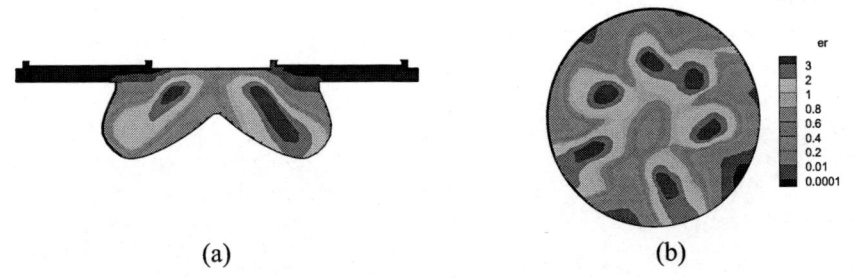

Fig. 15 Fuel mass fraction distribution at axial central cross section and the cross section of the cylinder halfway between the piston and the cylinder head at 350 °CA in VMAX=0.4

Fig. 14 shows the in-cylinder volume distributions within four different equivalence ratio ranges at 350 °CA, The volume fraction with equivalence ratio less than 0.0001, which means the region there almost none fuel exists, are all very close 0.36 in the ten cases. Accordingly, there is about 64% volume being occupied by the mixture with diesel vapor. Because the volume of the combustion chamber is approximately 60% of the total in-cylinder volume at 350 °CA, therefore, the chamber is full of the diesel vapor for all cases. The further information can be found in Fig. 15 for the fuel mass distribution in EVC=-60°.

As the maximum valve lift is decreased, or IVO is retarded, the unmixed region shrinks, and the rich fuel region expands because of the high intake velocity and high turbulence intensity, which benefit for forming the homogeneous charge. But it should be noted that the difference between these ten cases in equivalence distribution is not significant, and the effect of VVA/VVT on mixing is not very critical in the cases studied here.

5 CONCLUSIONS

A three-dimensional modeling was applied to investigate the effects of negative valve overlap on the gas exchange processes and mixture and temperature distributions in a diesel HCCI engine with early in-cylinder injection. Firstly, four cases with different EVO and IVC were calculated to compare the effects of the begin time and the end time of the gas exchange process on HCCI engines with NVO. And then three NVO strategies with different maximum valve lift and valve timing were compared to make it clear that the effects of valve lift and valve timing on retaining residual and control of the in-cylinder temperature, and their further influences on air motion and mixture preparation process.

It should be noted that there is not significant difference in the magnitude of swirl ratio, tumble ratio and turbulence intensity for all the ten cases at SOI, furthermore, the effect of VVA/VVT on mixing is not very critical in the cases studied here. In order to discuss the influence of different NVO strategies on the mixture preparation, only one simple injection strategy was employed in this study. And it is believed the droplet distribution may be influenced by the in-cylinder air motion caused by NVO when the droplets with small Sauter mean diameter (SMD) are injected into the cylinder during the intake stroke. By taking advantage of the higher flowing motion from the longer NVO, the strategy realized by split injection or other special injectors with low injection pressure and small SMD during the induction stroke and early compression stroke could help form more homogeneous fuel-air mixture after the liquid fuel is vaporized, furthermore benefits the combustion process for the diesel HCCI combustion, especially for the cold start and low load conditions.

ACKNOWLEDGEMENTS

The financial support from the EPSRC and the Nuffield Foundation are gratefully acknowledged.

REFERENCE LIST

(1) M. Christensen, A. Hultqvist, B. Johansson. Demonstrating the Multi-Fuel Capability of a Homogeneous Charge Compression Ignition Engine with Variable Compression Ratio. SAE Paper 1999-01-3679, 1999.

(2) S.S. Morimoto, Y. Kawabata, T. Sakurai, T. Amano. Operating Characteristics of a Natural Gas-Fired Homogeneous Charge Compression Ignition Engine (Performance Improvement Using Egr). SAE Paper 2001-01-1034, 2001.

(3) M. Yao, Z. Bo. Experimental Study on the Effects of Egr and Octane Number on Hcci Operation. SAE Paper 2005-01-0174 2005.

(4) N. Milovanovic, R. Chen, J. Turner. Influence of the Variable Valve Timing Strategy on the Control of a Homogeneous Charge Compression (Hcci) Engine. SAE Paper 2004-01-1899, 2004.

(5) H. Yamashita, T. Suzuki, H. Matsuoka, M. Mashida, K. Kitano. Research of the Di Diesel Spray Characteristics at High Temperature and High Pressure Ambient. SAE Paper 2007-01-0665, 2007.

(6) G.B. Parvate-Patil, H.H.a.B. Gordon. Analysis of Variable Valve Timing Events and Their Effects on Single Cylinder Diesel Engine. SAE Paper 2004-01-2965, 2004.

(7) P.J. O'Rourke, A.A. Amsden. The Tab Method for Numerical Calculation of Spray Droplet Breakup. SAE Paper 872089, 1987.

(8) Z.W. Han, R.D. Reitz. Turbulence Modeling of Internal Combustion Engines Using Rng K-E Models. Combustion Science and Technology, 1995, 106:267-295.

(9) G. A., T. Lechner, J. Jacobs, C.A. Chryssakis, D.N. Assanis, R.M. Siewert. Evaluation of a Narrow Spray Cone Angle, Advanced Injection Timing Strategy to Achieve Partially Premixed Compression Ignition Combustion in a Diesel Engine. SAE Paper 2005-01-0167, 2005.

(10) M.Y. Kim, J.W. Kim, C.S. Lee, J.H. Lee. Effect of Compression Ratio and Spray Injection Angle on Hcci Combustion in a Small Di Diesel Engine. Energy and Fuels, 2006, 20(1):69-76.

(11) A. McLandress, R. Emerson, P. McDowell, C.J. Rutland. Intake and in-Cylinder Flow Modeling Characterization of Mixing and Comparison with Flow Bench Results. SAE Paper 960635, 1996.

Cyclic combustion variability in gasoline engines

R. Tily, C.J. Brace
Department of Mechanical Engineering, University of Bath, UK

ABSTRACT

This paper presents an analysis of combustion data from a spark ignition engine running at a variety of mixture strengths at low speed and load conditions. In each case, the heat release per-cycle is studied using statistical tools and nonlinear time-series analysis techniques to illustrate the systematic variation of the combustion. Measures are presented here which aim to detect these variations at less severe operating conditions than have been previously reported, and which, in turn, enables their use in regions more representative of those experienced in production calibrations. It is shown that as mixture strength reduces from an A/F of 14.5 to 20.5 the degradation in combustion increases, as expected. More importantly, as the variability increases it begins to exhibit a systematic structure, which contains repeated sequences that suggest the presence of deterministic behaviour.

NOTATION

A/F   Air-fuel ratio (or AFR)
CV   Cyclic variability (in combustion)
EVO   Exhaust valve open
GDI   Gasoline direct injection
IVC   Inlet valve closed
KS   Kolmogorov-Smirnov test
NTSA Nonlinear time-series analysis
SI   Spark-ignition

1. INTRODUCTION

Cyclic variability (CV) in combustion has been studied for a great many years with researchers adopting various approaches. In more recent years, several researchers have approached the subject using some of the analytical tools developed in the field of chaos theory, which is the study of nonlinear dynamical systems. A chaotic system is defined as one that is bounded, aperiodic, deterministic, and is particularly sensitive to initial conditions. Recently, efforts have been made to confirm, whether

© IMechE 2007

or not, there is an element of determinism associated with CV and, in turn, show that it is not a purely stochastic based behaviour.

Ascertaining whether an actual physical system behaves in a chaotic manner (in the deterministic chaotic sense) normally involves analysing a time-series obtained from experimental data. This is, in itself, often an extremely difficult task. The difficulty arises from the fact that unlike purely mathematical systems (i.e. a known set of differential or difference equations) experimental data has no known mathematical form. It is also often affected by noise, which complicates the analysis. In the literature (1, 2) one can find repeated warnings of the dangers of misapplying nonlinear time-series analysis (NTSA) techniques or misinterpreting their results. It also has to be born in mind that the field of analysis of chaotic experimental data is far less advanced than that for chaotic mathematical systems consisting of differential or difference equations (2).

Previously, researchers (4, 6) have found some evidence of determinism only at extremely lean fuelling conditions. Such conditions are outside the normal operating envelope of production engines. These findings have also been obtained from measurements taken from a single cylinder. The research described here is concerned with more realistic levels of fuelling, and also considers multiple cylinders and any possible inter-cylinder dynamics that may be present. During this research, a number of metrics were investigated: maximum in-cylinder pressure, its corresponding crank-angle, and heat release per cycle. However, it was found that the heat release per cycle data appeared to provide a fuller picture of the dynamics. For example, phase-lag plots produced using maximum in-cylinder pressure data did not appear to show as much structure as the corresponding plots using heat release data. Hence, only the findings determined from heat release data are reported here.

2. EXPERIMENTAL WORK

The data used, during the research, consisted primarily of in-cylinder pressure data from the right-hand bank of a V8 GDI SI engine with a bore of approximately 85mm, and a compression ratio of 12.0. Tests were conducted with the engine running at a steady 2000 RPM and 2 bar BMEP. This particular engine speed/load point was chosen to be representative of normal driving conditions. Most of the previous research in this area has, by contrast, been based on engines running at idle speed. In-cylinder pressure data was gathered under four fuelling conditions for all four cylinders. These being air-fuel ratios of: 14.50, 18.05, 19.84, and 20.47. Coolant temperature was maintained at 88 °C.

3. DATA ANALYSIS AND RESULTS

3.1 Initial analysis
Initially, the data was smoothed using a simple moving average filter, before being further processed. This was to lessen the possibility of any noise introduced by the instrumentation affecting the subsequent analysis. Net heat release per cycle was then calculated using the method described by Stone (3, equation 13.57 pp. 513).

The resultant sets of time-series were plotted for the individual cylinders, and the combined cylinders, in order to develop an initial understanding. Figure 1 shows time-series for the heat released per cycle from the four cylinders combined (in firing order sequence) under the various fuelling conditions. It can be seen that as the fuelling becomes increasingly lean, the heat release values become more varied, particularly at A/Fs of 19.84 and 20.47.

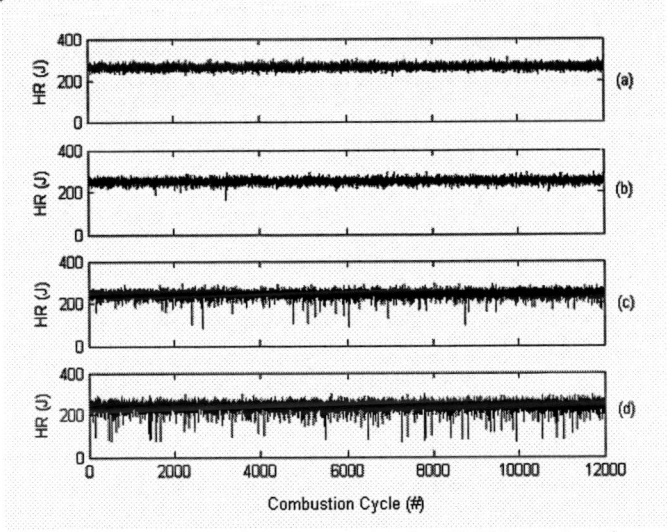

Figure 1. Heat release values time-series for 12000 combustion cycles for A/F ratios of (a) 14.50, (b) 18.05, (c) 19.84, and (d) 20.47.

3.2 Individual cylinders

To further examine the behaviour of the individual cylinders, the distributions of heat release values, at the various fuelling levels, were plotted. As the fuelling became increasingly lean the resultant distributions became more spread, indicating more variation between the individual combustion cycles. Figure 2 shows the distributions under stoichiometric fuelling (A/F 14.50) and the lean fuelling level of A/F 20.47 as an illustration of this behaviour. Note that there is also some variation between the individual cylinders.

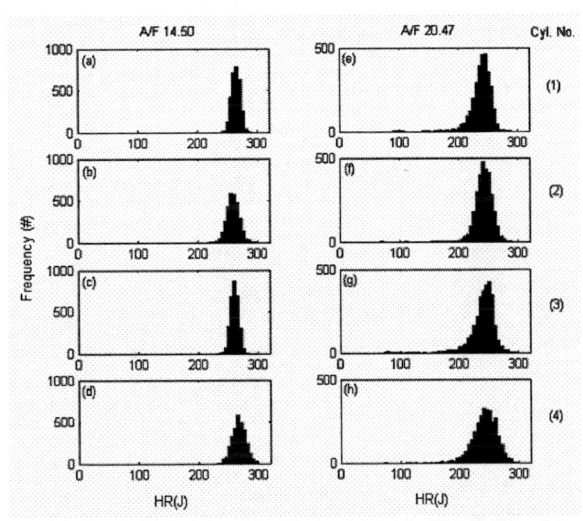

Figure 2. Distribution histograms of individual cylinder heat release values for fuelling levels of 14.50 (stoichiometric) and 20.47.

How the distributions compared with a normal distribution was revealed using the Kolmolgorov-Smirnov (KS) test, which is a non-parametric test for the null hypothesis that a random sample has been drawn from a specific distribution. In this case, a normal distribution.

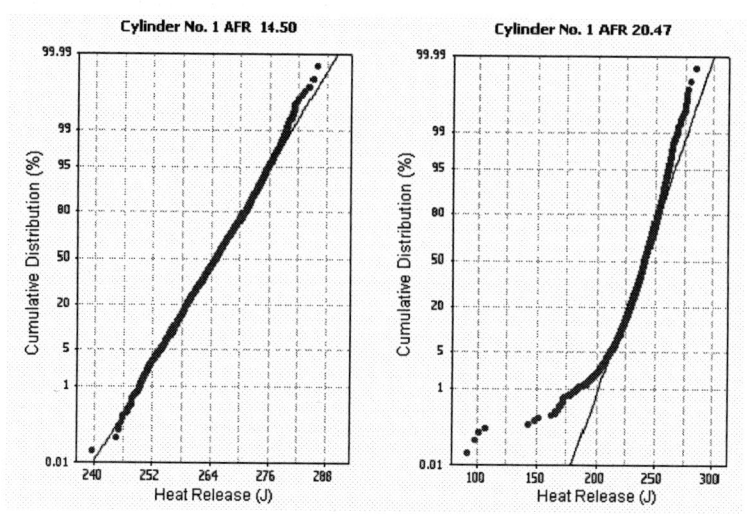

Figure 3. Kolmogorov-Smirnov normality test plots.

The KS statistic is the maximum distance between the curves shown in figure 3. The line represents the cumulative normal distribution while the points show the cumulative distribution of the test data. Overall, the results obtained showed a gradual increase in the value of KS as A/F increased.

Phase-lag plots which are often, wrongly according to Williams (5), referred to as return plots, are constructed by plotting a time-series against itself. Along the abscissa, values at time t are plotted against those at time t + lag on the ordinate. The resulting pattern provides some insight to the dynamical behaviour of the system under investigation. Figure 4 shows phase-lag plots with a lag of one cycle, for heat release values for each cylinder under the various fuelling levels. At A/F 14.50, the plots consist of a tight cluster of points in the upper right-hand corner. This reflects the fact that, under stoichiometric fuelling, there is little variation between cycles, and that there is strong combustion for each cycle. Note, however, that cylinders 2 and 4 show a slightly larger dispersion than cylinders 1 and 3. As the fuelling becomes increasing lean, the pattern of the plots changes from a disc shape to an angular or "boomerang" shape. This is particularly noticeable when the A/F is 20.47 where there are an increasing number of strong combustion cycles that are followed by a corresponding weak cycle. The cycles alternate between strong and weak combustion events, or vice versa, although not on a regular basis. This may to be due to the effect of unburnt fuel from a weak cycle enhancing the following cycle which produces a full burn. Running very lean, i.e. with too little fuel for complete combustion, produces the situation where there tends to be only sufficient fuel for each alternate cycle to achieve a relatively strong combustion.

The characteristic "boomerang" shape in the phase-lag plots for lean fuelling conditions has been previously reported in Daw et al. (4) and others. What has not been reported, apart from by Scholl and Russ (6), is that phase-lag plots can be misleading. Scholl and Russ found that similar plots can be produced from the same time-series but with its order randomly shuffled. Figure 5 shows the same data as Figure 4, but this time the plots have been produced using shuffled data.

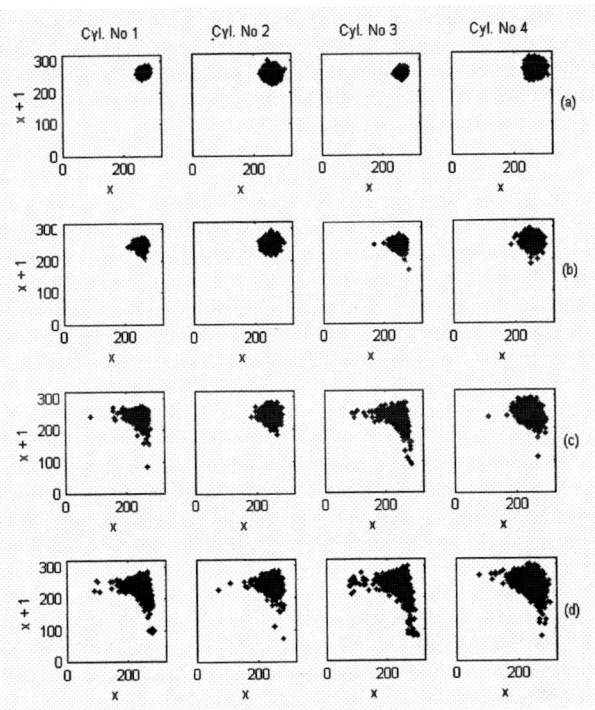

Figure 4. Phase-lag plots for heat release values at (a) 14.50, (b) 18.05, (c) 19.84, and (d) 20.47 air-fuel ratio levels for each cylinder.

Although both figures are similar, there are some distinctions. The "boomerang" shapes are not so well formed and there is a larger spread of points. However, there is still the danger that these plots could be misinterpreted. Scholl and Russ believe that phase-lag plots of skewed distributions can result in the angular appearance seen in many of the plots in Figure 5. Scholl and Russ offered a possible solution to this problem and propose converting the various time-series into quantile form before plotting. Quantiles are similar to percentiles, in that they divide the total frequency of a sample into a given number of equal proportions. Scholl and Russ describe the process of calculating the quantiles as follows. For a data set with n values, replace the smallest value with $1/n$, the second smallest value with $2/n$ and so on up to the last value which is replaced with n/n (this process is also in effect a normalization process). This process will convert a skewed distribution into a uniform distribution where all the values are evenly distributed between 0 and 1. The rank and sequence of the data are unchanged. With data processed this way, an uncorrelated time-series will appear uniformly distributed over a phase-lag plot, whereas a correlated time-series will appear with areas of higher or lower densities of points.

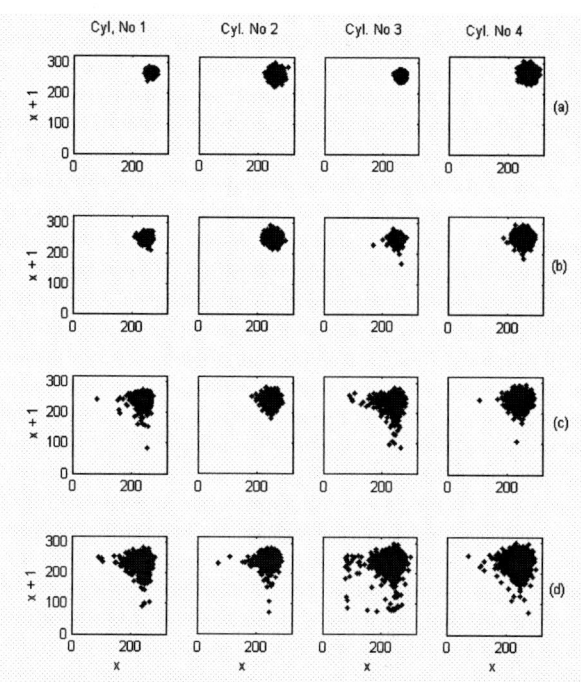

Figure 5. Phase-lag plots for heat release values at (a) 14.50, (b) 18.05, (c) 19.84, and (d) 20.47 air-fuel ratio levels for each cylinder. These plots have been produced using shuffled data. Compare with Figure 4.

Rather than rely on just a subjective interpretation of the phase-lag plots with the quantilized data, Scholl and Russ present an alternative quantitative method, which involves comparing the number of points within certain areas of the phase-lag plot to the numbers that would be expected for a uniform distribution.

Although Scholl and Russ's approach seems reasonable, it was felt that too much structure and information present in the relevant time-series was lost. Phase-lag plots produced with quantilized time-series took on a uniform point distribution (similar to those reported in Scholl and Russ's own paper).

Another potential reason for misleading phase-lag plots could be of the effect of a non-stationary time-series, and therefore that too should be checked. Such behaviour could be caused by any number of incidental factors, such as a periodic variation in one of the fluid temperature control circuits, which affects to some degree the variable being studied (heat release in this case). In fact, stationarity checks should always be made on data, particularly those that will be subjected to various NTSA techniques. Initially, a simple check for stationarity was performed, as described by Sprott (2), which involves dividing a time-series in half and checking their respective means and variance values and their differences using a standard error check for significance. However as the heat release time-series are quite long (at least 3000 points) it was felt that there was a need to break the data into shorter sequences. For the heat release data sets for the individual cylinders, it was decided to adopt the following method:

For each heat release time-series of 3000 points:
- Divide the data into ten 300 point windows;
- Calculate mean, variance, skew and kurtosis for each window;
- Carry out an ordinary least squares fit through the various moments;
- Note the gradient of the straight line fits.

If any of the gradients are greater than a chosen threshold value, then that particular series is not stationary. Plots of the fits allowed a visual check (Figure 6). In essence, if the line is not flat or nearly flat then the series is not stationary, and is changing over time; in fact there is a trend present. There are two types of stationarity. Weak stationarity is present when only the mean and variance stay constant, Strong stationarity is indicated when the process has higher-order moments, skew and kurtosis, that are constant.

The results for the individual cylinders, showed that the fits through the means produced straight lines with small gradients, i.e. the lines were nearly flat, and that the means were unchanging, whereas the variances' fits had some large gradients and therefore they must have changed over time, so the corresponding time-series was just weakly stationary at best and non-stationary in many cases. Figure 6 shows how the two measures varied over the complete time-series.

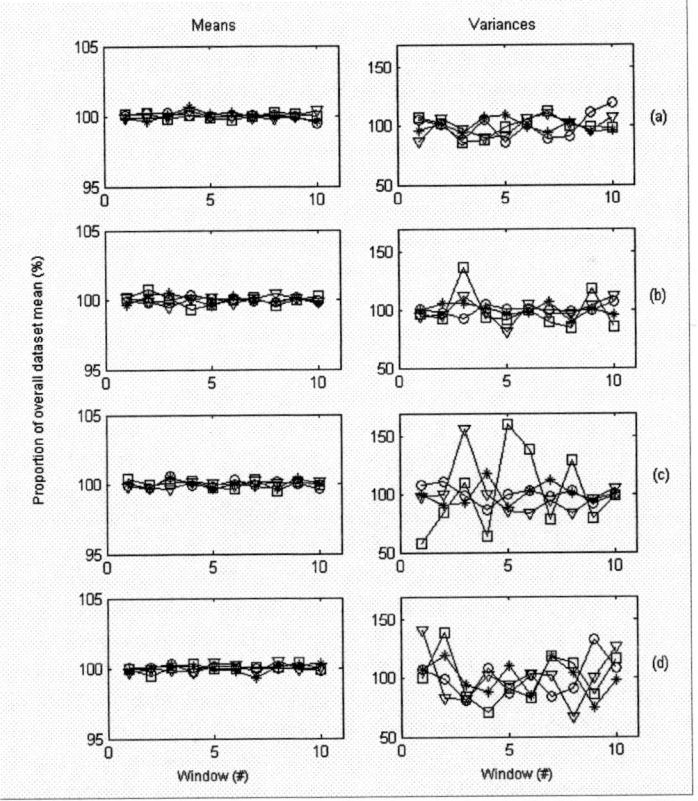

Figure 6. Stationary checks for individual cylinders: (1) ∇, (2) o, (3) □, and (4) *, for air-fuel ratios of: (a) 14.50, (b) 18,05, (c) 19.84, and (d) 20.47.

The results for the combined cylinders showed that the data could generally be considered stationary, as the longer time-series (12000 points, partitioned into 1000 point windows or segments) and the greater window length had a smoothing effect.

It is felt that the localised non-stationarity of the heat release time-series needs to be considered, and that this is something future research should be concerned with. As Sprott (2) remarks, in the real-world, stationarity is actually the exception not the rule.

3.3 Combined cylinders

In an attempt to gain some understanding of interrelationships between the individual cylinders, phase-lag plots were again employed. Using the combined cylinder data, increased lags were used so that one cylinder could be compared to its "neighbour". In this instance, a phase lag of 1 results in individual cylinders being compared to the next cylinder in the firing order while a phase lag of 4 would compare subsequent cycles on the same cylinder. Figure 7 shows the relationships between cylinder 1 with cylinder 4, 3, 2 and itself as the lag is increased from 1 to 4 respectively. The engine's firing order was 1, 4, 3, 2.

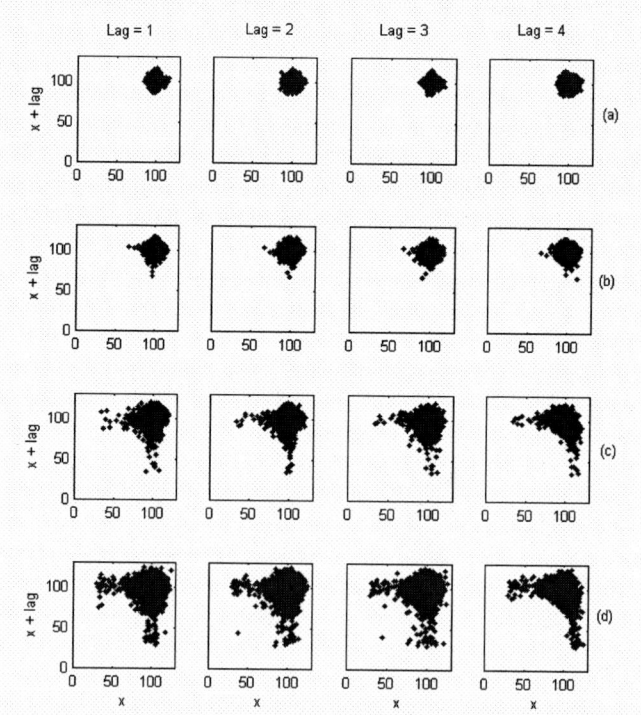

Figure 7. Phase-lag plots for combined cylinders with lags of 1, 2, 3, and 4 cycles for air-fuel ratios of (a) 14.50, (b) 18.05, (c) 19.84, and (d) 20.47. The data have been normalized. The engine's firing order is 1, 4, 3, 2.

Under a stoichiometric fuelling level, with an A/F of 14.5, there is little variation between the individual cylinders and only a slight increase in variation at an A/F of 18.05. However, there are noticeable variations at A/F 19.84 which increase further under the lean fuelling level of A/F 20.47. Compared to Figure 4, at these lean fuelling levels, the points form a less angular asymmetrical "boomerang" shape and have a larger circular cluster of points within the overall angular shape. The plots indicate that there are a considerable number of strong-weak or weak-strong combustion event combinations among related cylinders. This is similar behaviour as that observed for individual cylinders as shown in Figure 4. Although a brief investigation was taken to study the patterns of combustion performance, for the combined cylinders, at a more fine and detailed level, it did not lead to any deeper insights; however it is intended to further this study in the future. One observation is that at phase lags of 1, 2 and 3 cycles there is a general diffusion of the characteristic "boomerang" form of the plot for lean A/Fs. This diffusion is less evident when the phase lag is set to 4 cycles, perhaps due to this case being a comparison of adjacent combustion events on the same cylinder and hence being less influenced by unrelated or stochastic variations on neighbouring cylinders.

4. CONCLUSIONS

NTSA has been successfully applied to metrics describing combustion in one bank of an automotive V8 SI engine. Of the metrics studied, total net heat release appears to offer the greatest insight into deterministic CV. Phase-lag plots show a structure to the data suggestive of the existence of deterministic CV when the A/F rises to 19.84 and above. Some evidence of deterministic behaviour was also evident at A/F 18.05. It is possible that the somewhat non-stationary nature of the data contributed to the impression that deterministic behaviour was present. This needs to be investigated further. It is also possible that skewed distributions of the heat release data at lean conditions contributed to the features observed in the phase-lag plots. Plotting randomly shuffled data shows somewhat similar features but with some significant differences.

The combined cylinder data did not show appreciable non-stationarity and therefore the results obtained from them can be viewed with more confidence than those of the individual cylinders. Analysis of both sets of data showed clearly that there is an underlying structure to the dynamical behaviour of consecutive combustion events, which became more marked with increased lean fuelling conditions. However, even at the moderately lean fuelling level of A/F 18.05 a structure begins to appear. The conclusions drawn from this indicate that continuing efforts need to undertaken to understand the deterministic causes of CV as it seems to affect performance even under moderately lean fuelling.

It must be emphasized that the research described above is very much an initial investigation, and that it is intended that this research be extended. Although phase-lag plots are extremely useful tools, it is felt that there is a need for additional tools that can show a visual representation of the underlying structure of CV. A visual representation of the variations in combustion that enabled the viewer to see repeated patterns or particular events (e.g. misfires) in a long sequence more easily would be most welcome. With this in mind, further research needs be directed towards methods that aid the interpretation of long time-series. Techniques from the fields of data mining, visualization, sequence matching, classification, clustering, and anomaly detection may well prove useful. There definitely appears to be a need to move on and explore techniques that have so far been ignored by automotive researchers.

REFERENCE LIST

(1) Kantz, H., Schreiber, T. (2004) Nonlinear Time Series Analysis, Cambridge University Press, Cambridge.

(2) Sprott, J. C. (2003) Chaos and Time-Series Analysis, Oxford University Press, Oxford.

(3) Stone, R. (1992) Introduction to Internal Combustion Engines 2nd Edition, Macmillan Press Ltd, London.

(4) Daw, C.S., Finney, C.E.A., Green, J.B. Jr, Kennel, M.B., Thomas, J.F, Connolly, F.T. (1996) A Simple Model for Cyclic Variations in a Spark-Ignition Engine. SAE Paper No. 962086.

(5) Williams, G. P. (1997) Chaos Theory Tamed, Taylor & Francis, London.

(6) Scholl, D., Russ, S. (1999) Air-Fuel Ratio Dependence of Random and Deterministic Cyclic Variability in a Spark-Ignited Engine, SAE Technical Paper No. 1999-01-3513

EMISSIONS

Characteristics of catalytic oxidation for carbon black simulating diesel particulate matter over promoted Pt/Al$_2$O$_3$ catalysts assembled in two stage

Jong-Woo Jeong, Byungchul Choi, MyungTaeck Lim
School of Mechanical Systems Engineering, Chonnam National University, Korea

ABSTRACT

Catalytic oxidation activity for Carbon black simulating the soot of diesel particulate matters to CO_2 over Pt/Al$_2$O$_3$, Pt-Mn/Al$_2$O$_3$ and Pt/Ba-Al$_2$O$_3$ catalysts has been investigated with model gases of diesel emission. Two stages of catalysts with the strategies of promoting the soot and NOx reduction, simultaneously, was composed of the oxidation catalyst for carbon black and the other functional catalyst. The catalytic oxidation for carbon black to CO_2 was accelerated by activated oxidants and some exothermic reaction over the catalyst which lies in upstream of two stages. But the two stages of catalysts with carbon black oxidation catalyst sited in upstream showed the initiation of catalytic oxidation activity at lower temperature than other case. Two stage of catalysts composed of Pt/Al$_2$O$_3$ in upstream and Pt-Mn/Al$_2$O$_3$ with carbon black in downstream showed high catalytic oxidation activity for carbon black to CO_2 with 80% consumption rate of carbon black to total loaded carbon black in the range of 350-500 ℃ in the test process made of 4 ℃/min heating rate.

1 INTRODUCTION

There are all applications in which diesel engines have been used for many years, due to the inherent fuel efficiency and the excellent durability of such engines. The diesel engine is being increasingly used to power passenger cars and vans. However, like the gasoline engine, the diesel engine also emits carbon monoxide (CO), hydrocarbons (HC) and oxides of nitrogen (NOx). Diesel engines also produce significant levels of particulate matter (PM) which consists mostly of carbonaceous soot and a volatile organic fraction (VOF) of hydrocarbons that have condensed on the soot (1). There is growing concern about these PM emissions, since emerging evidence potentially links PM with acute health effects (2, 3). In particular, concern has grown over the effects of fine particulate material which may be linked to increased instances of asthma, respiratory and cardiovascular complaints. It is, therefore, a challenging topic that the reduction of emissions by improved fuel combustion processes and advanced exhaust after-treatment systems. With a heat-regeneration control strategy, PM filtering is probably one of the acceptable way to achieve sufficiently low emission of PM but the thermal shock cracking of the filter with a high fuel penalty is to be solved. Catalytic filtering resulting in collection and oxidation of soot that will ensure acceptable oxidation rates of PM at low temperature is preferred. This technique presents a lesser energy required for regeneration, a lower recycle frequency, a smaller chance of filter failure and a more simple regeneration strategy (4).

Catalyzed diesel particulate filter (CDPF) in order to oxidize diesel soot also provide significant reduction in the CO and HC emissions. In general, the catalysts are coated on the walls of the channels of a honeycomb monolithic filter with a wall-flow made of cordierite or silicon carbide. The interior surfaces of this high surface area substrate are coated with catalytic precious metals, such as platinum or palladium. Although the particle mass decreases when oxidation catalysts were used, smaller particles remain unburned, rendering the particle number the same (5-7). The degree of physical contact between the catalyst and the soot is an important factor on the reactivity of catalyst for soot oxidation along with the intrinsic properties of the catalyst. It was shown in published results that the catalytic activity of a metal oxide catalyst is different by the sample preparation procedures, which are the methods of mixing catalyst with soot. In general, the samples prepared with shaking in a bottle with a spatula ('loose contact') have a low reactivity, while samples prepared by mechanical mills and mortar ('tight contact') have a high reactivity (8-10). Nitrogen dioxide, NO_2, is far more reactive than oxygen, and can combust PM at a reasonable rate at temperatures as low as 250-300℃. The use of NO_2 to oxide soot is the basis of the passive regeneration strategy, since the NO_2-carbon reaction occurs at temperatures which are frequently accessed during the normal operation of many Diesel vehicles. The majority of the NOx emissions from diesel engines are in the form of NO, not NO_2, so some of this exhausted NO must be oxidized into NO_2 to enable passive regeneration (11, 12). This can be done using an oxidation catalyst, which can either be located upstream of the DPF, or can be coated onto the DPF.

In the present work, the Pt based catalysts added with transition metal oxides, Mn and Ba, are prepared and tested their activities in carbon black oxidation profiles further including the effects of oxidation catalyst located upstream or downstream of the CDPF. The two stages catalytic system composed of DOC and CDPF are also investigated for the catalytic activity of soot and reactants.

2 EXPERIMENTAL

2.1 Sample preparation
The sample of platinum 3 wt.% ('3Pt') was prepared by incipient wetness impregnation of the γ-Al_2O_3 with an aqueous solution of Pt precursor. The samples

added with Manganese ('3Pt5Mn') and barium ('3Pt/30Ba') were prepared by addition of the manganese and barium precursor dissolved in an aqueous solution to the already prepared $3Pt/\gamma-Al_2O_3$ aqueous solution. Therefore, the amount of Pt loaded with the samples, 3Pt5Mn and 3Pt/30Ba, are substantially 2.9 and 2.3 wt.% Pt, respectively. After drying in a rotating evaporator, and overnight drying in air at 120℃, the samples were also calcined in air at 700℃ for 4 h. The carbon black sample studied in this work (Printex-U^{TM} by Degussa) represents a synthetic diesel soot (13, 14). This carbon black was chosen to avoid any interference due to the presence of adsorbed HCs, sulfates, or fly ash present in real diesel soot. A proper and reproducible mixing between catalyst and carbon black was obtained by gently shaking the two counterparts in a vessel and a sonification using ultrasonic cleaner. These mixing conditions are somewhat different from the normally known "loose contact" condition with a spatula because they allow for better reproducibility due to added sonification process. An initial amount of mixed sample is 150mg carbon black and 50mg catalyst in a ratio of 3:1.

2.2 Catalytic activity assessment

The activity of the metal oxides and the effects of reactant gases on the temperature programmed reaction (TPR) were determined in a fixed-bed flow reactor under atmospheric pressure (Fig. 1). The carbon black-catalyst mixture (0.2g) was placed in a U-shaped quartz-glass reactor, ca. 12 mm i.d., and upheld by quartz-wool layer, whereas the tip of a K-type thermocouple was located well inside the bed itself with 20mm above the sample. After holding at 100℃ for 30 min with an inert gas N_2, TPR was started at a heating rate of 4℃/min. The reaction gases with various partial pressure of O_2, NO, and C_3H_8 were obtained by adequately blending gas flow channels. The controlled flow of artificial reaction gases (1.1 cm^3/min) were passed through the reactor with 800ppm C_3H_8, 1000ppm NO, 7.5% O_2 and N_2 balance gas. The outlet gas was analyzed continuously by a FT-ir (Midac) with 16sec/point data acquisition rate. Carbon monoxide and dioxide were the oxidation products of the soot and C_3H_8 in every case.

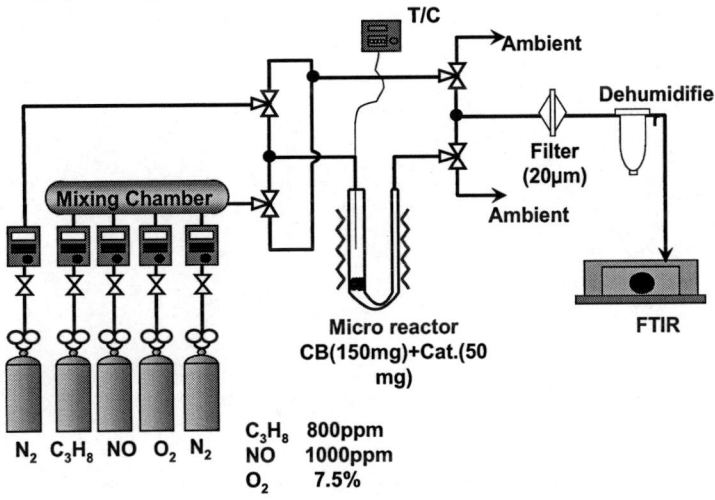

Fig. 1. Schematics of experimental apparatus.

3 RESULTS AND DISCUSSION

3.1 Reactivity of soot oxidation

Carbon black oxidation in the catalyst-carbon black mixed sample was performed by temperature programmed oxidation experiments, in order to understand the reaction of catalysts against soot oxidation. Fig. 2 shows a formation of CO_2 and CO during a typical soot combustion experiments obtained from TPO experiments using the 3 wt% Pt based catalysts loaded metal oxides, 5 wt.% Mn and 30 wt.% Ba, under 7.5% O_2. The first peak of CO_2 curves corresponding to 3Pt-5Mn/Al_2O_3 and 3Pt/30Ba-Al_2O_3 catalyst are observed at low temperature in comparison to 3Pt/Al_2O_3. The first peak temperatures for 3Pt and 3Pt5Mn catalyst are shown at 410 and 445℃, respectively. After the first peak, Carbon oxidation was continued and a few peaks of CO_2 are simultaneously observed with progress in the temperature, which is different from the other published results. Whereas only one peak of CO in terms of partially oxidized carbon black is obviously observed in exception of 3Pt catalyst at the temperature coextensively evolving in the first peak of the CO_2.

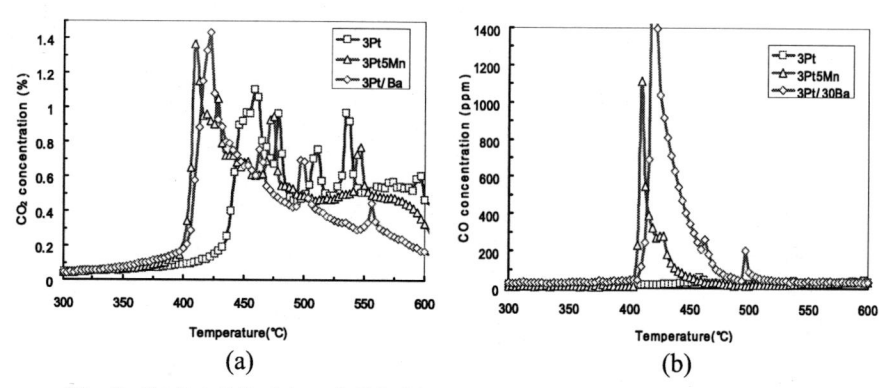

Fig. 2. Outlet CO_2 (a) and CO (b) concentrations during temperature programmed carbon-O_2 reaction over Pt based catalysts.

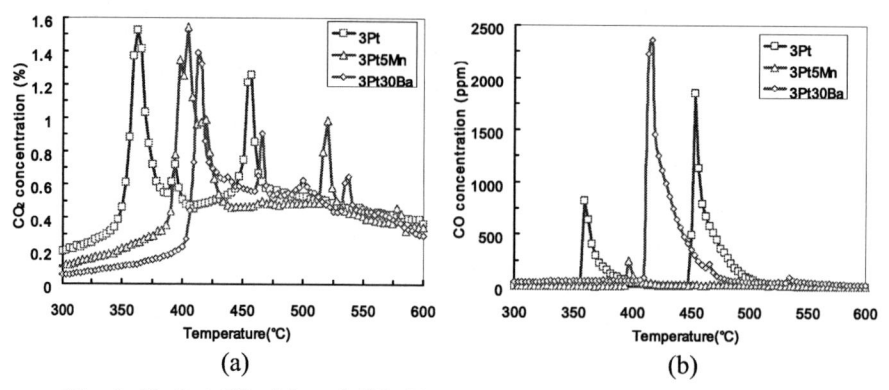

Fig. 3. Outlet CO_2 (a) and CO (b) concentrations during temperature programmed carbon-O_2 –NO-C_3H_8 reaction over Pt based catalysts.

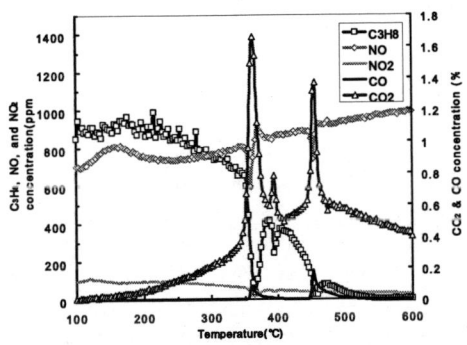

Fig. 4. Outlet gases concentrations during temperature programmed carbon-O$_2$–NO-C$_3$H$_8$ reaction over 3Pt/Al$_2$O$_3$ catalyst.

This may be a sign of the large amount of carbon black to catalyst (3/1, w/w) in mixture sample. Insufficient oxygen fed (7.5%, 1100 cm^3/min) in mass balance to carbon black in a point of sudden increase in amount of CO$_2$ is sequent to the partial oxidation using the lattice oxygen of the catalyst (7). And, Peaks of CO$_2$ after the first peak are attributed to the regional combustion of the carbon black, which is not in contact with catalyst particles. Self ignition of carbon black near the catalyst particles is initiated by heat transfer resulted from exothermic reaction on the near catalyst. This explanations related to high weight ratio of carbon black to catalyst in mixture sample may be acceptable in consideration of realistic CDPF exposed to 'soot cake'.

Temperature programmed reaction (TPR) experiments were performed using reaction gases, O$_2$, NO, N$_2$, and C$_3$H$_8$, in order to investigate the effect of reaction gases simulating diesel exhaust gas on the soot oxidation. Fig. 3 shows the evolution of CO$_2$ produced by oxidation of carbon black using the reaction gases on catalyst. Temperature programmed reaction (TPR) experiments were performed using reaction gases in order to investigate the effect of reaction gases simulating diesel exhaust gas on the soot oxidation. When NO, O$_2$, and C$_3$H$_8$ are used as feed gases for carbon black oxidation, oxidation reaction starts actively at lower temperature and is dramatically increased in comparison to the results under carbon-O$_2$ reaction (Fig. 2), especially in case of 3Pt/Al$_2$O$_3$. In 3Pt/Al$_2$O$_3$ sample, the presence of reaction gases, NO+O$_2$+C$_3$H$_8$, shifts the CO$_2$ profiles to significantly lower temperatures with the magnitude of about 90°C. This is due to the effect of NO, while NO easily oxidized to NO$_2$ with excess oxygen on catalyst is well known as a further possible oxidizing agent, in addition to the catalytic reaction of gaseous reactants each other. In 3Pt/30Ba-Al$_2$O$_3$ catalyst, the effect of reaction gases insignificant to oxidation reaction because 3Pt/30Ba-Al$_2$O$_3$ catalyst is loaded with a small amount of Pt compared to 3Pt/Al$_2$O$_3$. From the results of Fig. 4, the catalytic reaction of the samples under reaction gases is completed by correlative reaction of oxidation and reduction in direction of accelerating the reaction. This phenomenon was identified in all samples. The reactions of soot and propane gas using NO and O$_2$ as an oxidizing agent are occurred simultaneously. At a point of sudden oxidation of carbon, carbon oxidation was involved with the strong reaction of C$_3$H$_8$ and NO.

3.2 Two stages catalytic system

The effect of oxidation catalyst coupled with soot oxidation catalyst on a catalytic reaction is investigated using a Pt/Al$_2$O$_3$. In order to support the reactivity of soot oxidation and the sequent oxidation for the partial oxidized products, two stage catalytic systems were composed of oxidation catalyst (Pt/Al$_2$O$_3$) and soot oxidation catalyst. Fig. 5 shows the catalytic reaction in case of oxidation catalyst placed at the front or rear of soot-catalyst (3Pt/Al$_2$O$_3$) mixture. The first peak of CO$_2$ curve is observed at lower temperature in rear oxidation catalyst system (Fig. 5(b)) than in front oxidation catalyst system (Fig. 5(a)). This result is conflicted with general aspect that the effect of NO on soot oxidation is emphasized by adopted oxidation catalyst for NO to NO$_2$ at upstream of soot oxidation catalyst. In Fig. 5(a), oxidation reaction of C$_3$H$_8$ is above the point of CO$_2$ peak due to the fact that carbon oxidation is superior to that in Fig. 5(b). Under the circumstance of high activation energy for catalytic reaction, propane reaction gas is not competitive with downstream carbon black in terms of oxidation reaction, where it is easily oxidized by NO$_2$ resulted from oxidized NO on front oxidation catalyst.

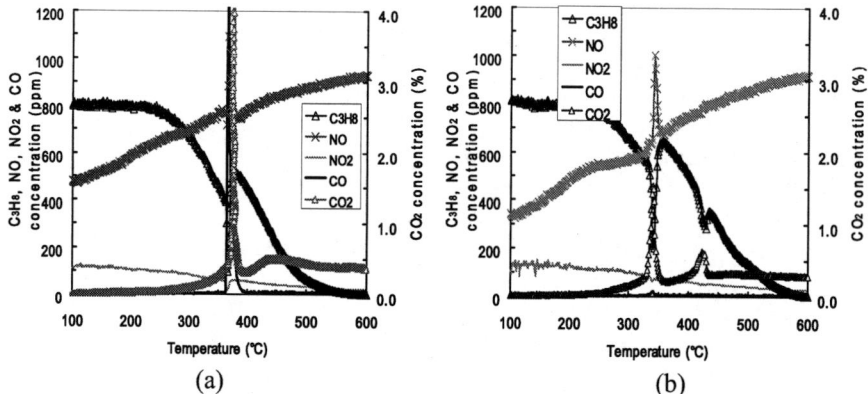

Fig. 5. Outlet gases concentrations during temperature programmed carbon-O$_2$–NO-C$_3$H$_8$ reaction over two stages systems of Pt+CB-3Pt/Al$_2$O$_3$ (a) and CB-3Pt/Al$_2$O$_3$ +Pt (b).

Table 1. Oxidized carbon black weight upto 500 ℃ within programmed temperature reaction.

Two stages systems		T$_{max}$	Oxidized carbon weight	
Front	Rear	(℃)	g	%(C$_{ox}$/C$_{in}$)
Pt	CB-3Pt	374	0.092	61
CB-3Pt	3Pt	341	0.082	55
3Pt	CB-3Pt-5Mn	384	0.131	87
CB-3Pt-5Mn	3Pt	379	0.142	95
3Pt	CB-3Pt/30Ba	406	0.083	55
CB-3Pt/30Ba	3Pt	423	0.059	39

T$_{max}$: The first peak temperature of CO$_2$ curves evolved by carbon oxidation on two stages catalyst sysems

According to NO-C$_3$H$_8$ reaction, the effect of NO as a strong oxidizing agent is not played directly on soot oxidation at downstream carbon-catalyst mixture. But the exothermic heat due to oxidation of reaction gases on front oxidation catalyst affects the reaction of soot-catalyst downstream to stable carbon black oxidation with high degree. As a matter of fact, the effect of NO upon soot oxidation under realistic circumstance in terms of DOC+CDPF system is insignificant in consideration of trade-off relation of HC and NOx each other and higher reactivity between gaseous reactants. On the other hand, the peak temperature of CO$_2$ curve in Fig. 5(b) is lower than in Fig. 5(a). This may be due to the chain-reaction on one stage of carbon-catalyst mixture while NO is oxidized to NO$_2$ and then serially played as a strong agent for carbon black oxidation, which is not completely understood. These results are ascertained by Table 1, which is evaluated by CO$_2$ concentration resulted in oxidized carbon weight ratio to initial fed carbon weight upto 500°C during the TPR experiment procedure. As comparing the oxidized carbon weight, two stages catalytic systems, the oxidation catalyst placed at upstream of carbon-catalyst mixture, shows the higher oxidized carbon ratio due to their support for the catalytic reaction of reaction gases than that in the reverse order of two stages systems in exception of 3Pt-5Mn/Al$_2$O$_3$ catalyst. Two stage system of 3Pt-5Mn/Al$_2$O$_3$ catalyst regardless of their order of the front or back shows superior reactivity of carbon oxidation to other catalytic systems.

4. CONCLUSIONS

Two stages catalytic system, as in the DOC+CDPF system, was prepared and tested for potential applications.
1. Reactivity of 3Pt-5Mn/Al2O3 for oxidizing the carbon black in TPO was began at 400°C evolving CO2.
2. Oxidation reaction for carbon black on 3Pt/Al2O3 was accelerated by catalytic reaction of reactant gases, as showing the activation temperature of 350°C.
3. Two stages systems composed of DOC (front) and soot oxidation catalyst (rear) was superior in catalytic oxidation rate of carbon black to that of the reverse.
4. Pt oxidation catalyst (front) + 3Pt5Mn/Al2O3, presents catalytic oxidation rate of carbon black with 95% within up to 500°C.

ACKNOWLEDGEMENT

This work is part of the project 'Development of Near Zero Emission Technology for Future Vehicle' and the authors are grateful for its financial support.

REFERENCES

(1) Walker, A.P. Controlling particulate emissions from diesel vehicles, Topics in Catalysis, 28 (2004), 165-170
(2) Maynard, R.L. and Howard, C.V. (1999). Particulate Matter: Properties and Effects upon Health, Oxford.
(3) Seaton, A., MacNee, W. Donaldson, K. and Godden, D. (1995). Particulate air pollution and acute health effects, Lancet, 345, 176-178
(4) Walsh, M., Global trends in diesel particulate control a 1995 update, SAE paper 950149
(5) Neeft, J.P.A., Makkee, M., Moulijn J.A., Fuel Process. Technology. 47 (1996) 1.
(6) http://www.dieselforum.org
(7) Uner, D., Demirkol, M.K., Dernaika, B., A novel catalyst for diesel soot oxidation, App. Cat. B: Environmental 61 (2005) 334-345
(8) Stanmore, B.R., Brilhac, J.F., Gilot, P., The oxidation of soot: a review of experiments, mechanisms and models, Carbon, 39 (2001), 2247-2268
(9) van Setten, B.A.A.L., Schouten, J.M., Kakkee, M., Moulijn J.A., Realistic contact for soot with an oxidation catalyst for laboratory studies, App Cat. B: Environmental 28 (2000) 253-257
(10) Fino, D., Russo, N., Saracco, G., Specchia, V., Catalytic removal of Nox and diesel soot over nanostructured spinel-type oxides, J. of Cat. 242 (2006) 38-47
(11) Timothy V. Johnson, Diesel emission control in review, SAE papers 2006-01-0030 (2006)
(12) Balle, P., Bockhorn, H., Geiger, B., Jan, N., Kureti, S., Reichert, D., Schroder, T., A novel laboratory bench for practical evaluation of catalysts useful for simultaneous conversion of NOx and soot in diesel exhaust, Chemical Engineering and Processing 45 (2006) 1065-1073
(13) Yezerets, A., Currier, N.W., Kim, H.H., Eadler, H.A., Epling, W.S., Peden, C.H.F., Appl. Catal. B Environ. 61 (2005) 120
(14) Mul, G., Zhu, W., Kapteijn, F., Moulijn J.A., The effect of NOx and CO on the rate of transition metal oxide catalyzed carbon black oxidation: An exploratory study, Appl. Catal. B Environ. 17 (1998) 205-220

Phenomenological NO model for conventional heavy-duty diesel engine combustion

R.S.G. Baert,[1,2] X.L.J. Seykens,[1]
[1] Faculty of Mechanical Engineering, Eindhoven University of Technology, The Netherlands
[2] TNO Automotive, Delft, The Netherlands

ABSTRACT

A phenomenological model is presented that predicts NO emissions from a conventional heavy-duty diesel combustion process. As the model is intented for application in the field of engine control, modeling activities have aimed at limiting the computational effort while maintaining a sound physical/chemical basis. The main inputs to the model are in-cylinder pressure data and trapped in-cylinder conditions together with basic fuel spray information. For its validation measured concentrations of respectively in-cylinder and exhaust gas NO from two different engines have been used. Relative changes in NO for a variation in fuel injection timing and for a variation in exhaust gas recirculation rate are well captured by the model. However, absolute emission levels are significantly overpredicted. Observed deviations are evaluated and possible explanations are suggested.

1. INTRODUCTION

The freedom to vary diesel engine operating conditions for a given speed and load requirement has significantly increased over the last decades. This flexibility comes at the cost of a more complex engine control system annex algorithms. Real time combustion models that predict heat release rate and emission would be a powerful and valuable tool for controller design, calibration and testing, reducing controller development time and costs. To limit computational effort, previous attempts at such models mentioned in the literature often had an empirical or semi-empirical character, e.g. [1, 2] and needed an extensive and expensive set of data for fitting/training of the model. Preferably a model should be based on actual physics and chemistry. Such phenomenological combustion models presented in the past (e.g. [3-5]) tend to be quite complex and too time-consuming for the purpose of this study. However, a simple NO_x prediction model was presented already more than 30 years ago by Shahed et al. [6]. From a heat release analysis they determined the amounts of fuel that would burn at a given crank angle value (within a step of 1 degree crank angle). They then postulated that these fuel amounts reacted with (within that time frame available) charge air at stoichiometric equivalence ratio. The

© Eindhoven University of Technology 2007

NO-formation rate in the resulting package of combustion products would - at any instant - be determined by the evolution in cylinder pressure and its effect on the temperature of this package. No mixing between packages was assumed. Overall engine-out NO was the sum of the contributions from all of the packages. Murayama et al. [9] extended this model by allowing for mixing of the products packages with surrounding air. For the mixing rate they postulated a Wiebe-like expression with a time constant of 0.5 ms. Focus in the research community then shifted further towards including more details. It took until 1983 before Ahmad and Plee [10] returned to the Shahed approach: they demonstrated a strong relation between the fuel specific NO_x-emissions of a direct injection diesel engine and the adiabatic flame temperature of a stoichiometric mixture of fuel and unburnt gas that would exist in the engine at maximum pressure conditions. Later [11], their colleagues Wu and Peterson returned to the path suggested by Murayama [9]. For the NO formation in a package that mixes with surrounding gas during a mixing time τ_{mix} they use:

$$dm_{NO} = M_{NO} \int_0^{\tau_{mix}} \frac{d[NO]}{dt} dt dV_{burnt} \qquad (1.1)$$

where M_{NO} is the molar mass of NO. Using an Arrhenius type expression for the initial NO-formation rate, they correlated measured flame temperatures with emissions for different – constant – mixing times. They considered different mixing times (between 4.4 μs resp. 2.4 ms) and different entrained gas compositions but could not identify a best approach for matching engine out emissions.

The background for this early work was the assumption that diesel sprays would burn in a manner similar to that of steady turbulent diffusion (spray) flames. Study of these flames indicated that NO formation would start in the regions with (near-) stoichiometric equivalence ratio (and therefore with highest local temperature). NO formation then continues as the combustion products are diluted with ambient gas. Recent laser diagnostic research on diesel spray combustion confirms this picture: shortly after the end of premixed burn, a quasi-steady situation is reached where final oxidation takes place in a relatively thin diffusion-burning sheet at the outer edge of the spray [7]. NO formation then starts in the diffusion flame and continues in the post-flame region [8].

More recently, Timoney et al. [12] and Dodge et al. [13] returned to the Shahed-approach (neglecting dilution and flame radiation losses), still using simple (thermal) NO kinetics only. Similarly, Chikahisha et al. [14] and Andersson et al. [15] have suggested variations to the approach suggested by Murayama. Both groups suggested semi-empirical mixing rate expressions but these expressions were not given a scientific basis. Only Andersson et al., like Wu and Petersen, considered the cooling effect of soot radiation on the temperature level of the reacting products, but they had to adjust this correction depending on engine load and speed to ensure an agreement with measured NOx emissions.

All of the above models focused on thermal NO formation. Only Mellor et al. [16] took a different approach: abandoning the concept of different fuel packages they postulated that the NOx formation would resemble a two stage process with high temperature NO formation at stoichiometric conditions (at start of injection pressure and temperature) being followed by simultaneous NO formation and decomposition reactions at lower overall equivalence ratio and at end of combustion pressure and temperature levels. Radiation losses were not considered.

The present study follows the approach of Murayama, but with attention to the full complexity of the NO kinetics, with attention to soot radiation effects and with an attempt at a less empirical model for package dilution. Furthermore it focuses on conventional heavy-duty diesel engine combustion.

2. MODEL DESCRIPTION

2.1 General model assumptions
With conventional heavy-duty diesel combustion – usually - the majority of the fuel is burned in the diffusion controlled combustion phase. Furthermore, the equivalence ratio of the premixed burnt mixture is estimated to be typically in the range of 2 tot 4 [8]. Such mixture strength is too rich for significant NO formation as confirmed experimentally by Dec [7]. Similarly, local in-cylinder Laser-Induced Fluorescence (LIF) NO measurements performed by Verbiezen et al. [17] showed very little NO up to the point of the occurrence of the diffusion flame. Hence, in this study, NO formation is considered only for the diffusion controlled combustion phase. This is contrary to most of the previous references [6,7,9,11,12,13] where contributions from (heat release during) both premixed and diffusive combustion are collected. New packages are formed with a fixed time interval of 1 crankangle degree. Heat release rate is calculated from the measured pressure trace as described in [18]. Blow-by is neglected and heat loss is computed using the well-known Woschni-correlation. The constants in this correlation are scaled such that gross cumulative heat release matches the amount of fuel injected.

2.2 Chemical kinetics
For conventional diesel combustion the extended Zeldovich reaction mechanism is the primary pathway for (thermal) NO formation [18]:

$$O + N_2 \Leftrightarrow NO + N \quad (1.2)$$
$$N + O_2 \Leftrightarrow NO + O \quad (1.3)$$
$$N + OH \Leftrightarrow NO + H \quad (1.4)$$

However, the trend towards combustion at lower temperatures (f.i. by using large amounts of EGR) has increased the importance of the N_2O-intermediate route [16]:

$$N_2O + O \Leftrightarrow 2NO \quad (1.5)$$
$$O + N_2 + M \Leftrightarrow N_2O + M \quad (1.6)$$

In this study both reaction pathways to NO have been included.

2.3 Initial temperature calculation

The initial temperature and composition of a package results from adiabatic combustion of a stoichiometric mixture of cylinder charge gas and fuel. The cylinder charge supposedly has undergone polytropic compression from trapped conditions. Fuel input is at fuel injection temperature. For this calculation the following comments are important:
1) Adiabatic flame temperature and equilibrium composition are computed using a chemical equilibrium solver that takes into account dissocation and its (lowering) effect on temperature.
2) Evaporation of liquid fuel is taken into account; this results in a significant drop of the temperature of the initial reactants and subsequently of the adiabatic flame temperature in the range of 50 – 80 K.
3) Based on the findings of Turns [19] adiabatic flame temperature is corrected for the cooling effects of gas radiation and of radiation from the soot that forms as the partially oxidised fuel molecules approach the diffusion reaction zone. Soot radiation was modelled as a fixed percentage of total heat loss.

2.4 Mixing model

As explained in the introduction, previous models treated mixing in a semi-empirical manner. Of course it is well known that in modern DI diesel engines fuel injection is the primary energy source driving the mixing process [20]. In this study the mixing process is therefore linked to the injection process: it is assumed that mixing of oxidizer into the combustion product zones is analogous to the mixing of oxidizer into the fuel spray. For a free, fully developed, steady spray Naber and Siebers [21] showed that the following expression holds for the cross sectional averaged equivalence ratio $\overline{\phi}$ along the spray axis coordinate:

$$\overline{\phi}(\tilde{x}) = \frac{AFR_{st}}{\dot{m}_{ox}(\tilde{x})/\dot{m}_f(\tilde{x})} = \frac{2 AFR_{st}}{\sqrt{1+16\tilde{x}^2}-1} \quad (1.7)$$

where $\dot{m}_f$ and $\dot{m}_{ox}$ are respectively the fuel and oxidizer mass flux, AFR_{st} is the stoichiometric air-fuel ratio and $\tilde{x}$ is the dimensionless axial coordinate:

$$\tilde{x} = \frac{x}{x^+} = \frac{x}{\left(d_f \sqrt{\overline{\rho}}/a\tan\left[\theta_{spray}/2\right]\right)} \quad (1.8)$$

with x the axial coordinate, d_f the effective nozzle hole diameter, $\overline{\rho}$ is the ratio between fuel and oxidizer density and θ_{spray} the spray cone angle. For this cone angle either (tabulated) measured cone angles can be used or correlations like the one presented by Naber and Siebers [21]. The parameter a is preferably tuned to measured spray penetration measurements (a values of 0.75 was suggested [22]).

Figure 1 gives a typical evolution of the equivalence ratio as a function of the spray coordinate. As mentioned before, new packages are generated at the stoichiometric position. It is assumed that the equivalence ratio of a package continues to follow this curve from this point onwards. In this way every package is given a spatial coordinate and therefore can also be referred to as a zone. Both package and zone are used in the remainder of the text. Of course this is a rather crude approximation and actual fuel air mixing is much more complex: this model neglects the three-dimensional nature of the diesel spray; mixing rates in regions of stoichiometric equivalence ratio will vary depending on their position in the spray. In Figure 1 also the position of the piston bowl is shown. Of course, mixing will be affected as the spray hits the bowl, an effect that is at this moment neglected. Finally, as the combustion product packages move within the reacting spray they can catch up with the slower tip region. When this happens, mixing rates will fall off and re-entrainment will occurs, again influencing NO formation in a manner not taken into account. Clearly, there is scope for further improvement of this model.

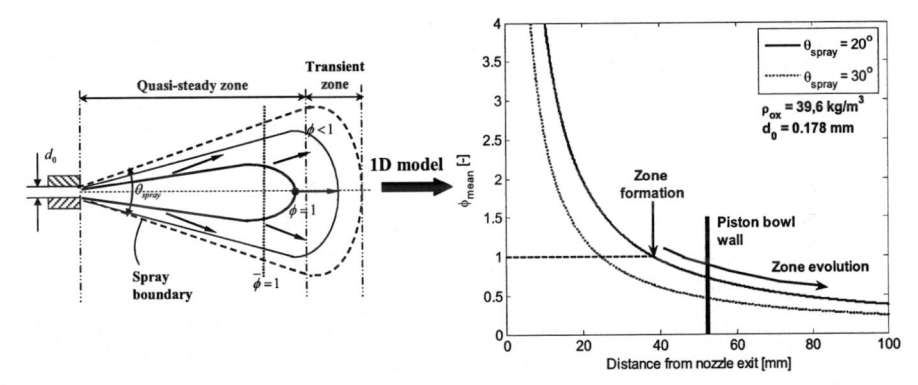

Figure 1. Evolution of 1-D quasi-steady spray according to model assumptions; θ_{spray} =20°, $\bar{\rho}$ = 39.6. Conditions as in Table 2, 18.1 bar imep condition.

Similarly, Desantes et al. [23] derived an expression for the cross sectional averaged axial mean flow velocity $U_{mean}(x)$:

$$U_{mean}(x) = 0.505 \cdot \frac{\dot{M}_0^{\frac{1}{2}}}{\rho_{ox}^{\frac{1}{2}} \cdot \left(\frac{\pi}{2\alpha}\right)^{\frac{1}{2}} \cdot (1-\exp[-2\alpha])^{\frac{1}{2}} \cdot x \cdot \tan\left[\frac{\theta_{spray}}{2}\right]} \quad (1.9)$$

with $\dot{M}_0$ the momentum flux at the nozzle hole exit and α the shape factor for the Gaussian radial velocity profile (to be consistent with eq. (1.7) α = 4.605 should be used). Contrary to equation (1.7), (1.9) is a function of injection velocity, and therefore of injection pressure. Combining equations (1.7), (1.8) and (1.9) – for a given engine speed – the change of equivalence ratio in a package can be determined. The mass of fresh charge m_{fresh} in the zone then follows from:

$$m_{fresh}(\theta) = dm_{fuel,burnt} \frac{AFR_{st}}{\phi_{zone}(\theta)} \quad (1.10)$$

where $dm_{fuel,burnt}$ is the burnt fuel mass corresponding to that zone. At this point it is worth mentioning that the diesel spray is a confined jet with a equivalence ratio value at large time/distance equal to the overall equivalence ratio (and not zero as with the free jet). Therefore, the expression in (1.7) is – at every instance – given an appropriate off-set.

3. EXPERIMENTS AND MODEL VALIDATION

3.1. In-cylinder NO measurements

Instantaneous local in-cylinder NO was measured using LIF on a single cylinder heavy-duty optical research engine [17]. Engine data and operating conditions are summarized in Table 1 / data set 1. Measured temporal NO-profiles were calibrated by assuming that at exhaust valve opening, these in-cylinder values would match the measured exhaust gas concentration values.

Table 1: Engine specifications and operating conditons, data set 1 and 2.

Engine	data set 1	data set 2
Bore, Stroke [mm]	130, 146	130, 158
Compression ratio [-]	15	16.0
Piston bowl shape	"Bath tub"	M-shaped
Diameter piston bowl [mm]	84	100
Swirl number [-]	1.8	-
Diameter nozzle holes [-]	0.128	0.178
Number of nozzle holes [-]	8	8
Engine speed [rpm]	1430	-
Imep [MPa]	0.5	-
Boost pressure [kPa]	140 (no EGR)	-
Fuel injected [mg]; inj. pres. [bar]	60; 1200	-

Figure 2 shows measured local NO concentrations as a function of crankangle for four different injection timings. Significant NO is measured only after the diffusion flame enters the PLIF probe volume (appearance of the diffusion flame was determined from the corresponding soot radiation). From 40 degree crankangle after TDC (that is after end of combustion for all injection timings) all NO concentration curves decay at the same rate. Measurement (using CO_2 absorption spectra) indicate that by this time local temperatures are well below 1800 K and consequently NO reactions are frozen. This implies that this concentration decay results from mixing of the combustion products with unburned gas. This process is identical for all four experimental conditions. Figure 3 shows the corresponding heat release rate profiles. As shown diffusive combustion at different timings is very similar. Start of diffusion combustion is taken at the moment when premix burn rate peaks. For the calculations, both premix and diffusion burn are fitted with a Wiebe function.

Clearly, from Figures 2 and 3, both NO and premixed burn change with timing. As explained by Musculus [24] the effect of a change in premix burn on NO is an indirect one (i.e. through its effect on the diffusive combustion phase). Furthermore, in Figure 2 the local in-cylinder NO concentration measurements are compared with two simulated NO concentration profiles. The dotted lines represent the limiting case of no mixing of fresh oxidizer into the combustion products zones. The solid line represents results using the mixing model proposed in this paper. The simulated curves are constructed from the predicted NO concentration values in the sequentially formed zones. From this it can be concluded that neglecting the mixing of fresh oxidizer into the combustion products results in a significant over prediction of NO concentration during the combustion cycle and also of the final exhaust gas concentration value, i.e. at exhaust valve opening. This over prediction becomes larger when injection is retarded. For these timings, temperatures are lower and closer to 1800 K, which is considered to be the limit for thermal NO formation. The time available for NO formation is then significantly influenced by the mixing rate.

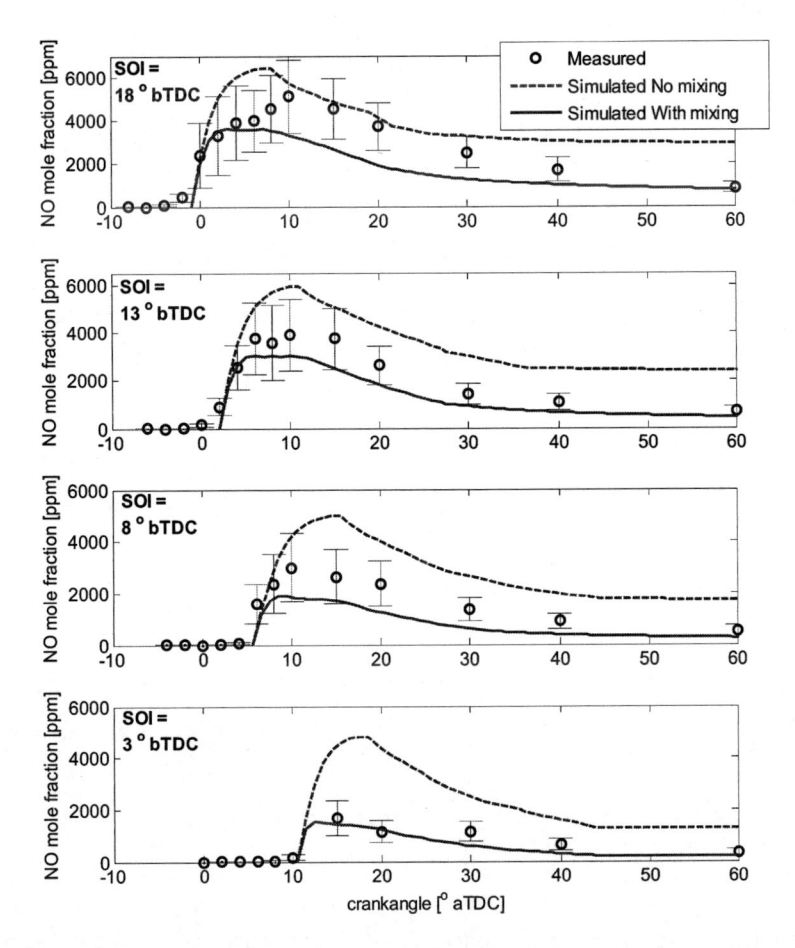

Figure 2. Measured and simulated in-cylinder NO concentration profiles for varying injection timings. Error bars contain cyclic variations as well as uncertainties in processing the LIF data [17].

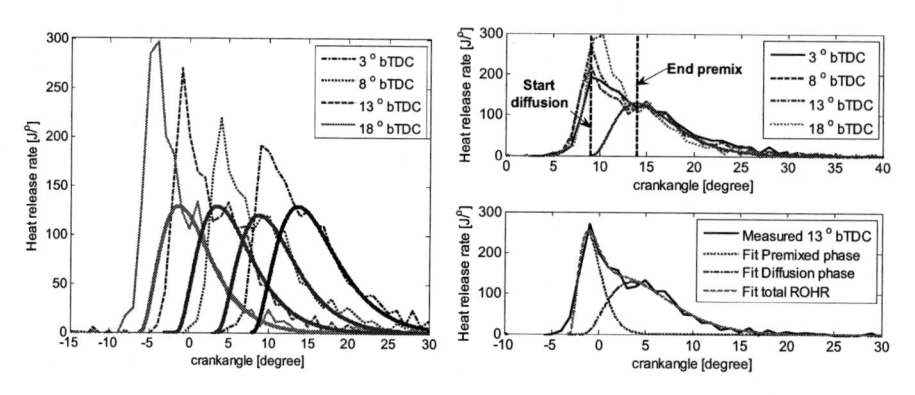

Figure 3. Heat release rate profiles for varying injection timings

3.2. Engine-out NO emissions with variation of load, SOI and EGR

A second set of data represents exhaust gas NO concentration measurements from a research type six-cylinder, heavy-duty diesel engine of which one cylinder is used for combustion analysis and exhaust gas concentration measurements. Boost pressure and EGR rate can be independently adjusted. Table 1 gives some details on the test engine and Table 2 on the different engine operating points.

Table 2: Global operating conditions; data set 2.

	EGR sweep	SOI sweep
Boost pressure	2.60 bar	1.20 bar
Start of Injection / EGR perc.	10° bTDC	0 %
Engine speed	1200 rpm	1200 rpm
Injection duration and pressure	3 ms; 1250 bar	1.7 ms; 750 bar
Imep	18.1 bar	7.7 bar

In the engine tests, EGR rates were varied while keeping boost pressure and fueling parameters fixed. Rate of heat release (ROHR) analysis showed that premixed burn was very limited and only small changes in ROHR with EGR were observed. Because of this, In the calculations, start of diffusion burn was taken at the start of combustion, and the diffusion burn rate was assumed constant until the end of premix burn.

Figure 4 presents measured and simulated NO emission indices EINO (defined as the ratio of total mass of NO produced and total mass of fuel burnt), normalized on the 0% EGR operating point. Table 3 gives an overview of the corresponding quantitative data. For control algorithm development, good qualitative agreement is essential. As can be concluded, the qualitative agreement between simulated NO emission and measured values is very good with an R-square value of 0.993. However, absolute calculated values exceed measured values with approximately a factor 1.6 to 2.3. Figure 4 also shows simulation results using the extended Zeldovich NO-reaction mechanism. As shown, omission of the N_2O-intermediate NO formation pathway results in an decrease in NO formation ranging from 3.8%

for 0 % EGR to 6.5% for 30% EGR. Clearly, the thermal NO formation mechanism is still the dominant NO formation pathway at these operating conditions. Imposing a lower thermal NO-formation rate or lowering the initial package temperature with 100 K brings the absolute NO levels closer to the measured ones, but the shape of the ensuing curve does not follow the measured trend as well.

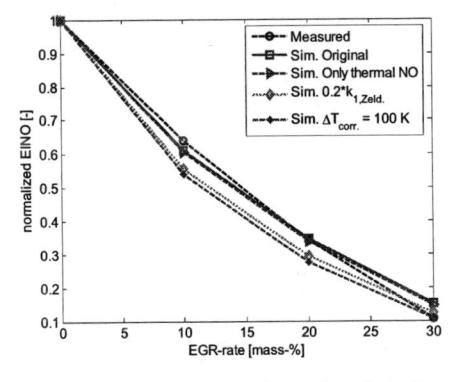

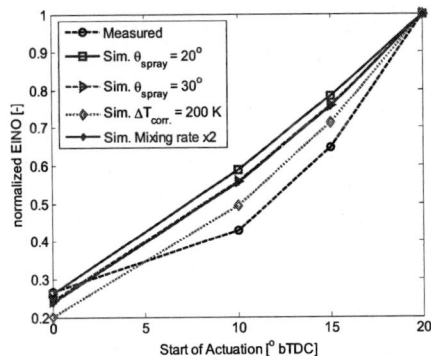

Figure 4. Measured vs. simulated EINO for varying EGR rate.

Figure 5. Measured vs. simulated EINO for varying injection timing.

Table 3: Measured vs. calculated NO (rEINO = relative NO emission index)

EGR sweep					
EGR-rate [%]	0	10	20	30	R^2
EINO measured	0.0464	0.0296	0.0158	0.0049	
rEINO original	1.65	1.57	1.67	2.34	0.993
rEINO thermal NO only	1.59	1.50	1.58	2.20	0.993
rEINO $0.2 \cdot k_{1,Zeld.}$	1.27	1.10	1.095	1.47	0.979
rEINO ΔT_{corr}= 100 K	1.082	-1.09	1.13	1.09	0.968
SOI sweep					
Timing [° bTDC]	0	10	15	20	R^2
EINO measured	0.0129	0.0207	0.0311	0.0483	
rEINO $\theta_{spray} = 20^0$	1.99	2.77	2.46	2.02	0.854
rEINO $\theta_{spray} = 30^0$	1.70	2.43	2.18	1.86	0.901
rEINO Mix. rate x2	1.57	2.27	2.06	1.75	0.905
rEINO ΔT_{corr}= 200 K	-1.32	1.04	1.03	1.10	0.956

In the same way, Figure 5 compares measured and simulated NO emission indices (normalized on the operating point with most advanced injection timing, i.e. most advanced start of injector actuation). Again, changes to the ROHR shape were not very large, and the same procedure to determine diffusion burn rate was used as with the EGR sweep results. Table 3 summarizes the results. Absolute emission indices are significantly over predicted with a factor of 2 to 2.8, and the trend with timing is only moderately well captured. This indicates that, although it has a phenomenological basis, the model does not capture all relevant physical and/or chemical dynamics. In addition, Figure 5 shows calculation results obtained with different values for the spray cone angle. From Schlieren observations of a reacting spray, an angle of 20 to 25 degree was determined (depending on ambient gas conditions). Given the difficulty to determine these outer spray edges, a range of 20 to 30 degree was considered for this comparison. Obviously there is an effect but not a pronounced one. Similarly, modifying mixing rate does not have a large effect. Imposing a significant additional T-drop when initialising a package would improve the correlation, but not the shape.

4. DISCUSSION

Absolute values of calculated and measured NO show differences in the order of a factor of two. There are a number of possible explanations for this:

1. Actual initial temperatures are lower than calculated values. In fact, direct measurements of soot radiation temperatures in direct injection diesel engines are up to 400 K lower than estimated adiabatic temperatures [25]. As our calculations show, radiative heat losses will only partly explain this difference. It is suggested that local flow field turbulence is partly responsible: this turbulence will put a strain on the reaction zones. Flamelet simulations of laminar diffusion flames [26] show that such strain reduces maximum temperatures. As shown in Table 3 a temperature correction in the order of 100 to 200 K significantly improves the agreement with measured NO values.
2. Because of turbulent fluctuations, average values of ϕ and of T correlate differently with average measured NO than in homogeneous reactor studies. This suggests that different reaction rate coefficients and activation energies should be used than those typically mentioned in the literature.
3. The NO-kinetic schemes presently used have been determined from experimental data obtained on relatively low pressure systems (typically << 10 bar). Their validity at higher pressures could be questioned. Miller et al. [27] for example use a pressure dependent correction factor on the reaction rate coefficient of the first Zeldovich reaction, eq. (1.2). For pressures above 35 bar this coefficient was decreased by a factor of 5. In Figures 4 and 5 and in Table 3 the effect of such a correction has been shown. Again, a considerable improvement on quantitative agreement results with only a marginal decrease in qualitative agreement.
4. The mixing model considers a free, quasi-steady spray and therefore disregards spray dynamics (i.e. transient spray behaviour), differences due to the confinement of multiple sprays, the effect of combustion on mixing and the likely influence on mixing of turbulence induced by in-cylinder flow. As shown in Figure 5 and Table 3 a twofold increase in mixing rate is not sufficient to improve the quantitative agreement to acceptable levels.

5. CONCLUSIONS

A simple NO emission formation model is presented that assumes that only fuel burnt during the diffusion combustion phase contributes to NO formation. This fuel reacts with surrounding charge air in stoichiometric proportions. The resulting packages of products are cooled by subsequent entrainment of fresh charge. The entrainment rate is based on a simplified 1-D quasi-steady spray model. Full NO chemistry is included as are the cooling effects of soot and gas radiation.

Comparison of measured and simulated in-cylinder NO shows that including entrainment has a strong effect on the absolute level of predicted NO. Furthermore, the reduction of engine-out NO with EGR is quantitatively well captured by the model. Similarly, the relative change in NO whith injection timing is well predicted. However, in both cases predicted absolute NO levels are significantly larger than the measured values. This indicates that there are still vital physical or chemical processes which are not fully captured. A correction on the package initial temperature was found to be a very effective measure to bring calculated and measured NO levels in line. The correction needed for matching is rather high, indicating that additional effects are involved. Adapting the chemical kinetics could be part of a further improvement. Future research will focus on these aspects to improve the NO emission formation prediction by the model. This will include a soot model for better estimation of radiation losses.

ACKNOWLEDGEMENTS

The authors want to thank TNO Automotive for its support.

REFERENCES

1. **Hohenberg, G., Gärtner, U., Daudel, H.** and **Oelschlegel, H.** A semi-empirical model for NOx simulation of HD Diesel engines, 8th Symp. „The working process of the Internal Combustion Engine", Graz, September 2001.
2. **Traver, M.L., Atkinson, R.J.** and **Atkinson, C.M.** Neural network-based diesel engine emissions prediction using in-cylinder combustion pressure. SAE paper 1999-01-1532.
3. **Khan, I.M., Greeves, G.** and **Wang, C.H.T.** Factors affecting smoke and gaseous emission from direct injection engines and a method of calculation. SAE paper 730169.
4. **Chiu, W.S., Shahed, S.M.** and **Lyn, W.T.** A transient spray mixing model for diesel combustion. SAE paper 760128.
5. **Heider, G.** Rechenmodell zur Vorausrechnung der NO-Emission von Dieselmotoren, PhD-thesis, Munich Technical University, 1996.
6. **Shahed, S.M., Chiu, W.S.** and **Yumlu V.S.** A preliminary model for the formation of nitric oxide in direct injection diesel engines and its application in parametric studies. SAE paper 730083.
7. **Dec, J.E.** and **Canaan, R.E.** PLIF imaging of NO formation in a DI diesel engine, SAE paper 980147.
8. **Flynn, P.F., Durrett, R.P., Hunter, G.L., zur Loye A.O., Akinyemi, O.C., Dec, J.E.** and **Westbrook, Ch.K.** Diesel combustion : an integrated view combining laser diagnostics, chemical kinetics, and empirical validation, SAE paper 1999-01-0509.

9. Murayama T., Miyamoto, N., Sasaki, S. and Kojima, N. The relation between nitirc oxide formation and combustion process in diesel engines. Proc. 12th Int. Congr. on Combustion Engines, Vol. B, p 1-23, CIMAC, Tokyo 1977.
10. Ahmad, T. and Plee, S.L. Application of flame temperature correlations to emissions from a direct-injection diesel engine, SAE paper 831734.
11. Wu, K.-J. and Petersen R.C. Correlation of nitric oxide emission from a diesel engine with measured temperature and burning rate, SAE paper 861566.
12. Timoney, D.J., Desantes, J.M., Hernandez, I. and Lyons, C.M. The development of a semi-empirical model for rapid Nox concentration evaluation using measured in-cylinder pressures in diesel engines. Proc. I.Mech.E., Vol. 129, Part D, Jl. Aut. Engng., p. 621-631, 2005.
13. Dodge, L.G., Leone, D.M., Naegeli, D.W., Dickey, D.W. and Swenson, K.R. A PC-based model for predicting NO_X reductions in diesel engines. SAE paper 962060.
14. Chikahisa, T., Konno, M. and Murayama T. Analysis of NO formation characteristics and control concepts in diesel engines from NO reaction-kinetic considerations, SAE paper 950215.
15. Andersson, M., Johansson, B., Hultqvist, A. and Nöhre A real time NOx model for conventional and partially premixed diesel combustion, SAE paper 2006-01-0195.
16. Mellor, A.M., Mello, J.P., Duffy, K.P., Easley, W.L. and Faulkner, J.C. Skeletal mechanism for NOx chemistry in diesel engines, SAE paper 981450.
17. Verbiezen, K Quantitative NO measurements in a diesel engine, PhD-thesis, Radboud University Nijmegen, The Netherlands, ISBN-10 90-9021435-6, 2007.
18. Heywood, J.B. Internal Combustion Engines Fundamentals, McGraw-Hill, London, 1988, ISBN 0-07-028638-8.
19. Turns, S.R. Understanding NO_x formation in nonpremixed flames: experiments and modeling, Prog. Energy Combust. Sci., Vol. 21, pp. 361 – 385, 1995.
20. Chmela, F.G. and Orthaber, G.C., Rate of heat release prediction for direct injection diesel engines based on a purely mixing controlled combustion. SAE paper 1999-01-0186.
21. Naber, J.D. and Siebers, D.L Effects of gas density and vaporization on penetration and dispersion of diesel sprays, SAE paper 960034.
22. Siebers, D.L, Higgins, B., and Pickett, L. Flame lift-off on direct-injection diesel fuel jets: oxygen concentration effects, SAE paper 2002-01-0890.
23. Desantes, J.M., Payri, R., Salvador, F.J. and Gil, A. Development and validation of a theoretical model for diesel spray penetration, Fuel 85, pp. 910 – 917, 2006.
24. Musculus, M. On the correlation between NO_x emissions and the diesel premixed burn, SAE paper 2004-01-1401.
25. Struwe, F.J. and Foster, D.E. In cylinder measurement of particulate radiant heat transfer in a direct injection diesel engine, SAE paper 2003-01-0072.
26. Evlampiev, A Numerical combustion modeling for complex reaction systems, PhD-thesis, Eindhoven University of Technology, 2007.
27. Miller, R., Davis, G., Lavoie, G., Newman, C. and Gardner, T. A super-extended Zeldovich mechanism for NO_x modeling and engine calibration, SAE paper 980781.

The porous medium approach applied to CFD modelling of SCR in an automotive exhaust with injection of urea droplets

S.F. Benjamin, C.A. Roberts
Automotive Engineering Applied Research Group, Coventry University, UK

ABSTRACT

The automotive industry is expected to adopt SCR after-treatment to control NOx emissions from Diesel passenger cars from 2010. Ammonia promotes NOx reduction and is introduced into the exhaust as a spray of aqueous urea droplets. This is a new aspect of CFD modelling of exhaust after-treatment. When modelling sprays the mesh must be 3D and so the porous medium approach is appropriate, circumventing the need for representative single channel modelling. The porous medium technique is well established for modelling three-way catalysis and its application to SCR is demonstrated in this paper, using a kinetic scheme available in the literature. Laboratory measurements of droplet diameters have been used to specify the input of aqueous urea to the CFD model. Representative droplet parcels are modelled using a Lagrangian model within the CFD code. In this way it is possible to fully model SCR in a 3D model of an automotive catalyst system.

NOTATION

A_V geometric surface area per unit reactor volume (m^2/m^3)
A_{PM} catalyst precious metal surface area per unit reactor volume (m^2/m^3)
C mass fraction
$C_{i\,g}$ mass fraction of species i in the gas phase
$C_{i\,sol}$ mass fraction of species i in the solid phase or washcoat pores
D species diffusivity (m^2/s)
D_d droplet diameter (microns)
K_{mi} mass transfer coefficient (m/s)
L channel length, monolith length (m)
M_i molar mass for species i (kg/mol)
ΔP pressure drop (Pa)
P_{O2} oxygen concentration mole fraction
q fit variable for Rosin-Rammler droplet size distribution

© IMechE 2007

Q	fraction of total volume in drops with diameter $< D_d$
R_i	rate of production of species i by reaction (mol /s /m³ reactor)
t	time (s)
T	temperature (K)
U	velocity (m/s)
U_C	velocity in substrate channel (m/s)
U_S	superficial velocity for the porous medium, $\varepsilon\, U_C$ (m/s)
V_w	solid phase pore volume per unit volume of reactor (m³ /m³ reactor)
x,y	coordinates
X	fit variable for Rosin-Rammler droplet size distribution
z	axial coordinate
ε	porosity of the substrate expressed as a volume fraction
μ_t	turbulent dynamic viscosity (kg/(m s))
ρ	density (kg/m³)
σ_s	turbulent Schmidt No.
θ	fraction of surface coverage by ammonia, ammonia storage

1 INTRODUCTION

In recent years various strategies have been rapidly developed for controlling the emissions from Diesel engine exhausts, and these were reviewed by Johnson (1). The technology was regarded as at an early stage of development (2) as recently as 2003, and Johnson (1) regarded selective catalytic reduction (SCR) as a pertinent technology only for heavy duty vehicles, with lean NOx trap (LNT) and Diesel particulate filter (DPF) technology earmarked for light duty, which includes the automotive market. It now looks likely that the preferred after-treatment technology for controlling NOx emissions in the exhaust of light duty diesel vehicles will be SCR rather than the alternative LNT system. This technology will be implemented in the period up to 2010. One disadvantage of the LNT system is the need to regenerate the NOx trap periodically by running the engine rich for short periods of about 3 seconds at intervals of typically one to two minutes. This causes some NOx slippage during the regeneration period and also a fuel penalty. This is discussed in some detail by Alimin (3). There is also the requirement to use ultra low sulphur Diesel fuel with the LNT. The SCR system typically consists of two bricks, a Diesel oxidation catalyst (DOC) followed by an SCR brick. The control system is simpler than the LNT in that SCR operates continuously without the need for regeneration, but an additional agent is required to promote the catalytic reactions, usually aqueous urea that is sprayed into the exhaust. The SCR system is therefore more complex in terms of hardware and ammonia slippage must be avoided. SCR technology has been developed over the last couple of years on heavy duty systems, for example in the work done by Gekas et al. (4). Both LNT and SCR systems may additionally require a DPF to be installed upstream to minimise soot particulate levels.

Although SCR has been proven as a technology for stationary engines and plant, the need for an additional reducing agent in a passenger car could be seen as a disadvantage of this system because of storage and filling issues. The foreseen problem with the infrastructure to supply the aqueous urea is, however, gradually being overcome, particularly in Europe where SCR technology is already in use on heavy duty vehicles. There have also been some questions about the safety of the principal catalyst chosen for SCR, vanadium. There are now, however, promising zeolite catalysts that can be used instead (5, 6).

Diesel exhaust lean after-treatment systems are expensive to develop and test and there are advantages in being able to model such systems to predict their performance. Computational fluid dynamics (CFD) modelling is particularly useful because the flow field, which in exhaust systems is complex because of packaging constraints and contrived exhaust component geometries, can be predicted. The chemistry of the catalytic reactions can be conveniently introduced into a CFD model if the porous medium approach is used to model the flow field. In this method each monolithic catalyst substrate with parallel flow channels is modelled as a porous medium that resists the flow (7). The particular challenge of modelling SCR systems is in modelling the introduction to the computational domain of the aqueous urea droplets. Making fundamental predictions of droplet formation within a nozzle is very difficult, and also dependent on the nozzle geometry. Therefore a more generally applicable semi-empirical approach is adopted. This requires the spray to be characterised by measurement of the droplet sizes and distributions just downstream of the nozzle. The drops are then modelled by the introduction of representative droplet parcels into the computational domain, where the parcels have known properties. This is similar to the work of Kim et al. (8) on a marine engine system.

In the work reported here, some measurements have been made to characterise the spray from a prototype automotive spray system, and a full 3D SCR model has been developed based on the porous medium approach. The CFD model mesh has the dimensions of an experimental exhaust system and predictions of emissions levels can be made. Emissions measurements for comparison with the model output will be available from an experimental test rig to validate the model at a later date. In this paper, the droplet parcel model is compared with spray measurements and the effect of droplet size on the emissions predictions from the full SCR system model is assessed. In this way a useful engineering tool has been devised by a new application of the porous medium modelling approach.

2 THEORY

2.1 SCR kinetic scheme
In order to model an SCR system with a CFD model it is necessary to have a kinetic scheme with rate constant values that can provide numerical values for reaction rates. Such a scheme is given by Chi et al. (9) and consists of rates for eight chemical processes. These are HCNO hydrolysis, ammonia adsorption and desorption, two alternative ammonia oxidation reactions, standard, fast and slow SCR reactions.

All the required rate constants for this scheme are given in the literature. The rates are calculated in mol/m^2/s units, so it is necessary to have an estimate of active catalyst area per m^3 of reactor in order to apply the scheme within the porous medium model. The scheme is for a vanadium catalyst. An alternative kinetic scheme for vanadium is also available in the literature (10, 11). One of the two ammonia oxidation reactions is

$$4NH_3 + 3O_2 \rightarrow 2N_2 + 6H_2O$$

The rate (mol/m^2/s) for this reaction (9) is $1.32E+07 \, \theta \exp(-15024/T)$

The rate (mol/m^3/s) for this reaction (10) is $1.1E+09 \, \theta (P_{O2}/0.02)^{0.27} \exp(-14193/T)$

where P_{O2} is oxygen mole fraction. The rates of ammonia oxidation are calculated from the two schemes (9, 10) for oxygen mole fraction of 0.07, ammonia surface coverage of 50 % and T at 550 K. If the two schemes are assumed equivalent then comparison of the calculated values for the specified conditions implies that about 533 m^2/m^3 is the catalyst active surface area. This is an unexpectedly small target area value when compared with A_v and A_{PM} surface values used in previous porous medium catalyst models (12), which are of order 10^3 and 10^4 respectively.

The standard SCR reaction is

$$4NH_3 + 4NO + O_2 \rightarrow 4N_2 + 6H_2O$$

The rate (mol/m^2/s) for this reaction (9) is
$(1/X) \, 2.36E+08 \, [NO] \, [NH_3] \exp(-7151/T)$

where X is $(1 + 0.0012042 \, [NH_3])$

Rate(mol/m^3/s) from (10) is
$2.2E+08 \, C_{NO}(\theta(1-\theta)/(1+7.2\theta))(P_{O2}/0.02)^{0.27}\exp(-6615/T)$

The rates of the standard SCR reaction can be calculated assuming [NO] concentration to be 0.002 mol/m^3 and [NH$_3$] concentration to be 0.002 mol/m^3. It is clear that the calculated rate from the Chi et al. scheme (9) is directly proportional to the ammonia concentration, whereas this dependence is not present in the Tronconi scheme (10). Comparison of the calculated values from the two schemes suggests an implicit factor of 100 m^2/m^3 for active catalyst surface area if the two schemes are equivalent. This value is even smaller than that deduced above for ammonia oxidation and suggests that the two schemes are not equivalent. The Chi et al. scheme (9) for vanadium can easily be substituted in the CFD model by an improved kinetic scheme or by kinetics specifically for zeolite. Such a scheme is now available, (13).

2.2 SCR CFD methodology
Modelling of SCR requires the introduction of the urea, which provides the ammonia for the reaction scheme, usually in the form of an aqueous spray. Spray

modelling using the Lagrangian approach is by definition transient and 3D. The porous medium approach (7) lends itself readily to full 3D modelling and can predict the flow field and conversion of species. The resistance of the porous medium to flow is described by the expression

$$\frac{\Delta P}{L} = -\alpha U_S^2 - \beta U_S \qquad [1]$$

where α and β are temperature dependent permeability coefficients for the porous medium. High values of α and β disallow flow at right angles to the axis of the porous medium block. A monolithic catalyst, where the flow is constrained within parallel channels, can therefore be modelled. Note that U_S is the superficial velocity for the porous medium such that $[\rho_{air} U_S]$ equals $[\varepsilon \rho_{air} U_C]$. The flow field is solved using the usual Reynolds averaged Navier Stokes methodology in the fluid upstream and downstream of the porous medium that represents the catalyst substrate.

For the porous medium approach the CFD model has a block of cells representing the fluid inlet. This is followed by the porous medium cells, which are used to model fluid flow through the catalyst monolith. A final block of cells models the fluid outlet. An extra block of cells with the same geometry as the porous medium cell block represents the solid properties of the monolith for heat transfer to the walls and conduction in the substrate. Enthalpy exchange between porous medium (fluid) cells and solid cells is described by source terms and appropriate heat transfer coefficients. The heat conduction equation is solved in the solid cell blocks and an effective radial thermal conductivity describes heat conduction transversely across the monolith. If the axial thermal conductivity is significantly greater than the radial value, then this can be accounted for between adjacent cells in user subroutines by an additional source term.

The general conservation equation for the transport of chemical species is shown in its full 3D version in [2] below, where the species source has units kg/m^3/s.

$$\frac{\partial(\rho C)}{\partial t} + \nabla \bullet (\rho U C) - \nabla \bullet \left[\left[\frac{\mu_t}{\sigma_s}\right] + \rho D\right] \nabla C \right] = \text{Source} \qquad [2]$$

The model applies modified forms of this equation to gas phase scalars in the fluid and to either gas or solid phase scalars in the porous medium. The transient term is always included. The convective term is excluded for solid phase scalars, but retained for gas phase scalars. The diffusion flux term only applies to gas phase scalars in the fluid. The diffusion flux along the porous medium is insignificant when compared with the convective flux. The source term applies to both gas and solid phase scalars in the porous medium. It describes the net effect of diffusion of species between the gas stream and the washcoat pores on the channel wall, i.e. the solid phase, as a mass transfer process. Hence the source term replaces the diffusion flux term between gas and wall. The source is calculated using a mass transfer coefficient derived from the thin film approximation. Further details of this approach are given in (7).

The modelling methodology requires that both the gas phase and solid phase species concentrations are properties of the porous fluid cells, since only the heat conduction equation with its heat transfer source term is solved for the solid cells. Generally, species sink from the gas phase and transfer to the solid phase. The net effect of the chemical reactions, however, is often that species sink from the solid phase in a catalyst system, although for some species the reactions provide a source in the solid phase. Equations [3] to [6] presented below are laid out to clarify the requirement for inclusion of the porous medium porosity, ε. A review article by Depcik et al. (14) mentions an apparent void fraction discrepancy in the history of catalyst modelling. In the CFD model here, the monolith axis is aligned in the z direction and there is no convection of species in the x and y directions in the porous medium. The equations below are presented in 1D form for clarity.

Equation [2] applied to an element of air in a monolith channel, with the diffusion flux term replaced by a source term, kg/s/m^3 air in channel, is written as [3].

$$\frac{\partial}{\partial t}[\rho_{air} C_{i\,g}] + \frac{\partial}{\partial z}[\rho_{air} U_C C_{i\,g}] = -K_{mi}\, \rho_{air}\, \frac{A_V}{\varepsilon}\, [C_{i\,g} - C_{i\,sol}] \qquad [3]$$

In [3] A_V/ε is active surface area per unit volume of air in the channel. For the gas phase species in the porous medium computational cell that represents the bulk monolith, equation [4] below is solved. The term on the RHS of [4] is the source term coded into the user subroutine.

$$\varepsilon \frac{\partial}{\partial t}[\rho_{air} C_{i\,g}] + \frac{\partial}{\partial z}[\rho_{air} U_S C_{i\,g}] = -K_{mi}\, \rho_{air}\, A_v\, [C_{i\,g} - C_{i\,sol}] \qquad [4]$$

Typically, the catalyst washcoat occupies about 10 % of the whole monolith reactor volume and the pores in the washcoat occupy about 50% of the washcoat volume. Thus only about 1/20 of the reactor volume is available in the solid phase. Equation [2] applied to an element of air inside a pore in the catalyst washcoat at the channel surface, with the diffusion flux and convective terms suppressed, is written as equation [5], where the source term has units kg/s/m^3 air in the washcoat pore.

$$\frac{\partial}{\partial t}[\rho_{air} C_{i\,sol}] = K_{mi}\, \rho_{air}\, \frac{A_V}{V_w}\, [C_{i\,g} - C_{i\,sol}] + M_i\, \frac{R_i}{V_w} \qquad [5]$$

In [5] A_V/V_w is active area per unit volume of air in the pore and R_i / V_w is the net reaction rate for species i per unit volume of the pore. Equation [6] below is solved by the CFD solver. This enables the species concentration in the pore, i.e. in the solid phase, to be obtained in the porous medium computational cell that represents the bulk monolith. The source term on the RHS of [6] is coded into the user subroutine.

$$\varepsilon \frac{\partial}{\partial t}[\rho_{air} C_{i\,sol}] = \{K_{mi}\, \rho_{air}\, A_v\, [C_{i\,g} - C_{i\,sol}] + M_i\, R_i\}\, [\varepsilon / V_w] \qquad [6]$$

With the source terms coded into the CFD model via user subroutines, the transport equations are solved to provide values for the mass fractions of species in the exhaust stream.

The 3D mesh for the SCR CFD model is shown in Figure 1 and has 110016 cells in total. Droplet parcels are injected at Z 27 mm. The slow, approximately 10 degrees, expansion cone replicates the geometry of an experimental exhaust on an engine test rig. The porous medium cells represent the SCR brick and the corresponding solid cells are a separate cylindrical block of 23040 cells. A block of fluid cells downstream of the porous medium completes the mesh. Another model consisting of a simple rectangular block mesh of 80 by 80 by 250 cells was used to separately investigate the functioning of the droplet spray model. The inlet duct is 50 mm in diameter and the catalyst brick is 118 mm in diameter. The catalyst brick is 182 mm in length.

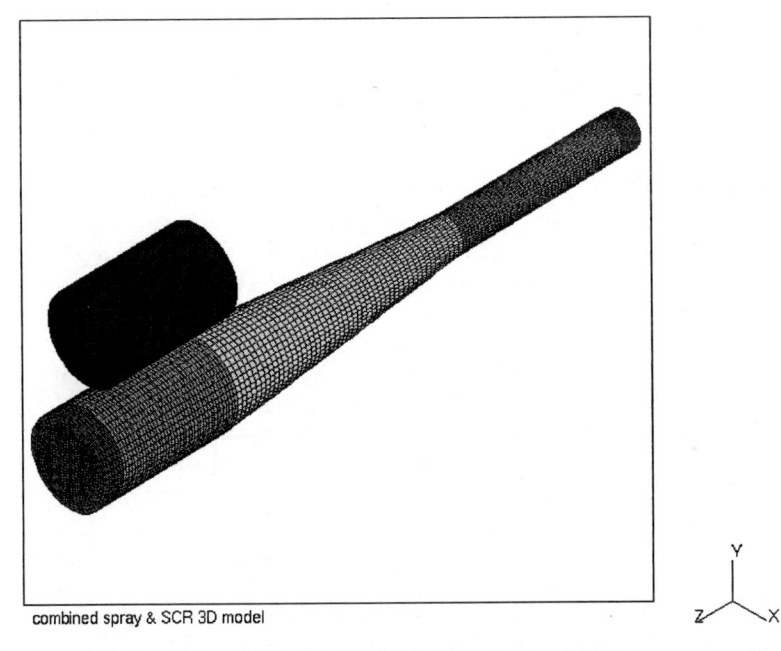

Figure 1: 3D mesh for SCR CFD model. The slow expansion cone replicates the geometry of a test rig and ducts the exhaust into the porous medium. The separate cylindrical block of cells models the solid properties of the porous medium cells.

The commercial CFD package Star-CD Version 3.26 was used for the studies described in this paper. The models were run on a 16 node Itanium-2 64-bit cluster under HP-UX. The rectangular block mesh for the droplet model studies was partitioned into 4 sets for parallel runs and the SCR model mesh, shown in Figure 1, was partitioned into 10 sets for parallel runs. When using the porous medium technique it is always necessary to keep together corresponding fluid and solid cells in the same partitioned set.

2.3 Droplet sub-model methodology

The urea spray droplets enter the domain. The water is driven off as the droplet temperature approaches 373 K, leaving solid or semi-molten urea spheres. At droplet temperatures of about 410 K these should sublime or vapourise to produce gaseous urea that rapidly dissociates.

$$CO(NH_2)_2 \rightarrow NH_3 + HCNO$$

The HCNO, produced by dissociation, hydrolyses with the excess of water present in the exhaust flow and spray; this is included in the kinetic scheme. Each HCNO mol reacts to produce one mol of ammonia so that each mol of urea ultimately supplies 2 mol of ammonia (15).

The urea is introduced into the domain in the rig with an air assisted spray unit. In order to avoid the need to fully model droplet formation in the nozzle, the model uses discrete parcel input. For example, 48 mg/s of urea spray can be modelled by 480,000 parcels/s in short runs < 0.1 second, or by 32,000 parcels per second in longer runs, for example 3 seconds. It is necessary to control the total number of droplet parcels to facilitate sufficiently rapid and convenient processing of the model output. Each parcel has mass in the range 0.1 to 1 µg but represents many individual droplets. CFD simulation time steps are arranged so that a few parcels enter per time step. Parcels are specified to the CFD package by size or size distribution, injection direction, mass flow rate and orifice size. The latter two parameters together fix the injection velocity. Also, the spray cone angle is specified. It has been found beneficial to partition the spray into five or six zones, see Figure 2, and to choose the parameter values for each zone to contour the spray to achieve predictions that compare favourably with measurements of a cold spray in a cold flow air. The real spray has a single orifice and the model has a single point of injection. Also, it was found necessary to a use a fluid source in the model to simulate the air assist to the prototype spray; it was not possible to model the detail of the behaviour of the spray by parcel injection alone. Although in reality the air assist air flow was choked, it was found that using a lower injection velocity but with the source having flux of the correct magnitude injected into a single mesh cell gave good agreement with measurements, as discussed in section 3.1. The cells in the vicinity of the air injection were refined but the dimension across the source cell was larger than the real nozzle orifice diameter.

The droplet model assumed rebound from the wall for any droplet impinging on the wall and neglected any effects of droplet collision. The geometry of the real system was such that the fairly narrow spray, 20 to 25 degrees angle, did not impinge on the wall of the duct within the computational domain when there was a surrounding air flow.

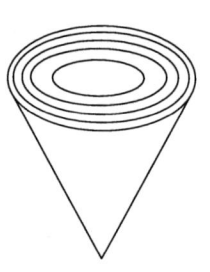

Figure 2: Schematic diagram of spray cone shown with five zones, with each annulus and the central circle having the same area.

2.4 Droplet measurements for input to CFD model

The droplets from the air-assisted prototype spray were measured by a TSI Phase Doppler Particle Anemometry (PDPA) system. This was operating in forward scatter mode. A typical droplet distribution is shown in Figure 3 for the spray discharging 48 mg/s. Data was for 30,000 valid droplets. The spray is characterised by large numbers of small droplets but small numbers of much larger droplets that contribute significantly to the volume flux. Rosin Rammler fit profiles are shown in Figure 4, indicating the uniformity of the spray when spraying into cold quiescent air. The parameters X and q given in Figure 4 are related by the standard expression

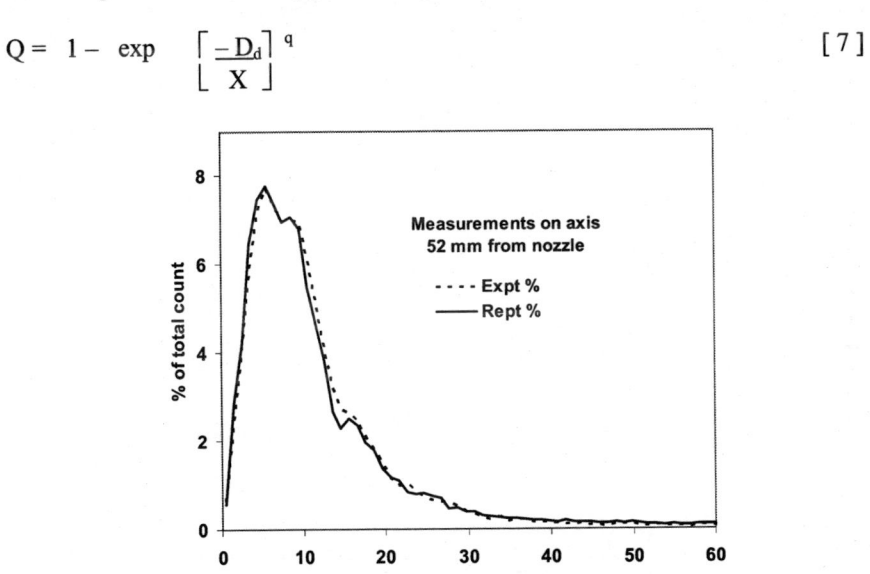

$$Q = 1 - \exp\left[\frac{-D_d}{X}\right]^q \qquad [7]$$

Figure 3: Typical droplet distribution. [Expt, experiment; Rept, repeat]

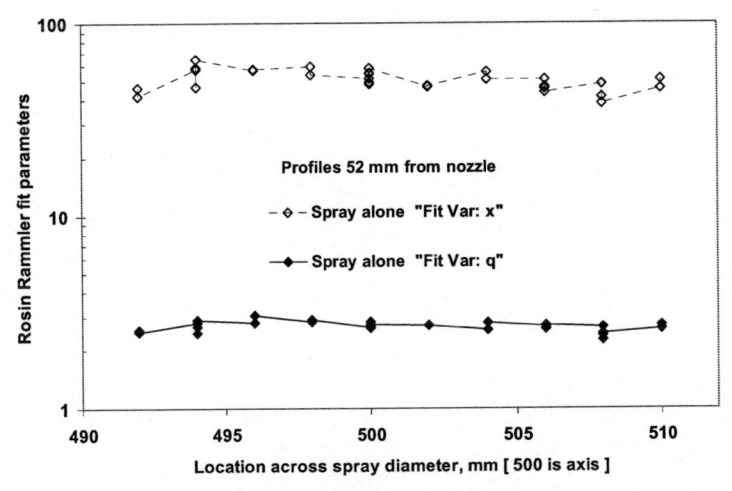

Figure 4: Profiles of Rosin Rammler fit parameters across spray spraying horizontally.

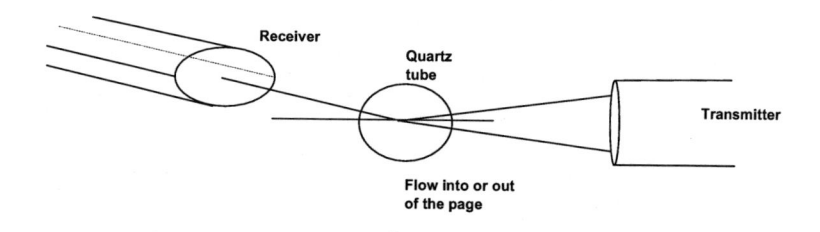

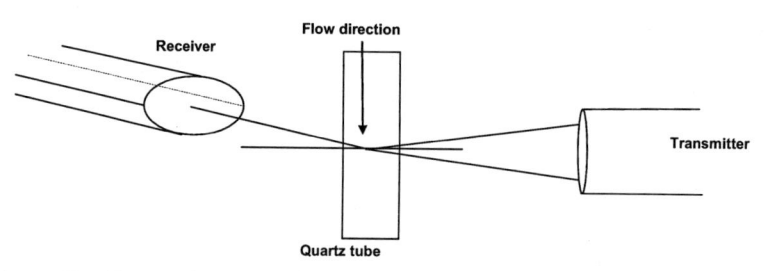

Figure 5: Alternative orientations for PDPA measurements of droplet diameters within quartz tube by forward scatter with receiver probe offset by 30 degrees.

The spray discharged through a duct of 50 mm diameter, a typical dimension for an automotive exhaust. The PDPA measurements of the spray could be made at the exit from different length tubes, either with or without additional air flow through the test rig. This technique was chosen because early measurements showed that the effects of measuring through a thin walled, 1.5 mm, quartz tube with the PDPA system were significant and attributable to refraction effects, rather than to the confining effect of tube. In the observations made by the present authors this effect could not be avoided by altering the orientation of the PDPA system relative to the tube, see Figure 5. Similar problems were anticipated even with angled plane windows. A similar effect was observed by Yimer et al. (16) who made PDPA measurements through two concentric quartz tubes of small diameters, 26 and 36 mm. Although they comment on refraction effects they seem to attribute the differences they observed in their measurements to the confining effect of the tube. In view of the uncertainty introduced by the quartz tube, the technique of using the short tubes was adopted here to circumvent the problem.

3 RESULTS FROM CFD MODELS

3.1 Injection of droplets into steady flow – comparison of CFD with data

Measurements made using the PDPA system of droplet diameters and velocities were compared with CFD predictions. The cold water spray was spraying into a steady cold air flow of about 11 m/s. The CFD simulation was run for a real time of 1.5 seconds with time steps of 0.00025 seconds. The droplet parcel injection rate to the model was 32,000 parcels/s.

Figure 6 shows fairly good agreement of predictions with the way that the droplet velocity profile is observed to decay in the experiments. The peak velocity falls with distance from the nozzle and the velocity falls across the spray to the surrounding air flow at the spray periphery. Figure 7 again shows the decay of the peak velocity along the axis and agreement between measurements and predictions is seen to be quite good when using a 75 m/s air injection in a single cell to model the air assist of the spray. Agreement between measurements and predictions is less good for droplet diameters in Figure 8, which shows D10 and D32 along the axis. The measurements show less change along the axis than shown by the CFD predictions. The profiles of the Sauter mean diameter in Figure 9 show good agreement on one side of the axis but fairly poor agreement on the other side. This is believed to be a feature of the prototype spray itself. The droplet CFD model developed from these simple measurements and CFD simulations was then applied to a full system SCR model.

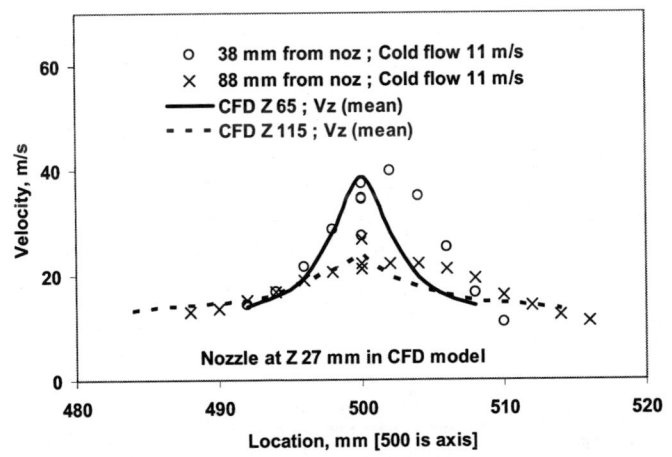

Figure 6: Droplet velocities measured and predicted across spray to show profiles.

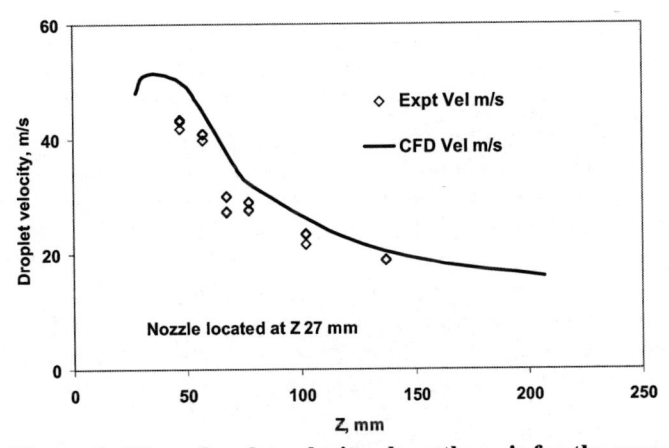

Figure 7: Mean droplet velocity along the axis for the spray through a tube with a surrounding 11 m/s air flow.

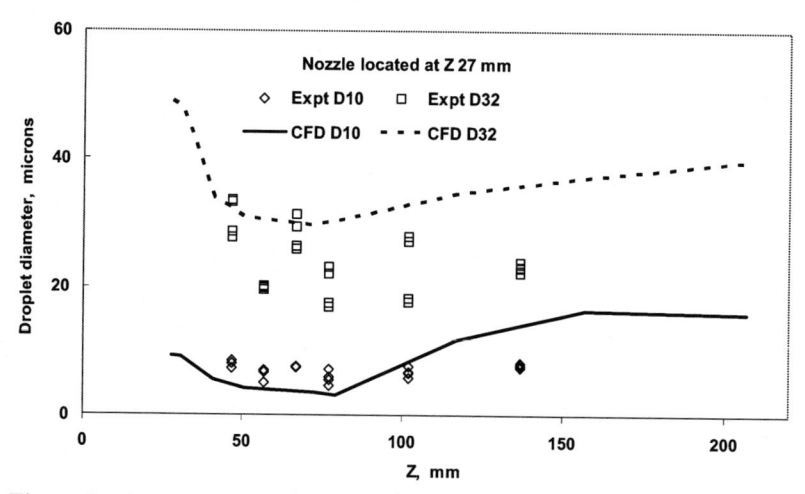

Figure 8: Comparison of drop diameter measurements with predictions.

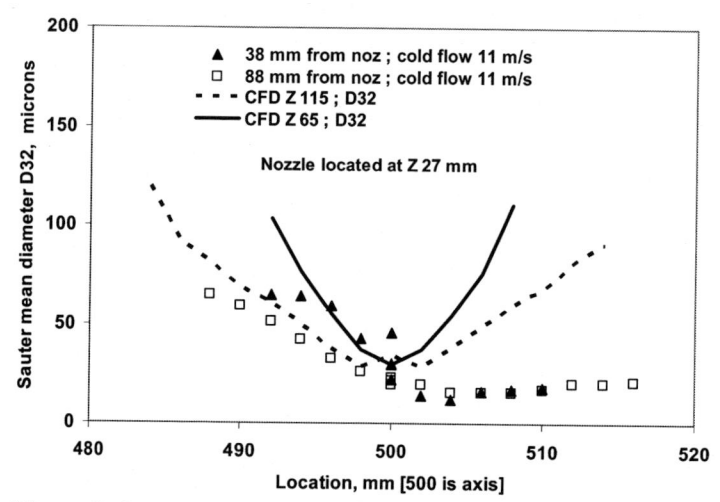

Figure 9: Sauter mean diameter profiles for droplets measured in 11 m/s cold flow, compared with CFD predictions.

3.2 Full model feasibility study

A full model simulation was carried out. This was a transient simulation that was run for 3 seconds of real time. The inlet flow was 25 m/s at 580 K. The inlet NO mass fraction was 0.00042 and the inlet NO_2 mass fraction was 0.00032. The inlet O_2 mass fraction was 9%. The inlet amount of aqueous urea was 48 mg/s entered as 20,000 droplet parcels per second. The Rosin Rammler fit parameters used to describe the spray were based on the measurements described in 3.1. The simulation took 50 hours of cpu time. The case was run in parallel across 10 processors.

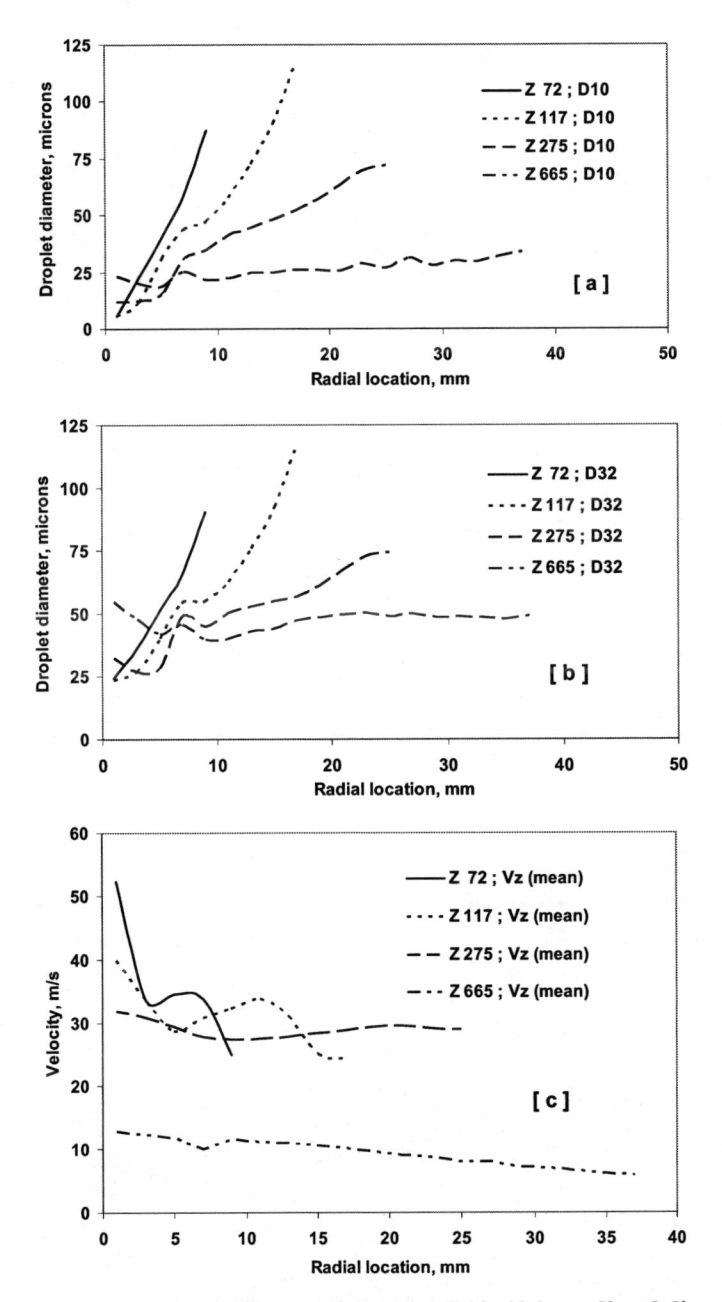

Figure 10: (a) Predicted droplet diameters, D10, (b) predicted diameters, D32, and (c) predicted velocities at four different axial locations in the full CFD model.

Figures 10 shows predictions for Z 72 and 117 mm, that is 45 and 90 mm from the droplet injection point, and also for Z 275 and 665 mm. These locations are just before the expansion cone and just before the catalyst. The model suggests fairly uniform droplet diameters and velocities at entry to catalyst, but the catalyst is 118 mm in diameter and no droplets are predicted to reach the periphery. Few droplets are found beyond 35 mm radius in the simulation. Figure 11 shows the predicted flow velocity at the exit from the SCR catalyst. The slow expansion cone, see Figure 1, has flattened the velocity profile but has not achieved a completely uniform profile.

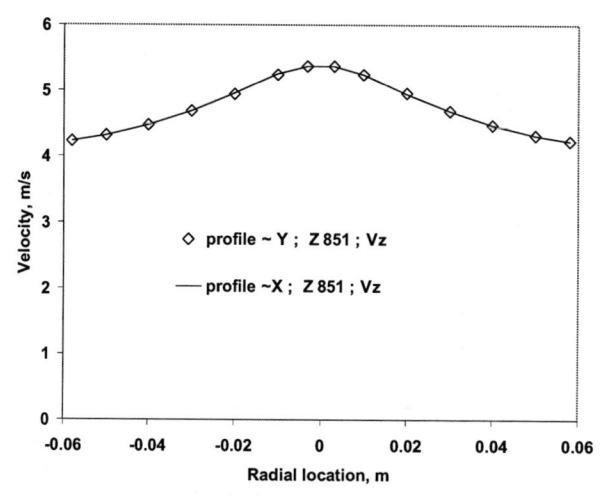

Figure 11: Predicted air velocity at Z 851 mm, the exit from the SCR catalyst in the full CFD model.

The consequence of the spray distribution is fairly poor NOx conversion at the periphery, see Figure 12. The NOx levels in Figure 12 are shown as mass fractions; hence the velocity profile in Figure 11 determines the NOx mass flow rate profiles. The ammonia distribution is also seen in Figure 12 to be non uniform at entry to the catalyst, with negligible levels at the exit from the brick. This suggests that the spray spatial distribution is not sufficiently uniform and that the amount of aqueous urea input to the model is too low.

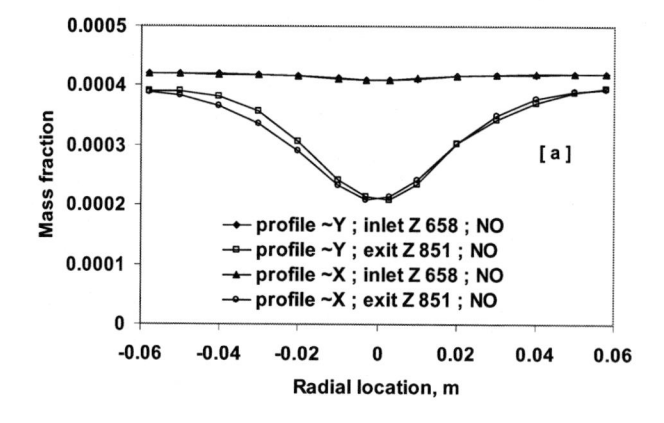

[a]

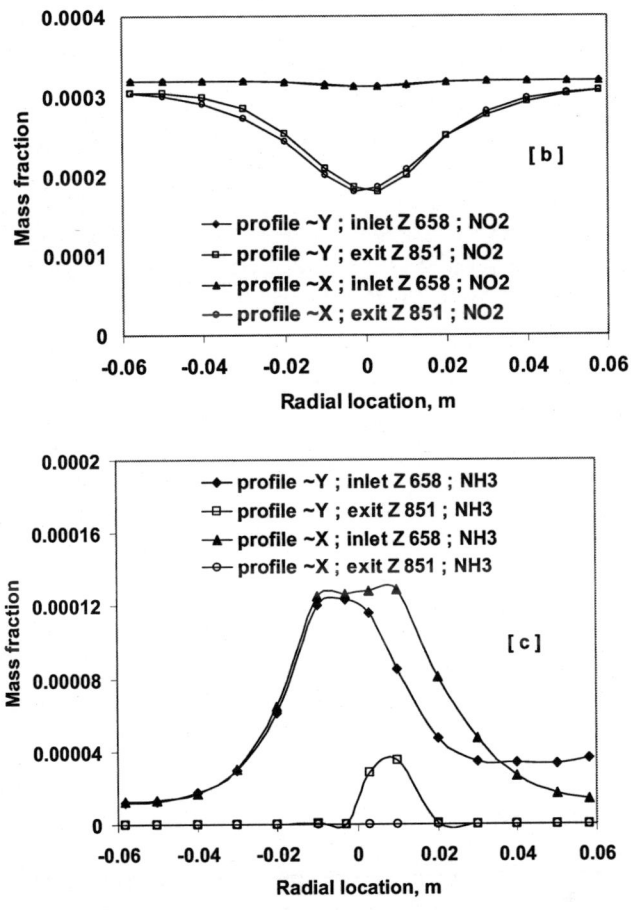

Figure 12: (a) Predicted NO, (b) predicted NO_2 and (c) predicted NH_3 profiles at inlet (Z 658 mm) and outlet (Z 851 mm) of SCR catalyst, along both X and Y axes.

4 SUMMARY OF RESULTS AND CONCLUSIONS

This paper has presented the results of droplet measurements using a phase Doppler particle anemometry system to measure both droplet diameter and velocity. The measurements were made at the exit from various length tubes to avoid the need to measure through quartz, which can introduce uncertainty into the measurements. The droplets were found to have mean diameter D10 less than 20 microns but the diameter varied across the spray profile so that larger droplets were at the edge of the spray field. The Sauter mean diameter D32 was about 30 microns on the axis but considerably larger, more than 60 microns, away from the axis. The CFD model was shown to be able to predict the droplet diameters and velocities at various distances from the nozzle. Both the change in droplet size along the axis and profiles at different distances from the axis were measured and predicted satisfactorily.

The porous medium approach can be used to model SCR catalysis in an automotive exhaust context. The spray model validated against PDPA measurements in cold flow studies was incorporated into the full SCR model. A kinetic scheme from the literature was evaluated and applied within the model. The results demonstrate the feasibility of the methodology, which incorporates a droplet sub-model into a model based on the porous medium approach. The output from the model indicates the limitations of the spray investigated here as a means of introduction of ammonia. They show that this spray is probably of too narrow an angle, and insufficiently well mixed and that the amount of urea injected is too low in the simulation. The spray angle and droplet size both influence how well the droplets convert to ammonia in the real exhaust system.

The work described here has been extended to an alternative spray and will ultimately be tested against species measurements in an engine exhaust. Matching the amount of urea injected to the levels of NOx in the exhaust is an ongoing challenge, particularly under transient conditions, but a validated CFD model is a useful tool in assisting further developments.

ACKNOWLEDGEMENTS

Technical and financial support from Jaguar-LandRover, EMCON Technologies and Johnson Matthey is gratefully acknowledged.

REFERENCE LIST

1. **Johnson, T. V.** Diesel emission control in review – the last 12 months. SAE paper No. 2003-01-0039. Also published in Diesel Exhaust Emissions Control, SAE SP-1754

2. **Twigg, M. V.** Automotive exhaust emissions control. Platinum metals Rev. Vol. 47 (4) (2003) pp 157 – 162

3. **Alimin, A. J.** Experimental investigation of a NOx trap using fast response emission analysers. PhD thesis, Coventry University, UK, 2006

4. **Gekas, I., Gabrielsson, P., Johansen, K.** Urea-SCR catalyst system selection for fuel and PM optimized engines and a demonstration of a novel urea injection system. SAE paper No. 2002-01-0289

5. **Baik, J. H., Yim, S. D., Nam, I-S., et al.** Control of NOx emissions from diesel engine by selective catalytic reduction (SCR) with urea. Topics in Catalysis. Vols 30/31 (July 2004) pp 37 – 41

6. **Li, G., Jones, C. A., Grassian, V. H., Larsen, S.C.** Selective catalytic reduction of NO2 with urea in nanocrystalline NaY zeolite. Journal of Catalysis Vol. 234 (2005) pp 401 – 413

7. **Benjamin, S. F., Roberts, C. A.** Three-dimensional modelling of NOx and particulate traps using CFD: A porous medium approach. Applied Mathematical Modelling Vol. 31 (2007) pp 2446 – 2460

8. **Kim, J. Y., Ryu, S. H., Ha, J. S.** Numerical prediction on the characteristics of spray-induced mixing and thermal decomposition of urea solution in SCR system. Proceedings of ICEF04, 2004 Fall Technical Conference of ASME Internal Combustion Engine Division. Paper No. ICEF2004-889

9. **Chi, J. N., Dacosta, H. F. M.** Modelling and control of a urea-SCR after treatment system. SAE Paper No. 2005-01-0966

10. **Chatterjee, D., Burkhardt, T., Bandl-Konrad, B., et al.** Numerical application of ammonia SCR catalytic converters: Model development and application. SAE Paper No. 2005-01-0965

11. **Tronconi, E., Nova, I., Ciardelli, C., et al.** Modelling of an SCR catalytic converter for diesel exhaust after-treatment: Dynamic effects at low temperature. Catalysis Today Vol. 105 (2005) pp 529 – 536

12. **Benjamin, S. F., Roberts, C. A.** Automotive catalyst warm up to light off by pulsating engine exhaust. Int J Eng Res Vol. 5 No. 2 (2004) pp 125 – 147

13. **Chatterjee, D., Burkhardt, T., Weibul, M., et al.** Numerical simulation of Zeolite and V-based SCR catalytic converters. SAE Paper No. 2007-01-1136

14. **Depcik, C., Assanis, D.** One-dimensional automotive catalyst modelling. Progress in Energy and Combustion Science. Vol. 31 (2005) pp 308 – 369

15. **Yim, S. D., Kim, S. J., Baik, J. H.** Decomposition of urea into NH3 for the SCR process. Ind Eng Chem Res Vol. 43 (2004) pp 4856 – 4863

16. **Yimer, I., Jiamg, L. Y., Campbell, I., et al.** Combustion noise reduction in a kerosene burner: Investigations in the spray characteristics of the fuel nozzle. National Research Council of Canada Report NRCC-47050. Also published in Combustion Institute Canadian Section Spring Technical Meeting, May 2004

Homogeneous diesel combustion challenges under vehicle compliant operating conditions

Roderich Otte, Jochen Müller,
Frank Weberbauer, Bastian Guggenberger
Robert Bosch GmbH, Corporate Research, Germany

ABSTRACT

To investigate the actual emission potential of homogeneous charge compression ignition (HCCI) a EU4 full-size engine was optimised for vehicle compliant operating conditions. Results from a single cylinder engine with fully variable valve train helped to identify operation strategies and hardware requirements. To reach both, sufficient amount of intake air and high EGR a 2-stage turbo charging system was obstructed. To fulfil dynamic demands transient operation was controlled using cylinder pressure sensors. Despite the great NO_x/soot-advantage HCCI combustion shows strongly increased HC/CO-emissions. HC species fractions were examined by means of mass spectroscopy. To improve HC/CO-after-treatment a pre-turbo-catalyst (PTC) upstream the turbine was tested.

1 INTRODUCTION

Modern Diesel engines allow strong road performance and low fuel consumption, resulting in a rising Diesel car fraction throughout Europe. At the same time stringent future emission legislation means a big challenge for Diesel engines. In Europe, for example, the expected 2014 nitrogen oxide threshold (Euro 6) will be approximately 30% of the actual Euro 4 value. To minimize tailpipe NO_x-emissions different attempts are pursued. Exhaust gas after-treatment with SCR (selective catalytic reduction) or NSC (NO_x-storage-catalyst) is one of the options. Since some time homogeneous charge compression ignition (HCCI) has been considered as a promising combustion concept to reduce engine raw emissions so that $DeNO_x$-systems could be dispensable. Additionally homogeneous combustion yields very low soot emissions.

On the other hand some issues make it difficult to utilise HCCI-combustion for vehicles in practice: complex transient operation, combustion noise and high HC/CO-emissions. Nevertheless homogeneous combustion reaches a good indicated efficiency and fuel consumptions at the level of current serial production engines due to short combustion duration and an optimal phasing of the burning process referring to top dead centre (TDC). The following paper identifies possible answers to the main challenges.

2 FUNDAMENTALS OF HCCI-COMBUSTION

Homogeneous self ignition with Diesel or gasoline engines has been of subject in industry and research facilities for some decades. Concepts for both fuels allow very low NO_x and soot emissions. Gasoline HCCI engines show improved fuel consumption compared to SI-mode rising from a dethrottled gas-exchange process.

Above 90 % of the nitrogen oxides emitted by combustion engines arise from thermal formation processes. In conventional Diesel engines high NO_x-production occurs on the very hot thin diffusion layer surrounding the main combustion zone. Soot formation occurs in the inner rich area of the combustion zone (1).

Homogeneous combustion, in contrast, avoids very high flame temperatures and local excess fuel. A high amount of exhaust gas is recirculated (EGR) to the cylinder to extend the ignition delay and therefore the period for mixture formation at the same time. This effect results in a homogeneous mixture of fuel, inert gas and air. Homogeneity causes low temperature combustion at relative A/F-ratios > 1 and results in low NO_x and soot emissions (Figure 1).

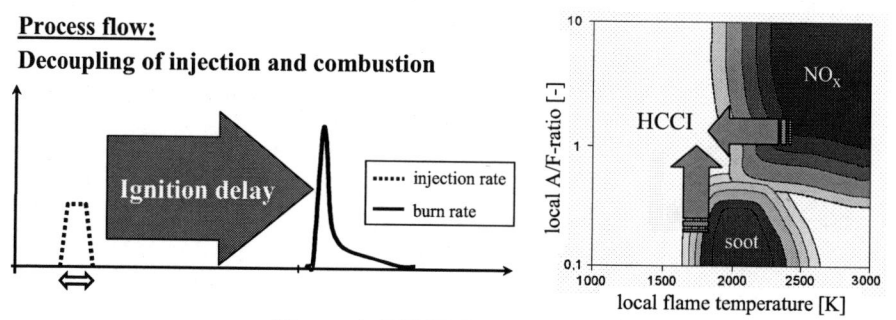

Process flow:
Decoupling of injection and combustion

Figure 1: HCCI-fundamentals

HCCI-theory describes that self ignition of each fuel molecule in the combustion chamber occurs at the same time – corresponding to the very efficient constant-volume-combustion process. Practically the process comes closer to this thermodynamic optimum than conventional diffusive combustion, resulting in high pressure gradients, short burning durations and increasing indicated efficiencies. The pressure gradient directly correlates with combustion noise and mechanical stress for the engine.

To take full advantage of homogeneous Diesel combustion it is essential to achieve high volumetric air efficiencies despite the demand for extensive exhaust gas recirculation. This requires high development effort for the turbo charger and the EGR-system and limits the possible engine load range.

Auto ignition of lean homogeneous fuel/air/exhaust-mixtures is a function of fuel properties, gas temperature, pressure and charge composition. Compared to the conventional Diesel engine, dynamic control of the HCCI process is much more difficult because usually direct access through injection timing is not given any longer. Nevertheless quick changes in load and/or speed are typical for vehicle operation and have to be performed with homogeneous combustion. Considering this fact HCCI research has changed within the last few years. Coming from early direct injection at the beginning of the intake stroke or even induction of a gas/fuel-mixture, modern concepts propagate later injection timings during the compression stroke. Under certain conditions this late injection provides for access to the start of combustion by injection parameters. Definitely it drastically reduces possible lubricant dilution through Diesel fuel condensing on the cylinder liner which is a severe problem of classic HCCI with early injection.

Regarding late homogenisation two different approaches are pursued, basically (2): relatively early injection (50...20°CA BTDC) combined with very high EGR and relatively late injection (after TDC) combined with EGR and high injection pressures. The latter allows higher loads because of the lower EGR but the early injection concept stands for higher fuel efficiency and thus mostly is preferred.

The low combustion temperature with HCCI causes high emissions of unburned and partially burned hydrocarbons. Oxidation of carbon monoxide is incomplete, in addition. HC/CO-after-treatment with a standard Diesel oxidation catalyst (DOC) is limited because of the low exhaust gas temperatures.

3   EXPERIMENTAL SETUP

Experiments were performed on a single-cylinder research Diesel engine with electro hydraulic valve actuation (Bosch-EHVS) and a modified four-cylinder engine, identical in construction (compare Table 1).

Table 1: Data of research engines

combustion process	four stroke Diesel
fuel injection equipment	Bosch CRS 2.1, $p_{Rail,max}$ = 1600 bars, solenoid inj.
injection nozzle	7 holes, 153°, ks = 1.5, hydr. flow = 310 cm^3/min @ 100bars (60 bars back pressure)
electronic engine control	EDC 16/DSpace control unit
number of cylinders	1 and 4
displacement/cylinder	538 cm^3
bore	88.0 mm
stroke	88.4 mm
compression ratio	15:1
intake valves	2 (swirl port/filling port)
exhaust valves	2
variable swirl	swirl flap closed: SN = 3, swirl flap open: SN = 2
valve train	1-cyl.: fully variable, electro-hydraulic valve actuation BOSCH EHVS, 4-cyl.: conv. cam-system
combustion chamber	ω-bowl
supercharging	1-cyl.: external compressor, 4-cyl.: 2-stage TC
EGR	cooled external EGR, (1-cyl. plus internal EGR)

The project started with the single-cylinder engine to identify operation strategies concerning fuel injection, internal/external EGR and controls. In addition hardware requirements for the later four cylinder engine and combustion properties for 1D-simulation have been derived from this. Reference car is a Mercedes E-Class (1800 kg)

4 MAIN ISSUES TO ENSURE HCCI VEHICLE CAPABILITY

As described before there are different challenges getting in the way of HCCI-commercialisation. These will be discussed in the following section.

4.1 Extension of HCCI load range

Compared to conventional Diesel combustion, the maximum load and speed range possible with HCCI concepts is clearly restricted. This is mainly caused by the high fraction of inert gas in order to extend the ignition delay. To ensure lean air/fuel mixtures at upper loads a high boost pressure has to be reached.

Furthermore increasing engine speed proportionally shortens the time for mixture formation and therefore the degree of homogeneity. This effect is partially compensated by intensified turbulence flow of the cylinder charge.

To enlarge the HCCI operation map different strategies were examined. As mentioned before, self ignition strongly depends on in-cylinder gas temperature during the compression stroke. Gas temperature, in turn, mainly depends on the temperatures of intake air, recirculated exhaust gas, the EGR-rate and on the compression ratio (CR). The first step was to lower the compression ratio from 18 to 15. With this, single-cylinder experiments showed that the relevant part-load map could be covered with HCCI performed with early injection during the compression stroke. Using CR = 15 due to lower temperatures less EGR is needed to delay self ignition for homogenisation which reduces necessary boost pressures to realistic values for a modern charging system.

First tests with the EU4-fullsize engine with lowered CR highlighted that the standard air-system represents a severe bottleneck for HCCI performance (Figure 2).

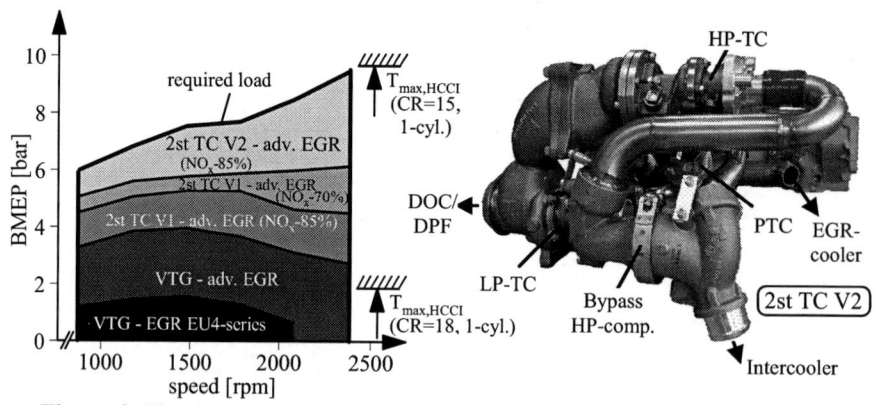

Figure 2: Hardware evolution from EU4-series to capable HCCI-system

The air system was redesigned using the 1D-simulation program GT-Power. The model comprised the whole engine including VTG-turbine, intercooler, DOC, DPF (Diesel particulate filter) etc. and was mainly established to enhance the charging system. To simulate a realistic gas exchange process burn-rates from single-cylinder experiments were implemented.

External EGR is driven by the pressure gradient between exhaust and intake systems whereas EGR values above 50 % entail low pressure gradients. The effective exhaust pressure upstream the EGR-valve is additionally minimised by losses in the narrow original EGR system. Consequences were a much bigger cross section of EGR piping/valve and, apart from this, a more effective EGR-cooling unit.

In order to increase boost pressure despite the very low exhaust gas mass flows with excess EGR a two-stage turbo charging system was developed. First a serial system, designed for similar reason for a conventional Diesel engine, was adapted to the engine and the new EGR-system. This concept already significantly helped to move the HCCI load limit upwards. Supported by GT-Power simulations a second two-stage turbo system was designed within the project. This version, pictured in Figure 2, has already approved its high performance on the test bed.

Besides possible maximum HCCI load depends on the certain emission target aimed, like exemplarily demonstrated in Figure 2 (85 % vs 70 % NO_x-reduction).

4.2 Controlled transient operation

Modern Diesel engines are operated with feed forward control through an electronic control unit (ECU) based on various parameter maps, actual sensor values (e.g. ambient temperature, boost pressure) and the drivers torque demands.

Measuring the in-cylinder pressure, on the other hand, allows a direct feedback combustion control through calculated characteristic attributes. This modification would drastically reduce ECU data volume and could include all those influencing variables not being considered so far.

Homogeneous combustion concepts depend on in-cylinder pressure based controls. Already very small deviations in EGR-rate lead to significant changes of NO_x-emission and combustion noise (1 bar/°CA per 1 % EGR). Since the ECU cannot determine EGR-values as exactly as needed, combustion has to be observed and continuously controlled via injection timing. This especially applies to transient operation where boost, exhaust and rail pressures and consequently cylinder charge condition are permanently alternating.

Limited load and speed range of HCCI and a possible catalyst heating strategy in conventional mode request a controlled combustion mode change to/from conventional combustion and back. In this case number of injections, injection timing, rail and boost pressure as well as EGR-rate have to be adapted to the new operating point as fast as possible. Figure 3 illustrates this problem.

Pressure based combustion control is necessary to minimize noticable steps in torque or noise which affect driving comfort. Besides it helps to avoid NO_x or HC peaks endangering a good overall emission result.

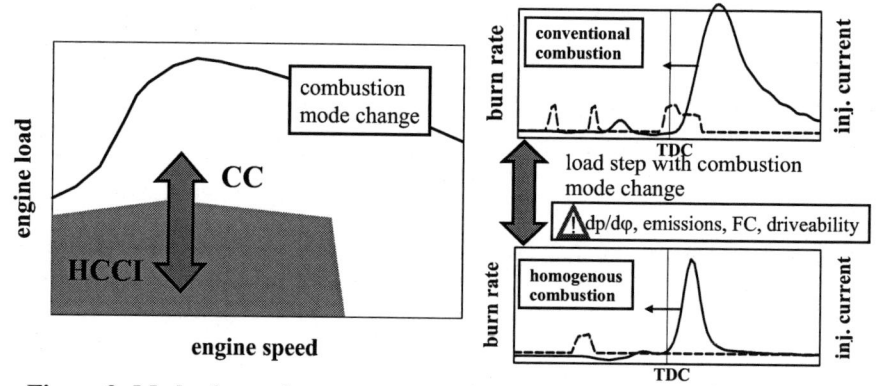

Figure 3: Mode change between conventional and homogeneous combustion

4.3 Combustion noise

As a matter of principle HCCI shows higher gradients of cylinder pressure than conventional combustion. Pressure gradients, and thus noise, depend on the reaction rate of the main combustion phase. To influence noise, the start of this combustion phase has to be shifted. With CR = 15 it is possible to change combustion phasing to "late" by means of retarded start of injection (SOI, Figure 4, left). NO_x emissions

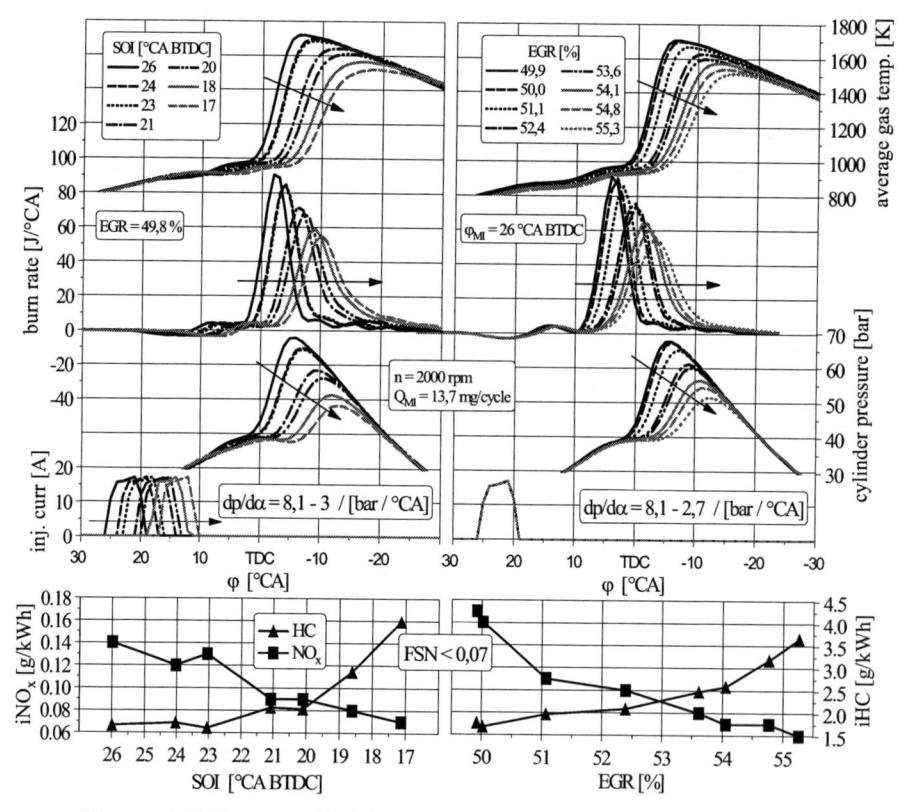

Figure 4: Influence of injection timing and EGR-rate on emissions and combustion noise

and noise decrease with later combustion phasing whereas HC-values drastically increase. In this regard higher EGR has an effect similar to retarded injection timing (Figure 4, right).

With HCCI Diesel combustion the NO_x/Soot trade-off known from conventional Diesel engines is replaced by a NO_x/Noise/HC conflict. To find the best application for each operating point exhaust gas temperature for save DOC-activity has to be considered, in addition. This section has thus demonstrated that there is a useful application range for this particular HCCI concept.

4.4 Emission of unburned hydrocarbons

The emission of unburned hydrocarbons and carbon monoxide is a severe problem for homogeneous Diesel combustion concepts. Basically the cold combustion avails HC and CO formation close to the even colder cylinder walls. This issue is tightened by the following circumstances: High EGR results in relatively low air/fuel-ratios, making the occurrence of local rich areas more probable, and low temperatures. Cold combustion combined with high EGR leads to low exhaust heat flux which seriously affects DOC-activity, especially at very low loads. Single-cylinder experiments showed that exhaust trapping with a fully variable valve train EHVS could lower HC raw-emissions by 38 % because of hot internal EGR and raise exhaust gas temperature by 45 K at the same time, which is valuable for HC-aftertreatment (3).

A widely unnoticed phenomenon is that composition of HC-emissions changes within the homogeneous application range. This clearly affects the conversion efficiency of the catalyst. Figure 5 shows the different HC-species for a variation of EGR-rate with constant pressure gradient, controlled by start of injection.

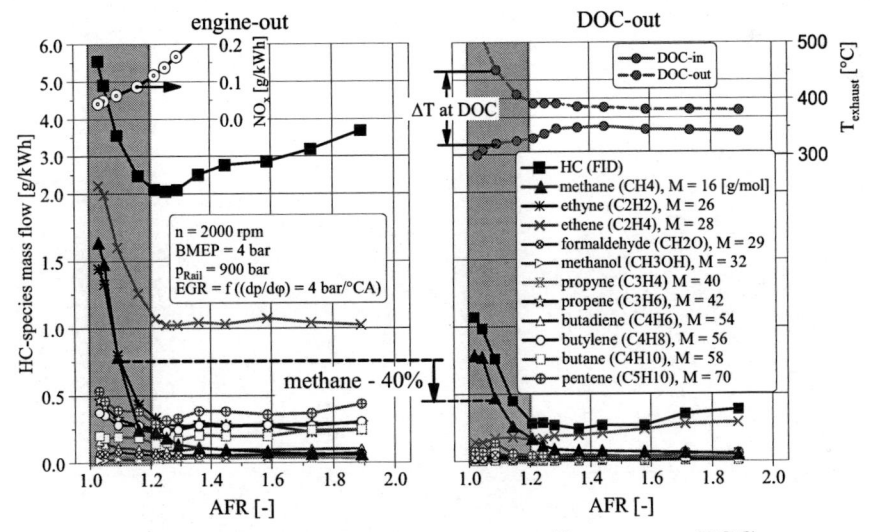

Figure 5: HC species fractions upstream/downstream DOC

With increasing EGR, characterised by decreasing AFR, SOI occurs earlier and ignition delay and degree of homogeneity rise. From AFR=1.3 to 1.1 the typical trade-off between HC and NO_x is clearly visible.

Hydrocarbon species were detected for molecular weights from 16 to 130 g/mol by a mass spectrometer. Total HC values come from parallel FID measurements. In contrast to Diesel fuel composition, molecules with more than five carbon atoms or molecular weights beyond 70 g/mol are hardly to find in the exhaust.

The biggest problem for after-treatment of HCCI exhaust obviously is the rising methane fraction below AFR = 1.2. Methane is a very stable and compact molecule that needs temperatures above 450 °C for oxidation. A methane conversion of only 40 % at AFR = 1.1 leads to the total methane fraction of 60 % downstream DOC. Methane oxidation at this point is fed by conversion of other hydrocarbons and carbon monoxide, indicated by a huge temperature rise of 130 K along the DOC.

During the urban part of the NEDC temperature peaks upstream the DOC never exceed 250 °C. Conclusion should be to avoid AFR < 1.2 and thus methane emission. Above all an air buffer of 20 % is very helpful to reach high dynamics during a positive load step. Nevertheless NO_x emission still loses 50 % of its value within this critical methane area.

Different measures are feasible to support DOC-activity. For low load operating points it is possible to increase exhaust gas temperature through bypassing the EGR-cooler and thus raising temperature of the intake gas mixture. Reduction of HC-raw emissions up to 30 % because of higher cylinder gas temperatures is more than a side-effect at this point. Compared to exhaust trapping with EHVS transient performance of this method is worse.

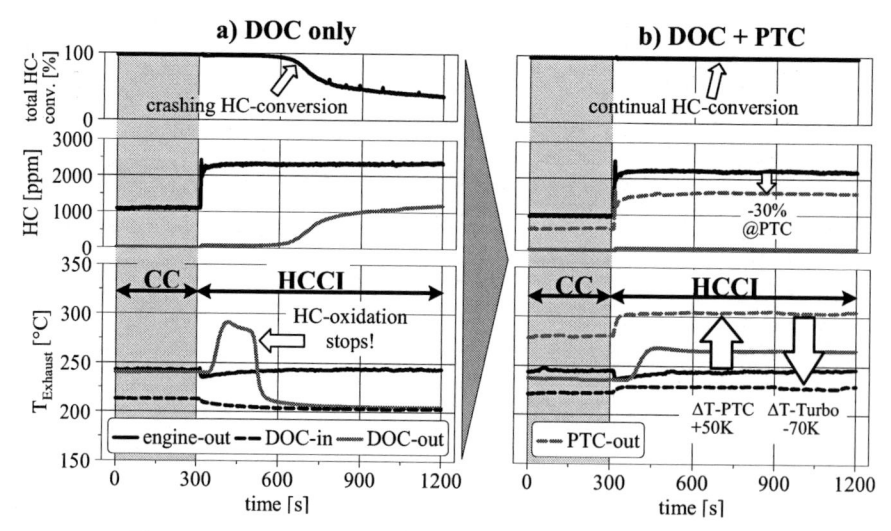

Figure 6: Effect of additional pre-turbo-catalyst (PTC) on total HC-conversion at HCCI (1500 rpm, IMEP = 3 bar)

By placing a small catalyst very close to the engine, it is possible to reach higher gas temperatures at the standard DOC inlet without any changes of engine parameters. Figure 6 shows the effect of an additional pre-turbo-catalyst (PTC) on HC-conversion in conventional combustion (CC) and HCCI mode at 1500 rpm.

At IMEP = 3 bar, a switch from CC to HCCI increases raw HC concentration and reduces the gas temperature upstream the DOC at the same time (a). That is, why HC-oxidation stops after a short while. Diesel oxidation catalysts are able to store HC because of their zeolite layer. Tailpipe HC (i.e. DOC-out) increases when the cumulated HC-amount exceeds the HC-storage capacity of the catalyst.

Implementing a PTC between the engine and the turbo charger, about 30 % of the total HC will be oxidised by this catalyst (b, centre). The temperature rise resulting from this exothermal reaction along the PTC ($\Delta T \approx 50$ K) supports the standard DOC and thus ensures continual and high HC-conversion all over the experiment. Sophisticated design, material and positioning of the PTC allows unaffected nominal power output. Right after an engine cold-start the PTC could support conversion during warm-up when the turbo chargers' heat capacity has the biggest effect on DOC inlet temperature.

5 PROGNOSIS OF TRANSIENT EMISSION POTENTIAL

Successful use of the identified combustion concept in a car depends on how the new components and control strategies interact at transient operation. Usually standard driving cycles like the NEDC are performed with a complete vehicle on an exhaust roller dynamometer test bench. With the applied modern dynamic engine test rig it is possible to simulate vehicle behaviour like damping or shifted gear to gain reliable emission results with an engine only.

It is also reasonable to predict driving cycle emission potential from steady-state operation results when full transient capability is not yet reached within the complex research process. This helps not to lose sight of the given target and to optimise the transient operation strategy. For this reason a modular simulation tool, developed at Robert Bosch (4), was adapted to the present application. In the cycle-module operating points from steady-state mapping are selected by calculating actual engine speed and torque based on vehicle mass/inertia, transmission ratio/efficiency as well as aerodynamic drag. To consider the higher mechanical friction during warm-up an indicated power characteristic based on coolant temperature is included. Calculated HCCI NO_x-emissions for the reference car with different engine warm-up strategies are benchmarked to the EU 4 baseline engine in Figure 7.

Although transient control effects are not included in the simulation, predicted NEDC emission for this baseline matches the test bed results very well. Total NO_x-emission mainly depends on the warm-up phase right after cold-start and the extra-urban phase during the last 400 s. If HCCI begins 200 s after engine-start the baseline NO_x-value is only halved with the charging system V1. Even with no CC warm-up this system is not sufficient. The most optimistic calculation with no CC in the beginning and with almost full HCCI acceleration (95 %) during the test gives very low NO_x because of highly improved extra-urban HCCI performance by 2-stage TC V2. For low HC-emissions with CR = 15 it will be essential to intensively glow during warm-up. With a glow-combustion-sensor (GCS) pressure based combustion control can be performed at the same time. Fuel consumption and soot emission are slightly below the EU4-application whereas HC/CO raw emissions are almost doubled and combustion noise is observably higher (not shown).

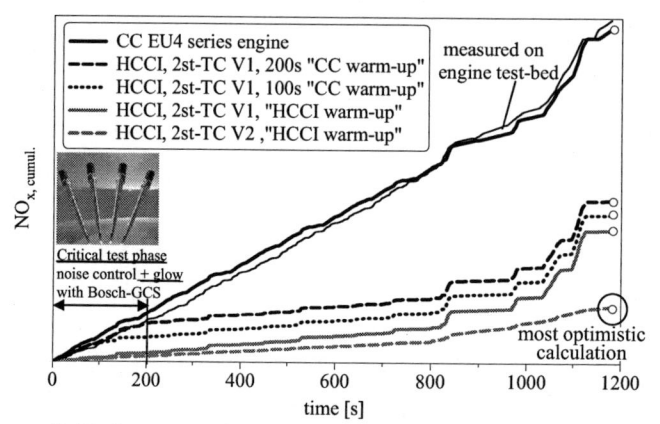

Figure 7: Estimated NO_x-Emission for different HCCI operation strategies during NEDC from steady-state experiments

6   CONCLUSIONS AND PROSPECTS

To utilise the well-known NO_x-potential of HCCI combustion concepts for vehicle-operation different issues have to be tackled. Fundamental hardware parameters and operation strategies can be identified with single-cylinder experiments. For reliable conclusions experiments with a full-size engine equipped with package-conform charging and exhaust after-treatment system are indispensable.

HCCI combustion needs a powerful charging system and a low-loss EGR system to provide for high volumetric air efficiency and high EGR-rates at the same time. For this reason the air-system was redesigned with the help of 1D-simulation. Emphasis was placed on a new 2-stage turbo charger to enable sufficient maximum loads for reasonable vehicle performance.

Homogeneous concepts need feedback combustion and noise control based on the measured cylinder gas pressure. This can be performed with a glow-plug sensor which allows high-temperature glow during engine warm-up at the same time. The latter is very important too because HCCI-success strongly depends on solving the severe HC/CO-problem. After-treatment of unburned HC with a Diesel oxidation catalyst (DOC) is limited because of low exhaust gas temperatures and partially high methane emission. A catalyst upstream the turbo system (PTC) shows significant potential to improve DOC-activity and thus tailpipe emissions.

Utilisation of HCCI in a vehicle depends on driveability and emission behaviour in transient operation. Extrapolation of cycle potential based on steady-state operation can give hints but cannot substitute the demonstration of real transient ability. Transient test cycle operation on the engine test bed will show residual NO_x-benefits and possible drawbacks. Finally it should not be unmentioned that conventional combustion has been brought to a very low NO_x-level in recent years whereas hardware requirements show a similar trend to what is discussed in this article.

7   REFERENCE LIST

(1) Flynn, P.
Diesels 2007, Promise & Problems
7th Diesel Engine Emissions Reduction Workshop, Chicago, 2001

(2) Weberbauer, F.; Rauscher, M.; Orlandini, I.; Kulzer, A.; Kopp, C.
Weiterentwicklung und thermodynamische Bewertung von Diesel-HCCI-Brennverfahren
Motorische Verbrennung, Haus der Technik, München, 2005

(3) Müller, J.; Otte, R.; Weberbauer, F.
Herausforderungen bei der Entwicklung von Diesel-HCCI-Brennverfahren
Motorische Verbrennung, Haus der Technik, München, 2007

(4) Kulzer, A.; Kufferath, A.; Christ, A.; Knopf, M.; Benninger, K.
CAI – Ein Brennverfahren auf Basis der BDE Technologie
15. Aachener Kolloquium Fahrzeug- und Motorentechnik, Aachen, 2006

Investigation of fuel injection strategies on a low emission heavy-duty diesel engine with high EGR rates

Andrew J. Nicol, Chris Such
Ricardo UK Ltd, UK

Ulla Sarnbratt
Volvo Technology Corp, Sweden

ABSTRACT

A test programme has been carried out to investigate fuel injection strategies on a single cylinder heavy-duty diesel engine. The programme was designed to use high rates of cooled Exhaust Gas Recirculation (EGR) to control NOx to below 1.0 g/kWh. A novel, flexible fuel injection system capable of up to 250 MPa maximum injection pressure was employed to explore the optimum trade-offs between NOx, soot emissions and fuel consumption. This paper addresses the potential of the system to control emissions using pilot and post injection. Post injections were found to reduce soot emissions at half load by a factor of two or more. With early pilot injection the soot-NO_x trade-off was improved at 25% load but with the drawbacks of high emissions of CO and fuel dilution of oil.

1 INTRODUCTION

The global trend to tighter emissions standards requires research into combustion and injection systems on heavy-duty diesel engines. A previous investigation focussed on the optimisation of injection parameters when using cooled exhaust gas recirculation (EGR) to achieve Euro IV emission limits (1). The results indicated that, for an electronic unit injector system with two solenoid valves, pilot and post injection strategies offered an improvement in soot emissions at part load.

This paper describes the investigation into the potential of alternative injection strategies at NOx levels corresponding to about one third of those measured in the previous project, that is, engine out NOx emissions of less than 1.0 g/kWh. The testing was carried out on a single cylinder 2 litre engine with a conventional, low swirl combustion system, high EGR rates and very high boost pressures. The fuel injection system comprised a combination of a cam-dependent EUI plus a lower pressure common rail, which provided a very flexible tool to investigate multiple injection strategies.

© Ricardo UK Ltd, 2007

2 OBJECTIVES

The objectives of the project were to investigate the potential of the combustion and fuel injection system to achieve the target emissions of 0.8 g/kWh NOx and 0.04 g/kWh soot emissions, with best possible fuel consumption.

3 TEST HARDWARE

3.1 Fuel Injection System

The fuel injection system used on this project was the prototype F1 system supplied by Delphi Diesel Systems Ltd, which consists of an E3 injector and a pressurised rail (2). The E3 injector has two solenoid valves integrated into the injector body. One solenoid valve controls the spill valve, the other solenoid controls the needle closing valve, which modulates fuel flow to the back of the needle, reinforcing the spring, and thereby increasing the nozzle opening and closing pressures. This enables very high mean effective injection pressures to be achieved to the benefit of low soot emissions. The E3 injector is already in production and is fitted to heavy-duty engines complying with Euro IV and US 2007 emissions legislation. In principle, injection pressures approaching 300 MPa are feasible with this system but, in the current test programme, the limit placed on the injection pressure was 250 MPa.

The rail connected by a short pipe to the injector. The spill control valve of the E3 injector was replaced by a rail control valve, which controlled the flow of pressurised fuel from the injector to the rail during those parts of the cycle when there was no fuel injection into the cylinder. As a result, no external or engine driven pump was needed to pressurise the rail.

The two valve injector allowed the possibility of pilot and post injections off the engine camshaft at relatively high injection pressures, and the rail allowed pilot and post injections at lower pressure levels. The limit placed on the rail pressure in this project was 80 MPa.

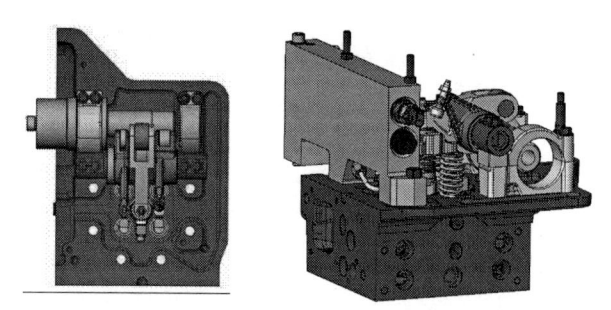

Figure 1: (a) EUI on Single Cylinder Test Engine; (b) Rail Installation

3.2 Single Cylinder Engine

The combustion system had an open combustion bowl and an inlet air swirl of virtually zero. Two levels of compression ratio were used: 15:1 and 16:1. The nozzles had 6 holes with a cone angle of 140° and a flow rate of 1.85 l/min at 10 MPa.

The testbed was equipped with a cooled EGR circuit. Boost pressure was simulated by means of compressed air, up to a limit of 6 bar absolute. Charge inlet air temperature, exhaust back pressure and EGR levels were selected on the basis of one-dimensional performance simulation, to be representative of the equivalent multi-cylinder engine, with a rating of 23.6 bar BMEP at maximum torque and specific power of 27.6 kW/litre. The settings at the key points tested are shown in Table 1.

Table 1: Single Cylinder Test Settings

ESC Test Mode		A100	A50	B50	B25	C100
Speed	rev/min	1200	1200	1500	1500	1800
Inlet manifold pressure	kPa abs	460	230	270	160	490
Exhaust pressure	kPa abs	500	260	300	190	580
Pressure differential	kPa abs	40	30	30	30	90
Inlet manifold temp	°C	66	51	60	54	67
EGR rate	%	32	33	39	40	27
EGR was varied about the mean rate shown in the Table.

Injector needle lift was measured with a Hall sensor. Injection pressure was calculated from the measurement of strain in the injector rocker. In addition, a pressure sensor was fitted to the rail. Soot emissions were calculated from the measurement of Filter Smoke Number (AVL 415 meter).

3.3 Test Methodology

Due to the prototype nature of the F1 system, a 'one factor at a time' approach was used initially. The first part of the test programme covered the investigation at three ESC part load modes (A50, B50 and B25), initially using a single injection. The variables were injection timing, EGR and injection pressure. For minimum soot emissions, injection pressures of 200 – 250 MPa were found to be necessary. After establishing the trade-off with a single injection, the experiment was extended to cover pilot and post injections off the cam.

Next, the combination of cam dependent injection and injection off the rail was explored, notably:

- A low pressure pilot injection off the rail, with timing varied from far pilot to close pilot, followed by a main injection off the cam

- A main injection off the cam, followed by a close post injection off the rail, with varied post injection quantities and intervals
- A main injection off the cam, followed by a close post injection off the cam (to increase post injection pressures) with varied post injection quantities and intervals

Turning to the two ESC full load modes (A100 and C100), the optimal injection strategy was a single injection at high injection pressures off the camshaft. In this case, after some initial investigations using "one factor a time tests", the Design of Experiment process known as Stochastic Process Modelling (1) was adopted, which enabled an advanced modelling and optimisation of the test results.

4 TEST RESULTS

4.1 Results at Part Load Test Modes

The trade-offs at A50 mode are shown in Figure 2. With the single injection, the target NOx level was feasible, but the soot emissions increased as the target NOx was approached.

The most effective way to improve the NOx/soot trade-off was found to be a post injection off the cam, equivalent to about 25% of the single fuel injection quantity. With post injection, the fuel consumption was slightly improved, compared with the baseline single injection.

Post injection at lower pressures off the rail was investigated but soot emissions were not improved, indicating the importance of sufficiently high energy in the post injected fuel sprays.

A combination of close pilot injection off the rail and post injection off the cam gave no benefit.

Early pilot off the rail slightly improved the NOx/soot emissions trade-off at the B50 mode, but with high fuel consumption, CO emissions and fuel dilution of the oil.

Analysis of the in-cylinder data at the A50 test point showed that there was a second peak in instantaneous heat release corresponding to the post injection, which caused the cumulative heat release to 'catch up' with the heat release of the single injection (Figure 3). As a result, the post injection gave little or no deterioration in fuel consumption. The effects were sensitive to the timing and quantity of post injection; in particular, sufficient separation between main injection and post injection was necessary. At the A50 test condition, the optimum timing for the post injection appeared to occur at a separation of about 5° crank, with a timing of the post injection of about 15° crank after TDC.

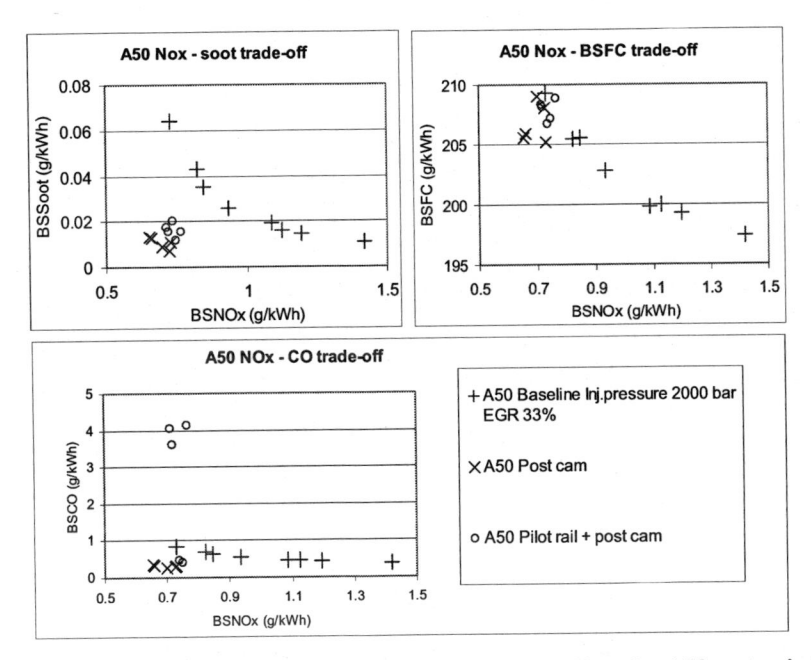

Figure 2: Emissions and fuel consumptions trade-off at the A50 test point

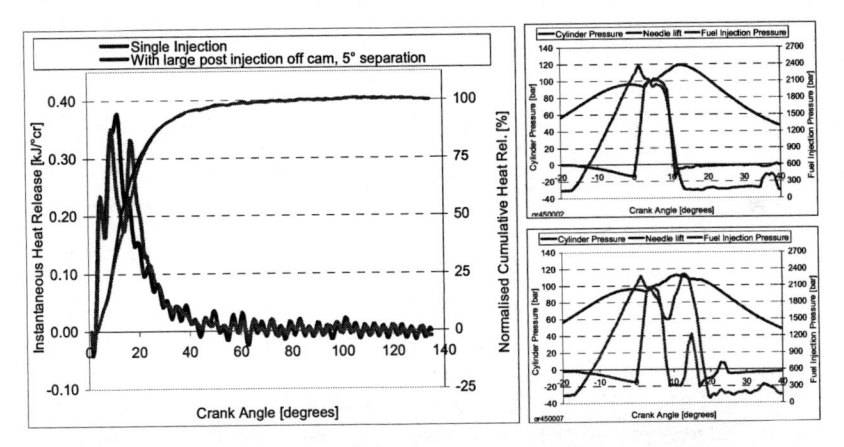

Figure 3: In cylinder diagrams showing the effect of post injection at the A50 test point

The emissions/fuel consumption trade-offs at the B25 test point are shown in Figure 4. Even at this low load condition, it was necessary to use maximum injection pressures of 200 MPa to control soot emissions. Investigation of alternative injection strategies showed that improvements in the NOx/soot trade-off could be achieved either by post off the rail with a rail pressure of about 600 bar, or by early pilot injection also off the rail. However, in both cases, the fuel consumption was increased, and early pilot injection resulted in very high CO emissions and increased oil dilution by fuel.

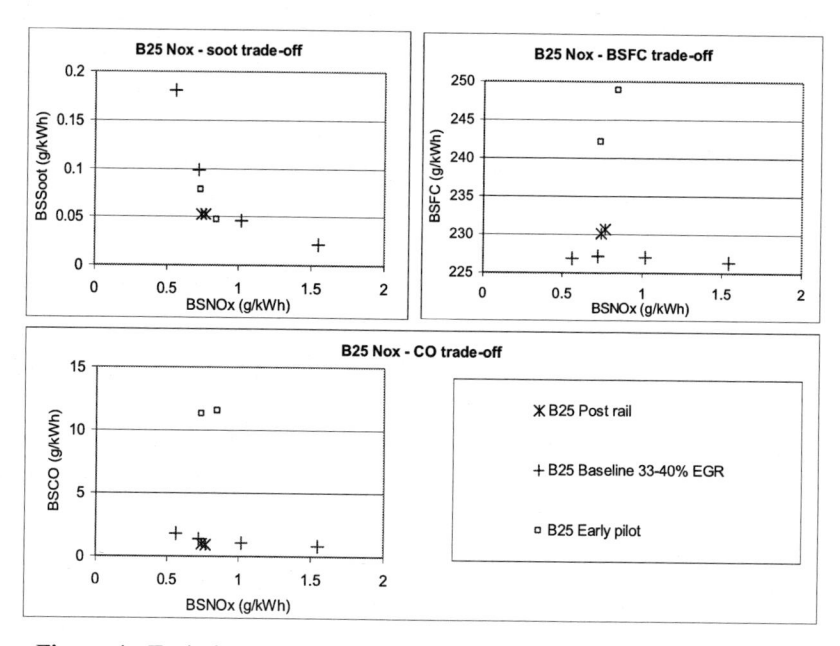

Figure 4: Emissions and fuel consumptions trade-off at B25 test point

An investigation of early pilot injection was carried out with the following results:

- Early pilot injection (at or before 60° before TDC) improved the NOx/soot trade-off at both B25 and B50 test points, with a penalty in fuel consumption, CO emissions and oil dilution.
- CO emissions and oil dilution were less of a problem at B50 than at B25, due to higher charge pressures and charge density.
- There was an optimal timing of early pilot injection for low soot. The optimal timing varied with charge pressure, rail pressure and pilot quantity. At B25, the optimum timing was typically 50° before TDC.

Previous studies have shown that very low emissions of NOx and soot can be achieved when the combustion is fully or partially pre-mixed (3,4). In general, this type of combustion is limited to low loads due to increased cycle-to-cycle variability and excessive rates of pressure rise at higher loads. The potential for achieving Homogeneous Charge Combustion Ignition (HCCI) was investigated at the B25 test point using a single injection at high pressure off the camshaft.

Figure 5 shows the effect of retarding the timing. As expected, the fuel consumption increased with retard but, when the start of injection became later than TDC, the emissions of smoke and NOx started to fall dramatically. At an injection timing of about 10° after TDC, the emissions were well within the target box but at the expense of an increase in fuel consumption of nearly 20%. Combustion at this point was fully pre-mixed.

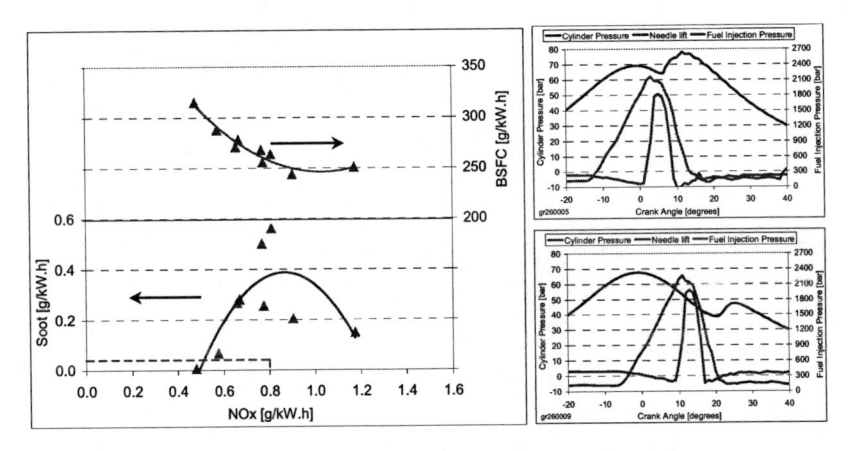

Figure 5: Effect of Retarding the Injection Timing at the B25 Test Point

4.2 Results at Full Load Modes

Previous studies have shown that high mean effective injection pressures are essential when using EGR at full load to control soot emissions, and these are best achieved by a single injection event of short duration (5).

In this programme, a variety of injection profiles were investigated, including single injection, post and pilot injections off the cam and rail. The screening tests confirmed that the best results were with a single injection at high pressure off the cam.

Figure 6 shows the initial screening tests with varying timing, varying nozzle opening pressure (NOP) and varying EGR. The optimal trade-off was predicted from the Design of Experiment. The validation test confirmed that the best measured result correlated well with the predicted result and was very close to the target emissions. As expected, the fuel consumption deteriorated as the NOx level approached the target.

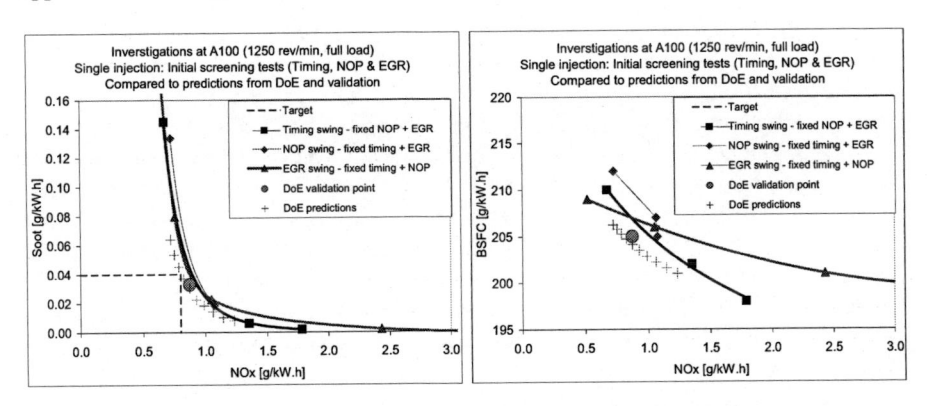

Figure 6: Emissions and fuel consumptions trade-off at B25 test point

5 DISCUSSION

The results of this investigation on the 2 litre single cylinder engine with prototype fuel injection system indicate that the target NOx levels of 0.8 g/kWh with soot levels of 0.04 g/kWh are feasible with internal engine changes alone, without exhaust aftertreatment.

To achieve such low NOx levels, EGR rates of 25% – 40% are required over most of the operating range. The use of high EGR levels at full load puts a premium on high boost pressures for highly rated engines. In this project, the boost pressures required for acceptable air/fuel ratios were between 4.6 bar – 4.9 bar (absolute) at full load and 2.3 – 2.7 bar (absolute) at half load, which drives the requirement for large single stage turbochargers, probably with variable geometry, or two stage turbochargers.

A pre-requisite for achieving low soot emissions when using high EGR rates is very high injection pressures, even at part load conditions. Compared with a single injection, the combination of a main injection having a maximum pressure in the range 200 – 250 MPa, and a post injection with a similar maximum pressure, showed significant improvements in soot at part load, with no increase in fuel consumption.

The benefit of post injection on soot emissions at part load conditions has been reported in other recent studies (such as 6,7), usually in cases where cam-independent common rail systems have been used, operating at lower injection pressures and with lower EGR rates. As an example, Payri et al (6) reported significantly reduced soot emissions with no increase in fuel consumption. The main reason for the soot reduction was given as the effect of the post injected fuel on the air/fuel mixing process during the later part of the diffusion combustion phase. The momentum energy of the fuel spray increases in-cylinder gas motion and stimulates air/fuel mixing, which then increases in-cylinder temperature and soot oxidation (8). Another explanation is offered by Han et al (9), based on results of Computational Flow Dynamic studies on a heavy-duty engine with quiescent combustion system, deep fuel sprays (125° cone angle) and 15:1 compression ratio. According to the computations, a split injection (with about 25% of fuel in the second injection) can reduce soot formation. This is due to the fact that soot producing rich regions at the spray tip are not replenished by fuel. During the dwell between the injections, the mixture becomes leaner. The subsequent injection takes place into a high temperature, oxygen-rich mixture, which reduces soot formation rates significantly. Irrespective of whichever mechanism is dominant, the soot-reducing effect is supported by the observations of the current study, at least in the case of high pressure post injection off the camshaft. Post injection at lower pressures off the rail was found to have no benefit at the part load condition, presumably due to a lack of injection energy and hence poorer mixing with the available oxygen.

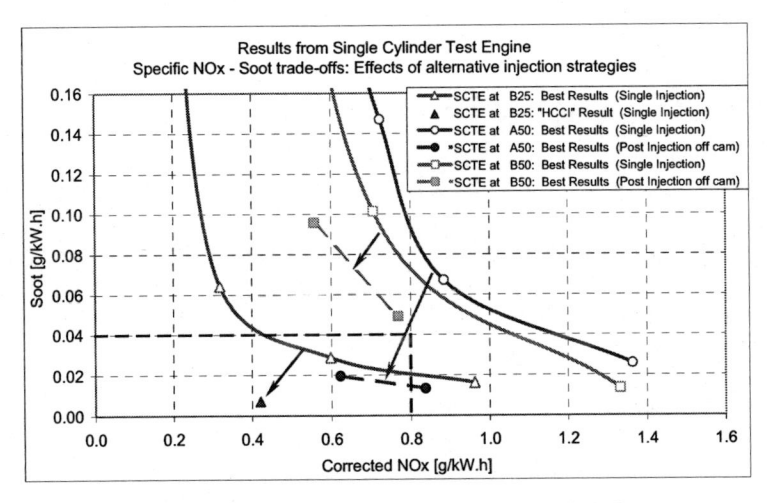

Figure 7: Emissions Trade-off at Part-Loads

At full load, post injection is not so effective because, with a single injection, fuel is still being injected during what would be the gap between the main injection and the post injection. Splitting the main injection at full load would result in a very late end of injection, which would increase fuel consumption and soot emissions.

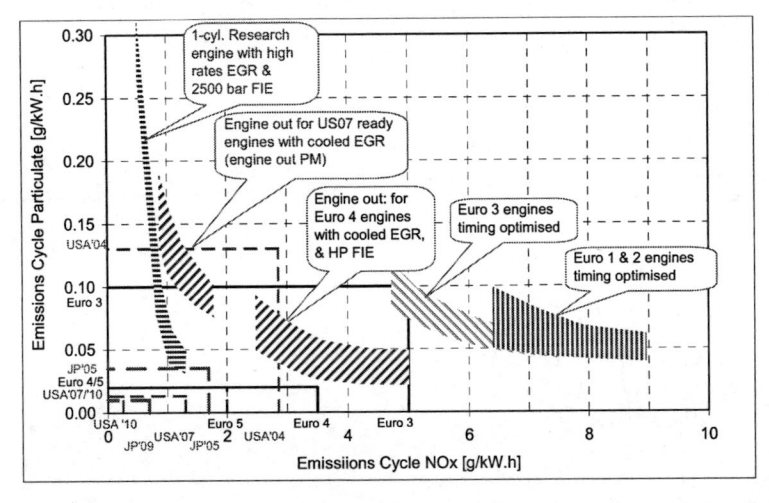

Figure 8: Summary trade-off over ESC test (shows engine-out results)

Figure 8 shows the predicted ESC trade-off for the single cylinder research engine, compared with data at other emission levels. In order to reach the US 2010 emissions limits, the expected Euro VI limits, and long term Japanese standards, exhaust aftertreatment will need to be applied to the corresponding multi-cylinder engine. The most likely combination is a full flow Diesel Particulate Filter (DPF) and Selective Catalytic Reduction (SCR), with consumption of urea solution (AdBlue) of about 2% of the diesel fuel flow.

6 CONCLUSIONS

A test programme has been carried out on a heavy-duty single cylinder engine equipped with high EGR rates and a flexible fuel injection system, consisting of a cam dependent EUI, capable of up to 250 MPa injection pressure and a rail capable of up to 80 MPa. The combustion system used was a conventional, low swirl type. The engine-out target for NOx was 0.8 g/kWh at each of the test modes, with a soot target of 0.04 g/kWh.

Maximum injection pressures in the range of 200 – 250 MPa were used to control soot emissions, with EGR rates of 27% - 40% depending on the test mode.

At the half load conditions (A50 and B50), post injection enabled significant reductions in soot emissions to be achieved with little or no increase in fuel consumption. In-cylinder data showed that the post injection increased the rate of heat release in the later part of combustion, and that the cumulative heat release period was no longer than with a single injection.

At the low load point (B25), Homogeneous Charge Compression Ignition (HCCI) was achieved by means of a single injection at very retarded timing. This resulted in extremely low emissions of NOx and soot, at the expense of higher fuel consumption. Early pilot injection also showed a benefit in the NOx/soot trade-off, but with higher CO emissions and fuel dilution of the oil.

At full load (A100 and C100), best results were achieved with a single injection off the camshaft.

The NOx and soot emission targets were achieved at each test mode, thereby showing that the concept is worth carrying forward.

ABBREVIATIONS AND DEFINITIONS

EGR	Exhaust Gas Recirculation
EUI	Electronic Unit Injector
NOx	Nitrogen Oxides
CO	Carbon Monoxide
HC	Hydrocarbons
BMEP	Brake Mean Effective Pressure
ESC	European Steady-State Cycle
TDC	Top Dead Centre
HCCI	Homogeneous Charge Compression Ignition
IVC	Intake Valve Closing
NOP	Nozzle Opening Pressure
DPF	Diesel Particulate Filter
SCR	Selective Catalytic Reduction

EGR rate is defined by the equation:
EGR (%) = ((CO_{2inlet} – $CO_{2ambient}$)/($CO_{2exhaust}$ – $CO_{2\ ambient}$)) x 100

Soot consists of dry, unburned carbon, which is a major constituent of particulate matter. The mass flow rate of soot is calculated from smoke meter readings (Filter Smoke Number) and the exhaust flow condition using an established correlation.

ACKNOWLEDGEMENTS

The authors wish to express appreciation to the European Union for supporting financially the GREEN project, under which the current investigation has been conducted.

The authors are grateful to the staff of Delphi Diesel Systems Ltd, particularly Mr Nathan Keeler, for their support during the project, to Mr Derek Beckman for supervision of single cylinder testing, and to Mr Nicholas Miche for carrying out engine tests.

REFERENCES

1. Rogers, B.J., Such, C.H., Best, C.H., "Investigation of a Two Valve Electronically Controlled Unit Injector on a Euro IV Heavy Duty Diesel Engine Using Design of Experiment Methods" Institution of Mechanical Engineers paper C610/005/2003
2. Tullis, S.D., Greeves, G., Barker, B, "Emissions Reduction with Advanced Two-Actuator for Heavy-Duty Diesel Engines" Institution of Mechanical Engineers paper C610/004/2003
3. Kimura, S., Aoki, O., Kitahara, Y., Aijoshizawa, E., "Ultra-clean Combustion Technology Combining a Low-Temperature and Premixed Combustion Concept for Meeting Future Emission Standards" SAE paper 2001-01-0200
4. Cooper, B.M., Penny, I.J., Beasley, M., Greaney, A., Crump, J.,"Advanced Diesel Technology to Achieve Tier 2 Bin 5 Emissions Compliance in US Light-Duty Diesel Applications" SAE paper 2006-01-1145
5. Such, C.H., Andersson, P.G.E., Needham, J.R., Freeman, H.D., "The Development of Technology for Heavy-Duty Euro 4 Engines" Aachen Colloquium 1998
6. Payri, F., Benajes, J., Novella, R., Dronniou, N., Lejeune, M., "Potential of a Two Stage Combustion Concept for Reducing Pollutant Emissions in a HD Diesel Engine" The Diesel Engine: The Low CO2 and Emission Reduction Challenge Conference, SIA Lyon, France May – June 2006
7. Hultqvist, A., Hildingsson, L., Chartier, C., Miles, P., "On Soot Reduction by Post Injection under dilute Low Temperature Diesel Combustion" DEER Conference, Michigan 2006
8. Chmela, F., Jager, P., Herzog, P., Wirbelheit, F., "Reducing Exhaust Emissions of Direct Injection Diesel Engines via Injection Rate Shaping" Motor Technische Zeitung September 1999
9. Han, Z.Y., Uludogan,A., Hampson, G.J., Reitz, R.D., "Mechanism of Soot and NOx Reduction Using Multiple-Injection in a Diesel Engine" SAE paper 960633

Recent developments in the control of particulate emissions from diesel vehicles

Martyn V. Twigg
Johnson Matthey plc, UK

ABSTRACT

High-speed turbocharged direct injection diesel engines in cars were introduced some fifteen years ago, and with continuous engine refinement they have become increasingly popular, so today 50% of new cars in Western Europe are diesels. Hydrocarbons, CO, NOx and particulate matter (PM -"soot") emissions from diesels were markedly reduced during this period. In keeping with the general lowering of exhaust emissions impending legislation will require fitment of efficient diesel particulate filters (DPFs) on all new diesel vehicles. Trapped PM in DPFs must be removed, and with availability of ultra low sulphur diesel (ULSD) many heavy-duty vehicles were retrofitted with low temperature passive catalytic DPFs that oxidise already present NO to NO_2 for PM combustion. This method is not suitable for cars, and high temperature oxygen regeneration is employed. The first DPFs on cars had an inorganic fuel additive to lower PM combustion temperatures, while second generation DPFs were catalysed and needed no fuel additive. The most recent DPFs have multiple catalytic functions and they are small enough to be mounted directly on the turbocharger. The nature of nano-sized PM, and how they are formed is being elucidated, and it appears the entire range of PM sizes can be prevented from entering the environment. DPF development requires much costly engine dynamometer or rolling road facilities to load DPFs with PM, and to alleviate this and facilitate fundamental research, a flexible laboratory diesel PM generator was developed.

1. INTRODUCTION

This article is about particulate matter (PM) emissions from diesel engines in vehicles. Large diesel engines are ideal for the efficient powering of trucks and buses, and they have been used in these applications over very many years. Several decades ago these vehicles produced much more exhaust pollutants than their modern counterparts, and what is especially evident now is the almost absence of visible black PM as they accelerate or climb inclines. As heavy-duty diesel engine technology steadily improved, there was a revolution in the use of diesel engines in

© IMechE 2007

passenger cars. No longer is a diesel engine in a car something of a rarity, at least in Western Europe, where there has been a huge increase in the production of diesel passenger cars, so today more than 50% of all new European cars have a diesel engine, as illustrated in Figure 1 [1].

Increased demand for diesel passenger cars came with the introduction of the turbocharged high-speed diesel engine. These engines have high torque at low speed, and good fuel economy compared to gasoline engines. They produce less PM than older diesel engines because of improved fuelling and better combustion.

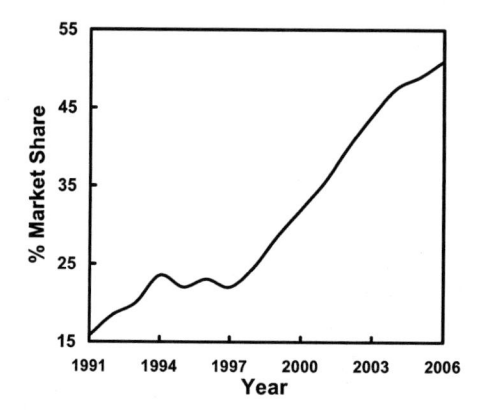

Figure 1. Growth in diesel powered European passenger cars over the last fifteen years. This was associated with the development of powerful turbocharged direct injection high speed diesel engines.

Very high-pressure systems enable fuel injection via several very fine nozzles directly into the cylinder and "common-rail" systems permit multiple, rather than single injections. In spite of improved PM emissions from diesel engines of all types there remain concerns about PM emissions into the environment, and legislation is being introduced that demands fitment of DPFs to all diesel engines. Already many heavy-duty diesel fleets have been retrofitted with DPF systems, and a growing number of new diesel passenger cars have filters, even though present legislation is not so demanding that they must be fitted.

2. ORIGIN AND CHARACTERISTICS OF PM

Conventional gasoline engines operate with a homogeneous charge of a stoichiometric mixture of air and fuel that is compressed and then ignited by a spark. The legislated gaseous exhaust emissions carbon monoxides (CO), nitrogen oxides (NOx) and hydrocarbons (HCs) are controlled by "three-way catalyst" (TWC) formulations coated on ceramic or metal foil monolithic honeycomb structures [2, 3]. The amount of PM formed is quite small compared to a diesel engine, unless for some reason the combustion mixture contains an excess of gasoline above the stoichiometric amount. A diesel engine operates in a completely different way. Air is compressed in the cylinder and is heated via the Joule-Thomson effect to around

700-800K. Then finely "atomised" fuel is injected at high pressure (up to 2000 bar) from fine nozzles directly into the hot gas. The atomised fuel droplets burn in the excess oxygen, perhaps via a pyrolysis process that strips hydrogen from the hydrocarbon leaving a carbon particle that burns in a shrinking core process. As fuel starts to burn the heat released causes remaining fuel to vapourise and burn rapidly, limited only by the rate at which oxygen can access the flame front. It is here where most of the PM is formed from combustion intermediates.

The fine PM in the exhaust gas agglomerate during passage into the exhaust manifold and associated pipes as the gas cools, and having a very high surface area the aggregates absorb HCs, partially combusted products as well as sulphur oxides and nitrogen oxides formed during combustion in the cylinder. The sulphur oxides originate from sulphur compounds present in the fuel and lubrication oil, while nitrogen oxides result from reactions of nitrogen and oxygen at the very high temperatures at the flame front. Other very small amounts of inorganic materials are formed from species present in the fuel and oil such as calcium, zinc, magnesium and phosphorous oxides [4]. These fine inorganic PM can be important nucleation centres for the growth other more volatile species. Figure 2 shows schematically how small carbonaceous particles agglomerate to form larger particles that adsorb gaseous species from the gas phase.

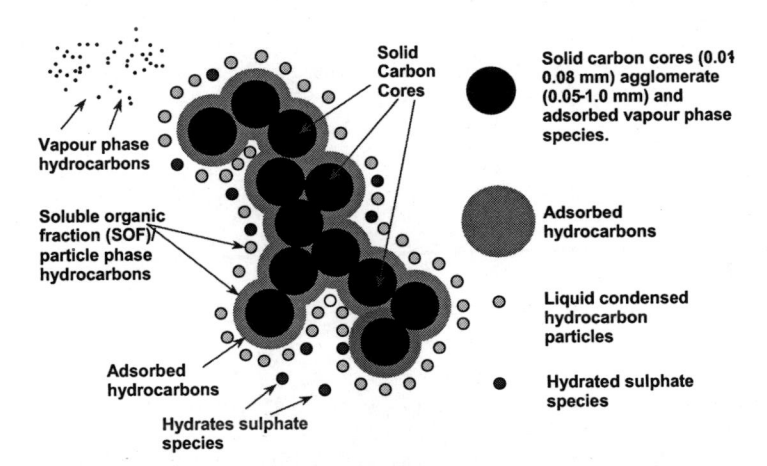

Figure 2. Schematic representation of diesel exhaust PM. Fine particles of carbon agglomerate forming larger particles that adsorb components from the gas phase.

PM present in diesel engine exhaust gas can cover a very wide size range, and this is illustrated in Figure 3. The micron sized PM is typically material that was entrained in the exhaust system and later became dislodged, while the accumulation mode is representative of material illustrated in Figure 2. The nucleation mode comprises very tiny particles less than say, 50 nm. These contain only a very small fraction of the total mass of PM, but because they are so very small they contribute very significantly to the total number of particles, and they are discussed further below.

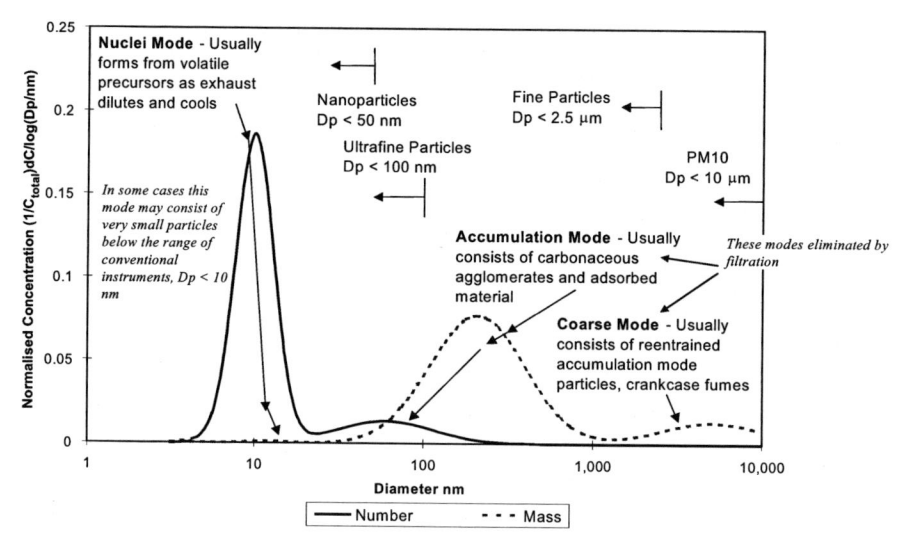

Figure 3. Sizes of PM in diesel exhaust. A DPF removes coarse and accumulation modes. The nucleation mode comprises almost no mass but contributes significantly to the total number of particles present. (Courtesy of D. B. Kittelson)

There are increasing concerns about health effects associated with nano-sized PM in diesel exhaust gas. Because they are so small they are able to pass almost unheeded into the lungs and thence even into the blood stream. This is now one of the main environmental concerns about diesel PM emissions.

3. LEGISLATION

Figure 4 shows how UK diesel PM emissions standards (mass/km) for passenger cars have decreased since 1983. The test cycles have changed, but it is clear the limits decreased considerably - by an order of magnitude. The proposed passenger car PM emissions for the next level of European legislation, Stage 5, due to be phased-in during 2009 can only be achieved by fitment of DPFs and similar demanding legislation will be introduced for heavy-duty vehicles at about the same time, so fitment of filters will almost certainly become necessary on all new European diesel vehicles. Related legislation in other parts of the world will also mean DPFs will be fitted to diesels elsewhere around world in the future. These standards are based on PM mass, and later some countries, most notably those in the Europe Union, will adopt particle number standards that are likely to be more demanding than those based on mass.

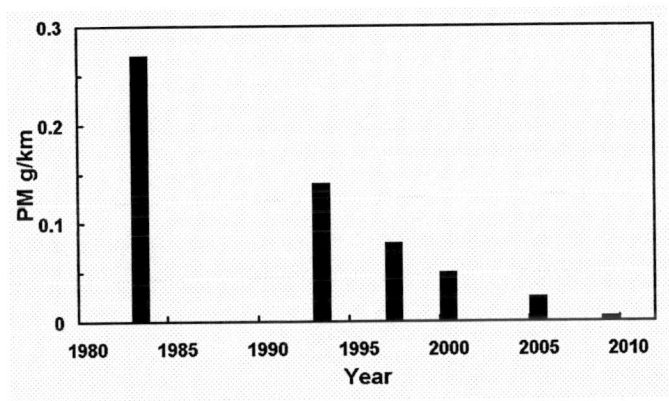

Figure 4. Reduction in light duty diesel PM standards in the UK. By 2009 they will have decreased more than an order of magnitude over three decades.

4. DPF TYPES

Several ceramic and sintered metal DPFs are available. The most commonly used commercially, are porous ceramic wall-flow filters shown schematically in Figure 5. Materials used for their fabrication include cordierite, silicon carbide and aluminium titanate. Alternate channels are plugged so the exhaust gas is forced to go through the channel walls. The PM does not pass through the walls and it is trapped in the DPF, the amount retained in the DPF, increases with time and causes an increasing backpressure across the DPF. If nothing is done to limit the amount of PM in a DPF the backpressure will become so high the engine will stop working. It is therefore essential the backpressure across a DPF is kept within defined limits, and that PM is removed to prevent excess backpressure build up. The best way of removing trapped PM is to oxidise it to CO_2 and water in a process called regeneration. There are two well-established ways of regenerating DPFs that are discussed in the next Section.

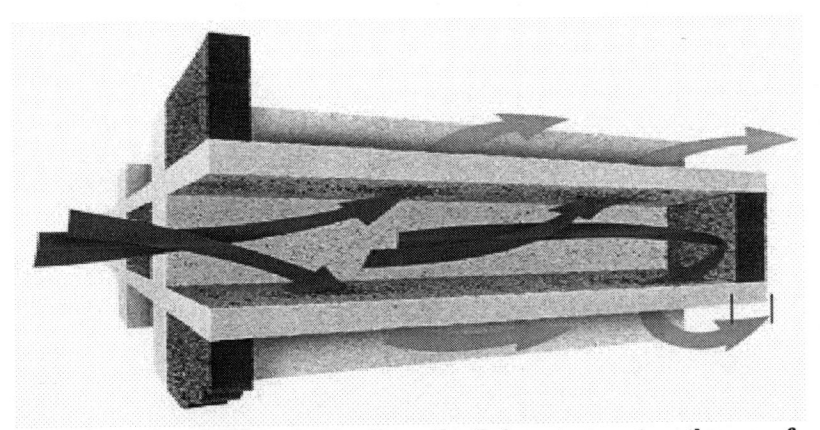

Figure 5. Ceramic wall flow DPF schematic. Exhaust gas enters the open front face channels and flows through the walls because the ends of alternate channels are plugged. Pores in the walls allow the passage of gas but not PM.

5. DPF REGENERATION

The backpressure across a clean DPF increases as the amount of retained PM increases, as illustrated in Figure 6. At first the backpressure rises rapidly as the temperature of the gas increases and PM particles enter the small pores in the DPF walls. Quite soon a layer of PM is formed on the walls and then the backpressure increase becomes less rapid, and it then increases linearly as the layer of PM becomes thicker. To prevent the backpressure becoming too high PM in the filter must be removed, and this is done in a process called regeneration.

It is critical regeneration is done in a tightly controlled procedure. In the past when the technology was not available to do this many DPFs were melted during regeneration, both on light- and heavy-duty applications. Today there are two well proven ways of doing this that are discussed in the following Sections.

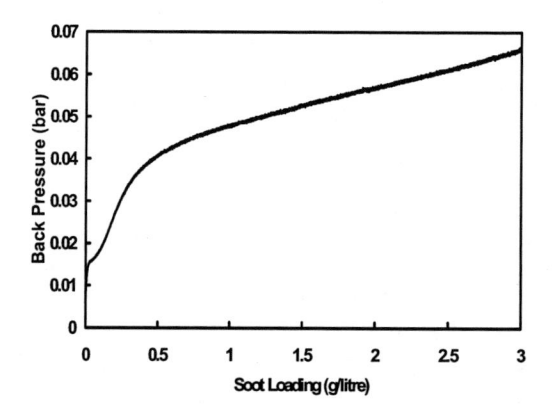

Figure 6. Back pressure across a DPF with PM loading. The initial steep rise is due to temperature increasing and PM entering the DPF wall pores. The linear region is due to an increasingly thicker PM cake on the filter walls

5.1 Passive continuous regeneration

On heavy-duty vehicle applications the engine often operates at high load with exhaust temperatures in the range 250-400°C. Under these conditions the already present NO in the exhaust can be used in a process that continuously removes PM in the DPF while the engine is operating. An oxidation catalyst upstream of the filter oxidises any HCs and CO in the exhaust gas to CO_2 and water. The catalyst also converts NO to the strong oxidant NO_2 that oxidises trapped PM in the filter at the normal operating temperatures, as in Equations (1) and (2), where PM is represented by "CH". On many applications this system operates continuously and the filter is kept clean all of the time, so the system was given [5] the name Continuously Regenerating Trap CRT®.

$$2NO + O_2 \rightarrow 2NO_2 \qquad (1)$$

$$5NO_2 + 2\text{"CH"} \rightarrow 5NO + 2CO_2 + H_2O \qquad (2)$$

The CRT® arrangement of an oxidation catalyst upstream of a DPF is shown in Figure 7. The critical catalytic oxidation of NO to NO_2 is inhibited by SO_2, and so it could not be introduced widely until ultra low sulphur diesel (ULSD) fuel became available. At first heavy duty fleets equipped with CRTs® made provision to have special supplies of ULSD, then countries such as Sweden had ULSD fuel generally available, and many other countries followed, so the CRT® became more widely introduced. Within the UK ULSD became generally available in the late 1990s, and it is now readily available and many thousands of CRTs® are in service on buses, trucks and delivery vehicles around the world [6, 7].

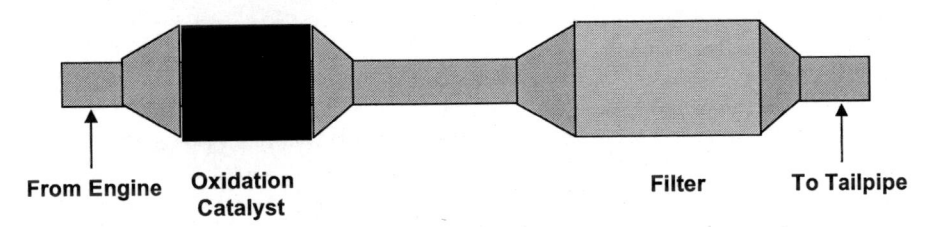

Figure 7. One arrangement for a CRT® with an oxidation catalyst positioned some distance upstream of the DPF.

By having catalyst in the walls of the DPF NO reformed from NO_2 during PM combustion, Reaction (2), can be reconverted to NO_2 that in turn can react with more PM so improving the overall PM removal efficiency. This system is called a "Catalysed Continuously Regenerating Trap" (CCRT®) and it is used [8] where the duty cycle does not always provide appropriate conditions for proper functioning of a CRT® itself.

The practical appeal of the CRT® system is it is passive, requiring no attention during use. Periodically it is however necessary to remove inorganic ash from the filter that accumulates after driving very long distances.

5.2 Active regeneration
It may not always be possible to use the passive DPF regeneration strategy using NO_2 to burn PM retained in a DPF because a vehicle's duty cycle does not provide sufficient temperature, or perhaps the NOx/PM ratio is not high enough to keep the DPF clean. In these situations it is necessary to use an alternative regeneration procedure known as active regeneration. Here the retained PM in the DPF is allowed to build-up to a predetermined level, and then the temperature of the exhaust gas entering the DPF is elevated to above 550°C when typical diesel PM starts to burn in oxygen.

The increased temperature from the engine can be achieved by injecting fuel after TDC (top dead centre) and oxidising remaining hydrocarbons and partially combusted products in the exhaust gas over a catalyst upstream of the DPF. Careful control of the process is necessary because once started the burning PM can easily release sufficient heat to result in temperatures considerably above 1000°C in the DPF.

To control the combustion of PM in a DPF it is necessary to have full control of the engine operating parameters. In particular, fuel injection timing, so additional fuel can be provided after TDC, ideally with feedback control from temperature sensors, and a means of controlling the oxygen levels in the exhaust gas. Here an oxygen sensor is needed together with a means of throttling the engine to restrict the amount of air taken into the engine. This is normally augmented by high exhaust gas recycle (EGR) to help restrict the amount of oxygen available for burning the PM, and so control the combustion kinetics. It is also necessary to have an algorithm that predicts when the DPF should be regenerated, and a backpressure sensor to overrule the algorithm when diving under unusual circumstances.

6. PASSENGER CAR DPF SYSTEMS

In contrast to heavy-duty diesel vehicles, the temperature of a diesel passenger car exhaust gas rarely exceeds 250°C during town driving, so active regeneration must be employed that periodically burns PM in the DPF with oxygen. This takes place every 400-2000 km depending on the actual engine and the DPF used, and driving conditions.

Three commercial DPF systems have been implemented on cars using active regeneration strategies, and these are illustrated in Figure 9. The first has a platinum oxidation catalyst upstream of a DPF to control HCs and CO emissions, and to convert NO to NO_2 for low temperature PM combustion when conditions permit this to take place. The catalyst also oxidises extra fuel when it is injected into the engine to raise the exhaust temperature for PM combustion in oxygen [9]. This system was introduced in 1999 [10] in combination with a base-metal fuel additive that lowers the temperature for PM combustion. This system works well although fuel additive residues are retained in the filter as inorganic ash (see below) and this contributes to an increasingly higher backpressure across the DPF. The second system, shown in Figure 8, also has an upstream oxidation catalyst, and the DPF is catalysed to promote PM combustion so a fuel additive is not needed. Many passenger cars use this arrangement.

The third, and most recent system, comprises a single small catalysed DPF with the oxidation catalyst functionality to oxidise HCs and CO during normal driving, and periodically to oxidise extra partially burnt fuel to give the temperature necessary to combust PM with oxygen. The catalyst also oxidises NO to NO_2 to provide some passive PM removal during high speed driving. It fits directly on the turbocharger and is thermally very efficient because during active regeneration there is only the filter to heat so heat losses are minimised, and the oxidation reactions used to boost the temperature actually take place in the DPF creating heat where the PM is trapped [11, 12].

- Fuel additive type

- Oxidation catalyst and catalysed filter

- Catalysed filter only (integrated oxidation catalyst)

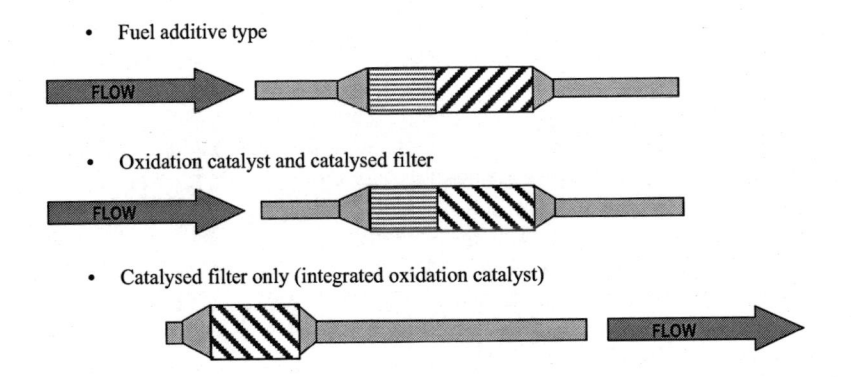

Figure 8. Three commercial DPF systems used on passenger cars in Europe.

7. NANO SIZED PM

DPFs remove coarse micron sized PM and "accumulation mode" particles of about 100 nm in size that together account for almost all the PM mass in Diesel exhaust. Attention is now being directed towards the extremely small particles about 10 nm and even smaller in sizes that are associated with potential health hazards. These are nanoparticles, and although collectively they have extremely little mass, they can be present in huge numbers. When they are inhaled nanoparticles are so tiny they can pass straight into the lungs, through the bronchial tissues and enter the bloodstream, and so go to all parts of the body.

Recent research [13] indicates most of the "nucleation mode" PM comes from volatile organic or inorganic precursors that are formed as the exhaust gas cools. It is difficult to mimic in the laboratory the very large exhaust gas dilution that takes place when driving so a mobile laboratory has been used to study real world nanoparticle formation. The results on heavy duty diesel engines showed under appropriate conditions an oxidation catalyst can remove effectively all the HCs in the exhaust gas, and that then most of the nucleation mode PM is "sulphate", probably sulphuric acid, ammonium sulphate ((NH_4)$_2SO_4$) or ammonium hydrogen sulphate (NH_4HSO_4) derived from sulphur compounds originally present in the fuel and lubrication oil. In agreement with this they are not formed if the sulphur level in the fuel and oil is reduced below a critical level. The lifetime of these nanoparticles is expected to be short because they will coalesce and undergo processes that take them out of the air [14]. Recently we showed careful chemical design of DPF systems can control nanoparticle emissions as well as the coarser types of PM [15, 16].

8. INORGANIC ASH

Inorganic compounds in lubrication oils provide anti-wear and antioxidant properties and keep PM in suspension. Common additives contain P, Ca, Zn, Mg and S, and because small amounts of oil are burnt in the cylinder these elements enter the exhaust gas, and are retained in the DPF. Inorganics derived from fuel PM combustion aids are also trapped in the DPF, and these include Ce, Sr, and Fe compounds. Because of the very high temperatures during combustion, thermodynamics will generally determine the nature of the final species [4]. What is deposited in the DPF on a heavy duty vehicle without fuel additive is typically $Zn_3(PO_4)_2$ and $CaSO_4$ together with material from engine wear, but the elements present depends on the actual oil used. Ash accumulation in a DPF is gradual and it increases backpressure that is not reduced by high temperature regenerations used to remove carbonaceous PM. So gradually the time between regenerations has to decrease, and to minimise this effect the levels of oil and fuel inorganics will have to be reduced. Other measures to reduce backpressure caused by ash are to use a larger filter, if this is possible, and to use DPFs with asymmetric channel structures that provide a larger inlet volume compared to that in the outlet side. The latter may cause slightly higher fresh backpressure compared to a symmetrical one, but this relative difference decreases as ash accumulates in the DPF and soon the asymmetric structure has the lower backpressure [17].

9. LABORATORY GENERATED PM

Loading DPFs with PM requires costly facilities, an engine dynamometer or a vehicle on a rolling road, but PM from a diesel engine is often poorly repeatable due to variations in conditions such as exhaust gas recirculation (EGR) rate, especially when a DPF backpressure varies during loading. With the need for repeatable DPF load characteristics and limited availability of engine-base test facilities we decided to develop a laboratory soot generator for studies on full size DPFs.

To have a similar chemical composition as diesel exhaust PM a diesel fuelled burner was used. To maintain constant combustion conditions regardless of the DPF backpressure, the gas flow through the entire system was drawn by a vacuum blower on the outlet of the DPF. The burner adopted was an ambient pressure swirl-type, and its operating conditions were optimised to produce particulates similar to those in diesel engine exhaust. This required a rich primary combustion followed by dilution and quenching with a secondary air flow. A schematic drawing of the apparatus is shown in Figure 9, and it is being commercialised by Cambustion Ltd under the name Diesel Particulate Generator (DPG).

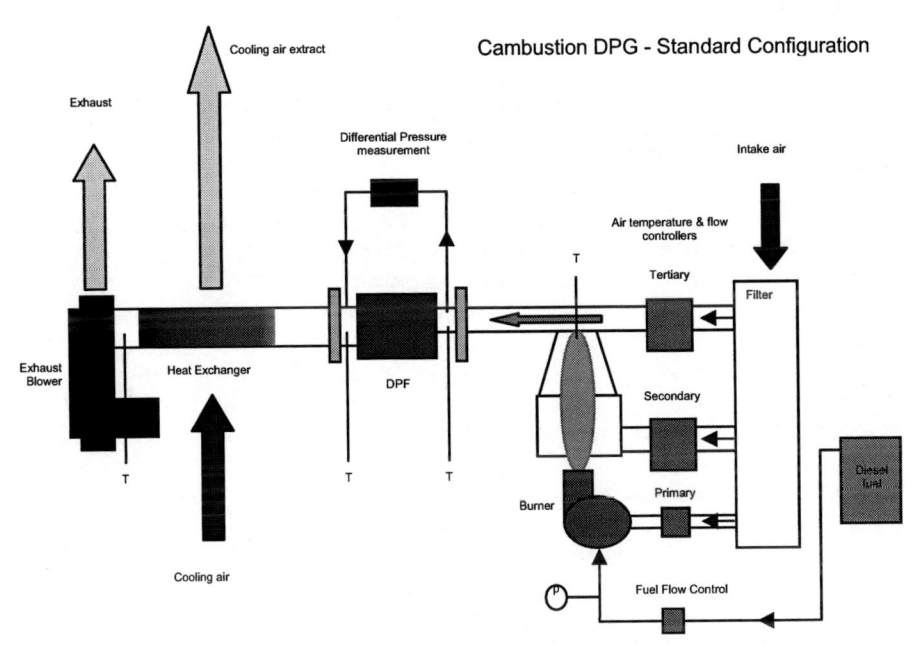

Figure 9. Schematic of laboratory diesel particulate generator (DPG) that loads full sized DPFs with PM in a similar way as an engine.

10. THE FUTURE

Diesel engines have better fuel economy than gasoline engines and provided their exhaust gas emissions are well controlled their use can provide environmental benefits in terms of CO_2 emissions. Fitment of filters to diesel engines is environmentally important, and future legislation will demand their use in Europe and elsewhere around the world. The overall trend is of increasing complexity in diesel exhaust emissions control systems. Initially oxidation catalysts were used on diesel cars to control just HCs and CO emissions [18]. More recently PM filters were introduced and the types used have evolved so now all of the oxidation required and filtration functions can be incorporated in a single small DPF. By analogy with gasoline "three-way catalysts", they are "three-way" systems.

In the future control of NOx emissions from diesel engines will be necessary, and two processes will be used to do this. In the first NOx is converted to nitrate species within a catalyst and the nitrate is reduced to nitrogen (N_2) by frequent pulses of enriched exhaust gas obtained by late injection of fuel into the engine. This has the advantage of having the reductant (diesel fuel) already available on the vehicle.

The second method uses ammonia derived from urea that is injected into the hot exhaust gas. Over a special catalyst the ammonia selectively reduces NOx to N_2 (a process known as selective catalytic reduction or SCR). To be cost and space effective some of these functions will be combined in single components, and already this process has started. It may be expected in the future combined systems will be fitted to diesel passenger cars so four emissions will be controlled with a special catalysed DPF and perhaps a small upstream flow-through catalyst. However, much work has to be done to realise such a "four-way" system.

11. CONCLUSIONS

Sophisticated emissions control systems are being developed for fuel-efficient (lower CO_2) modern diesel engines in passenger cars and heavy-duty applications. For several years catalysts have been fitted to oxidise CO and HCs, and the spotlight is now on preventing particulate matter from entering into the atmosphere. This is done with wall-flow DPFs and there are two methods for preventing build-up of PM in them. The first makes use of a catalytic process in which already present NO is oxidised to NO_2 that reacts with PM to reform NO. This can work on heavy-duty applications that operate at higher temperatures than are available on passenger cars where active regeneration is necessary. Here catalysts oxidise extra fuel from the engine to achieve the temperatures needed to burn PM with oxygen. DPF systems are capable of eliminating coarse and accumulation mode PM from diesel exhaust. Nanoparticles from diesel engines are the subject of much research and ways of controlling them is now understood. In the future NOx reduction systems will be needed to meet legislative requirements, and will involve NOx-trapping technology or SCR using ammonia derived from an aqueous solution of urea. Once these approaches have been developed it is likely multi-function catalytic systems will be developed in the same way three-way catalysts are used on gasoline engines today.

12. REFERENCES

(1) Over recent years there has been a major increase in the sales of diesel passenger cars in Europe. At present more than 50% of the cars sold have diesel engines. AID, Schmidt's auto Publications, 2006, 2 August, 1.
(2) M. V. Twigg, *Applied Catalysis B: Environmental*, 2007, **70**, 2; N. R. Collins and M. V. Twigg, Topics in Catalysis, 2007, **42-43**, 232.
(3) M. V. Twigg, Phil. Trans. Roy. Soc. A, 2005, **363**, 1013.
(4) A. N. Hayhurst, D. B. Kittelson, J. T. Gidney and M. V. Twigg, "Chemistry of Inorganic Additives to Fuel and Oil: In-Cylinder Reactions and Effects on Emissions Control Systems", IMechE Conference Tribology 2006: Surface Engineering and Tribology for Future Engines and Drivelines", 12-13 July 2006, London.
(5) B. J. Cooper and J. Thoss, *Soc. Automotive Engineers Technical Paper*, 1989, 890404.
(6) P. N. Hawker, *Platinum Metals Rev.*, 1995, **39**, 1.
(7) A. P. E. York, J. P. Cox, T. C Watling, A. P. Walker, D. Bergeal, R. Allansson and M. Lavenius, *Soc. Automotive Engineers Technical Paper*, 2005, 2005-01-0954.

(8) R. Allansson, C. Görsmann, M. Lavenius, P. R. Phillips, A. J. Uusimaki and A. P. Walker, *Soc. Automotive Engineers Technical Paper*, 2004, 2004-01-0072.

(9) G. Merkel, W. Cutler, T. Tao, A. Chiffey, P. R. Phillips, M. V. Twigg and A. P. Walker, "New Cordierite Diesel Particulate Filters for Catalyzed and Non-Catalyzed Applications", 9th Diesel Engines Emissions Reduction Workshop 2003, Newport, Rhode Island, 24 August 2003.

(10) O. Salvat, P. Marez and G. Belot, *Soc. Automotive Engineers, Technical Paper*, 2000, 2000-01-0473.

(11) P. G. Blakeman, A. F. Chiffey, P. R. Phillips, M. V. Twigg and A. P. Walker, *Soc. Automotive Engineers, Technical Paper*, 2003, 2003-01-3753.

(12) A. F. Chiffey, P. R. Phillips, D. Swallow, M. V. Twigg, W. A. Cutler, T. Boger, D. Rose and L. Kercher, "Performance of New Catalyzed Diesel Soot Filters Based on Advanced Oxide Filter Materials", 4th FAD Conference "Challenge – Exhaust Aftertreatment for Diesel Engines", Dresden 2006.

(13) M. Grose, H. Sakurai, J. Savstrom, M. R. Stolzenburg, W. F. Watts Jr., C. G. Morgan, I. P. Murray, M. V. Twigg, D. B. Kittelson and P. H. McMurry, *Environmental Science and Technology*, 2006, **40**, 5502.

(14) D. B. Kittelson, W. F. Watts, J. P. Johnson, C. J. Rowntree, S. P. Goodier, M. J. Payne, W. H. Preston, C. Ortiz, U. Zink, C. Görsmann, M. V. Twigg and A. P. Walker, *Soc. Automotive Engineers, Technical Paper*, 2006-01-0916.

(15) D. B. Kittelson, W. F. Watts, J. P. Johnson, C. J. Rowntree, S. P. Goodier, M. J. Payne, W. H. Preston, C. Ortiz, U. Zink, C. Görsmann, M. V. Twigg and A. P. Walker, *Soc. Automotive Engineers, Technical Paper*, 2006-01-0916.

(16) D. B. Kittelson, W. F. Watts, J. P. Johnson, C. Rowntree, M. Payne, S. Goodier, C. Warrens, H. Preston, U. Zink, M. Ortiz, C. Görsmann, M. V. Twigg, A. P. Walker and R. Caldow, *J. Aerosol Science*. In press.

(17) D. M. Young, D. Hickman, N. Gunasekaran and G. Bhatia, *Soc. Automotive Engineers Technical Paper*, 2004, 2004-01-0948.

(18) Launch of Volkswagen's "Umwelt Diesel", *Ward's Automotive Reports*, 1989, *18*, September 301.

Feasibility study of emission control of hydrogen fueled S.I. engine

Tetsuya Ohira [1], **Kenji Nakagawa** [2], **Kimikata Yamane** [2], **Hiroshi Kawanabe** [3], **Masahiro Shioji** [3]

[1] Suzuki Motor Corporation, Japan
[2] Musashi Institute of Technology, Japan
[3] Kyoto University, Japan

In an attempt to understand potential issues with a hydrogen direct injection lean burn engine with similar power output to a gasoline-fuelled engine, emission characteristics of a hydrogen engine were investigated. It was demonstrated that low NOx emission can be achievable without a catalytic converter. Two major issues, however, have been recognized, that is, combustion instability at low load conditions and too low exhaust gas temperature to get enough boost pressure. Heterogeneous hydrogen concentration of the mixture was focused in the CFD and visualization study. Hydrogen jet design of the injector could contribute to improvement of mixing.

1 INTRODUCTION

One of the concepts for future power sources for low CO_2 emission vehicles is a hydrogen fuelled lean burn S.I. engine with a boosting system. Possibilities to achieve the same power output of conventional gasoline-fuelled engine and low emissions without any catalytic converter are desired for small engines mounted on small vehicles that have a limited capacity for powertrain and fuel or energy storage.

Because gaseous hydrogen has low energy density relative to liquid fuel such as gasoline, a hydrogen-fueled engine as a power source for a small vehicle should be boosted. Application of direct injection to hydrogen engines has advantages in principle to trap more air mass than premixed or port injection and to avoid backfire. A direct injection S.I. engine with typical small engine's dimensions such as a 68mm bore was used in this fundamental study. Hydrogen was injected into the cylinder at 7MPa in the early compression stroke process in the direct injection mode. An external compressor was used to investigate boost pressure effects on the emissions without boost pressure fluctuations. Emission characteristics of the boosted hydrogen S.I. engine are discussed in this paper.

Key constituents of the emissions of the hydrogen engine are nitric oxide and nitrogen dioxide (NOx). Our target of the NOx concentration is below 10ppm in

© IMechE 2007

order to achieve low tailpipe emissions without any catalytic converter in the exhaust system.[1] The optimal excess air ratio (lambda) to keep low concentration of NOx was investigated. Even with different boost pressure or different air mass density in a fully throttled condition, NOx concentration showed similarity of exponentially increasing characteristics as the lambda decreased.[2] This formal characteristic can provide us with a potential methodology for emission control without any catalytic converter equipped on a vehicle.

Combustion instability was found in low load conditions of our previous study. A higher percentage of unburned hydrogen in the exhaust gas and a larger cyclic variation of combustion showing higher coefficient of variation (COV) of indicated mean effective pressure (IMEP) appeared when using direct injection comparing to that of premixed hydrogen fueling. Poor mixing of hydrogen injected into the cylinder and in-cylinder air could be the cause. Characteristics of mixing the hydrogen jet and ambient air are also discussed focusing on the relation of mixture condition to emissions and thermal efficiency. Unsteady behavior of the hydrogen jet and ambient gas entrainment characteristics was studied by a computer fluid dynamics (CFD) analysis and visualization research using a vessel chamber.[3]

In a high load condition, discussion has focused on a point if a hydrogen-fueled engine could have low emission even when highly boosted conditions cause higher combustion peak pressure. Air to fuel ratio influence on NOx and the other emissions has been studied in intake-boosted conditions. An external boosting system was successful in providing us stable intake pressure level and stable test conditions throughout this study.

The actual boosting device to get charge pressure at low engine speeds still remains to be considered. Utilizing lower temperature exhaust gas is very difficult for turbo charging. Mechanical supercharging would cause large friction loss and loose total thermal efficiency. A turbo charger with an electrically powered turbine being developed may be useful to raise its efficiency.

2. EXPERIMENTAL APPARATUS

2.1 Tested Engine
In order to confirm a concept of a hydrogen fuelled lean burn S.I. engine with a boosting system, a small direct injection engine, which was originally gasoline fuelled, was modified for the experimental works in this study. The major specifications are shown in Table 1.

Hydrogen was supplied from a cold liquid storage tank. For port fueling, hydrogen was drawn into the inlet plenum at 0.6MPa through a surge tank. For direct injection, hydrogen was injected at 7MPa through the injectors installed in the same configuration as the original gasoline engine as shown in Figure 1. This engine was originally developed to obtain good mixture with intense tumble flow through a CFD study.

Table 1. Engine Specifications

Item		Dimensions
Engine Type		Water-Cooled, 3-Cylinder,4-Stroke, DOHC, Direct Injection
Stroke Volume		658 (cc)
Bore x Stroke		68 x60.4 (mm)
Compression Ratio		9.1:1
Combustion Chamber		Pent Roof
Injection type		Electro-Magnetic Current controlled Single hole
Swirl Ratio		0
Tumble Ratio		1.2
Valve Timing	Intake V. Open	21deg.CA BTDC
	Intake V. Close	66deg.CA ABDC
	Exhaust V. Open	66deg.CA BBDC
	Exhaust V. Close	24deg.CA ATDC

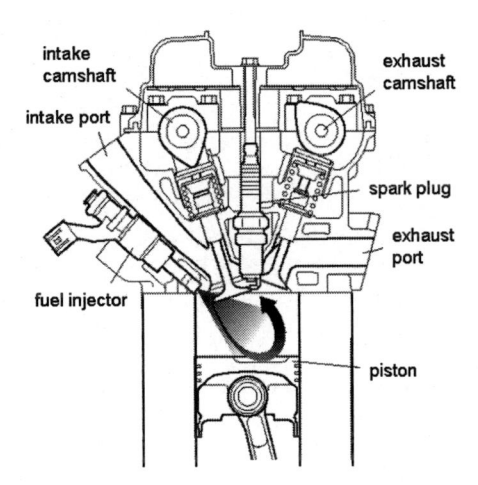

Figure 1. Injector Layout

Chutes-like bent intake port generated 1.2 averaged tumble ratio. A screw type compressor (Hitachi OPS-55UAR) boosted intake air to 0.8MPa and regulated pressure with a valve (CKD R4000) to desired levels. In order to avoid cyclic variation due to charge air mass fluctuation, the external boosting system was applied to obtain stable intake manifold pressure in the engine experiments.

2.2 Engine Conditions

The intake throttle was fully opened to maximize thermal efficiency during this study. Engine load was basically controlled by changing amount fuel injected into the cylinder. Operation conditions, coolant temperature and lubricant oil temperature were set at 80degC with 5degC tolerance.

Hydrogen was regulated to 7MPa by low-pressure valve after regulation by a high-pressure valve to 15MPa. Direct injection was driven by electric current control. Start of injection was fixed at 4 deg crank angle (CA) after intake valve close, i.e., 70 deg CA ABDC. The original electro-magnetic injectors had been modified for hydrogen fuelling together with fuel lines. Injection duration was calibrated to the hydrogen flow rate for each injector prior to the engine experiments.

Constant engine speed was at 2000min^{-1}. To investigate characteristics of combustion and emissions of the hydrogen engine along with loads, break mean effective pressure (BMEP) was changed from 0.1MPa to 1.7MPa. In boosted conditions exhaust backpressure was adjusted as same as intake manifold pressure (MAP) by a throttle valve in the exhaust passage. Bulk in-cylinder mixture concentration, lambda was swept from 1.3 to 4.7. Even in a leanest and lowest load condition, injection duration was sufficiently longer than its solid injection period. In the case of direct injection, backfire was not observed throughout all experiments.[2]

2.3 Engine Test Apparatus

Hydrogen consumption rate was measured by a calibrated mass flow meter (OVAL F-113K-A-15-11N). Air mass was measured by air flow meter (OVAL VXW1025-N12GL- 1114). Pressure transducer (AVL GM12D) was installed in a hole drilled in the cylinder head of the #3 cylinder to measure in-cylinder pressures and conduct combustion analysis using an analyzer (Ono Sokki DS-9100). Emission concentrations of NOx, CO, CO_2, O_2 and HC in exhaust gas were measured by an analyzer (Horiba MEXA-9100LE). Unburned hydrogen concentration was measured by a hydrogen analyzer (Horiba MEXA-1000WL). Exhaust gas was sampled for emission measurements out of the exhaust manifold at 20mm downstream from the exit of the exhaust port of the cylinder head. Exhaust gas temperature from each cylinder was monitored using K type thermocouples at 50mm downstream point from the exit of the exhaust port. The engine experimental schematic is shown in Figure 2.[2]

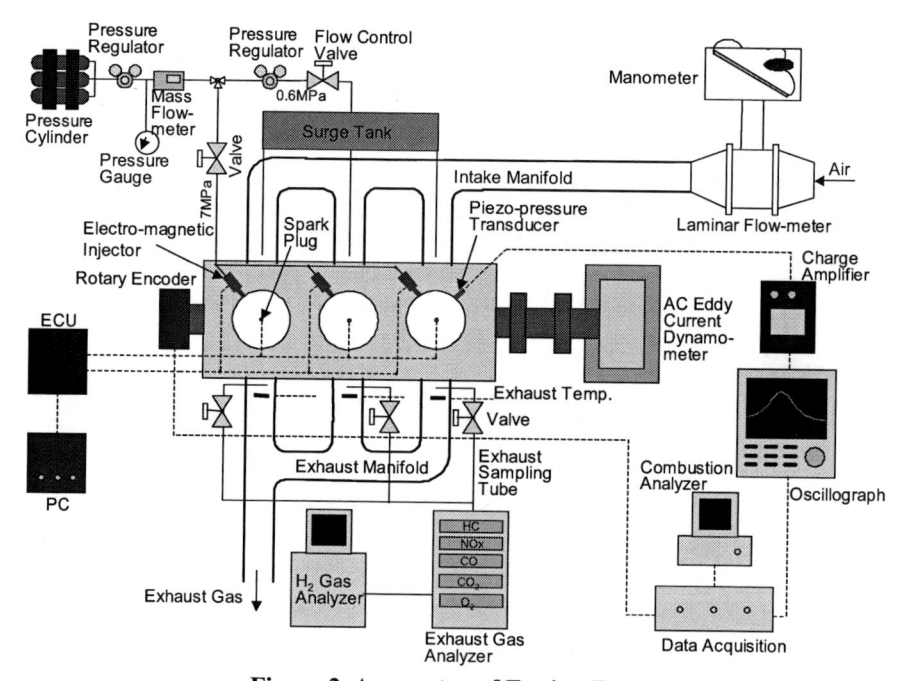

Figure 2. Apparatus of Engine Test

2.4 Visualization Apparatus

Visualization of hydrogen jet was conducted in a vessel chamber by laser sheet Mie-scattering and Particle Image Velocimetry (PIV) technique. A laser sheet with 0.5mm width was emitted by a Nd:YVO4 laser (Spectra Physics, Millennia wave length 532nm, rated power 5W). The visualization camera (Vision Research Inc., Phantom v7.0) used in this study had CMOS sensor composed of 304*600 effective pixels, 28μs exposure and 10000fps time response to capture a whole image of the hydrogen jet. When zooming up in PIV visualization, the camera settings were changed to 300*160 pixels, 5μs exposure and 34000fps. PIV signal was analyzed by commercialized software "Koncert".

A vessel chamber was used with 90mm diameter, 170mm height and 1080cm³ volume that could cover the hydrogen jet observation.[3] Visualization apparatus system is shown in Figure 3.

2.5 Injection condition for vessel tests

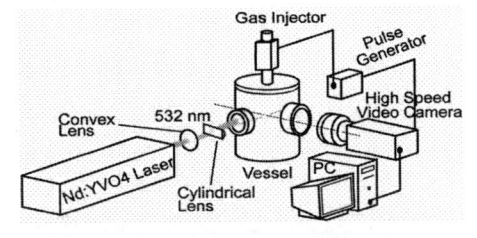

Figure 3. Apparatus of Vessel Test

Seeding particles (TiO₂ 0.2mm) were suspended and nitrogen was charged in the vessel at designated ambient pressures (Pa: 0.5MPa or 1MPa) and temperatures (Ta: 300K). Hydrogen or reference gas nitrogen pressured to 5MPa was injected downward into the vessel chamber through the injector mounted on its roof after the ambient gas became tranquil.[3]

3. RESULTS AND DISCUSSION

3.1 Engine Test Results

To understand basic emission characteristics of a hydrogen direct injection engine, measured data on a dynamometer with different boost pressures and lambdas were shown in Figure 4. Actual lambdas of the test condition were plotted on targeted lambdas in Figure 4 (a). Measured air flow rate was in proportion to boost pressure level regardless of lambda. BMEP is basically depending on the boost pressure level and lambda as shown in Figure 4 (c). BMEP at 160kPa could reach to the same level as a natural aspirated gasoline engine. Minimum advanced spark for best torque (MBT) ignition timing was not as related with boost pressure as lambda. Leaner mixture requires more advanced spark timing for MBT. Although in high load conditions at the end of injection timing was later than the ignition timing as shown in Figure 4 (e), combustion or emission data was stable and kept reasonable characteristics.

Break thermal efficiency showed close relationship with boost pressure level because intake manifold pressure would be dominant parameter to pumping loss. Higher BMEP, gave higher break thermal efficiency.

3.2 Combustion and Emission Feature

Combustion excitation of this hydrogen engine was not large as shown in Figure 5 (a) except lambda 1.3. At lambda 1.3 in-cylinder pressure rise (dp/dθ) was up to over 600kPa. It is presumed that combustion speed depended on mixture concentration in the combustion chamber.

In low load conditions, COV of IMEP, i.e., combustion instability, showed high levels especially at lean lambdas over 3.2. Unburned hydrogen level of the exhaust gas was synchronized to the combustion instability as shown in Figure 5 (c). Naturally, static hydrogen air mixture has wide ignitable lambda range from 0.14 to 9.98. Ignitable limit of actual hydrogen mixture in a engine normally reaches to 5 or higher. It is considered that actual charged gas in the cylinder had heterogeneous distribution of hydrogen concentration including over lambda 5.

(a) Actual Lambda

(b) Air Flow Rate

(c) BMEP vs MAP

(d) MBT Ignition Timing

(e) Injection & Ignition Timing

(f) Break Thermal Efficiency

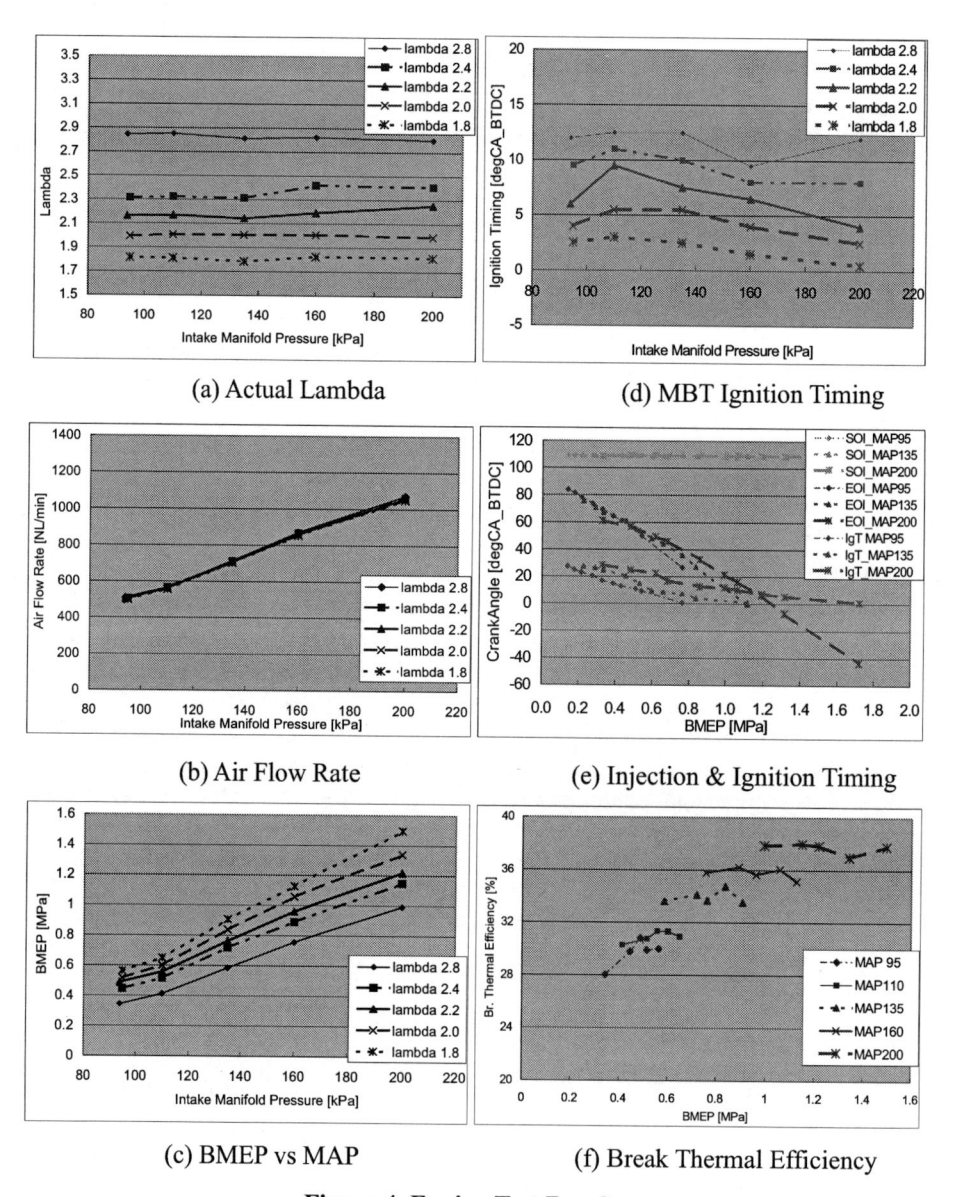

Figure 4. Engine Test Results

NOx level was dependent on lambda, too. NOx emission level seemed inversely proportional to lambda for all boost pressure levels or BMEPs. NOx emission from an engine without any exhaust-gas after-treatment system should be 10ppm level to meet the world lowest class emission regulation. From the observed data of the engine test, the hydrogen engine should keep lambda 2.4 in all load conditions. If we could have a lambda sensor to detect linearity in the 2.4 range for a hydrogen-fuelled engine, it would be feasible to control the major emission NOx and achieve converter free hydrogen engines for vehicles.

(a) Maximum Rise of In-Cylinder Pressure

(b) Maximum In-Cylinder Pressure

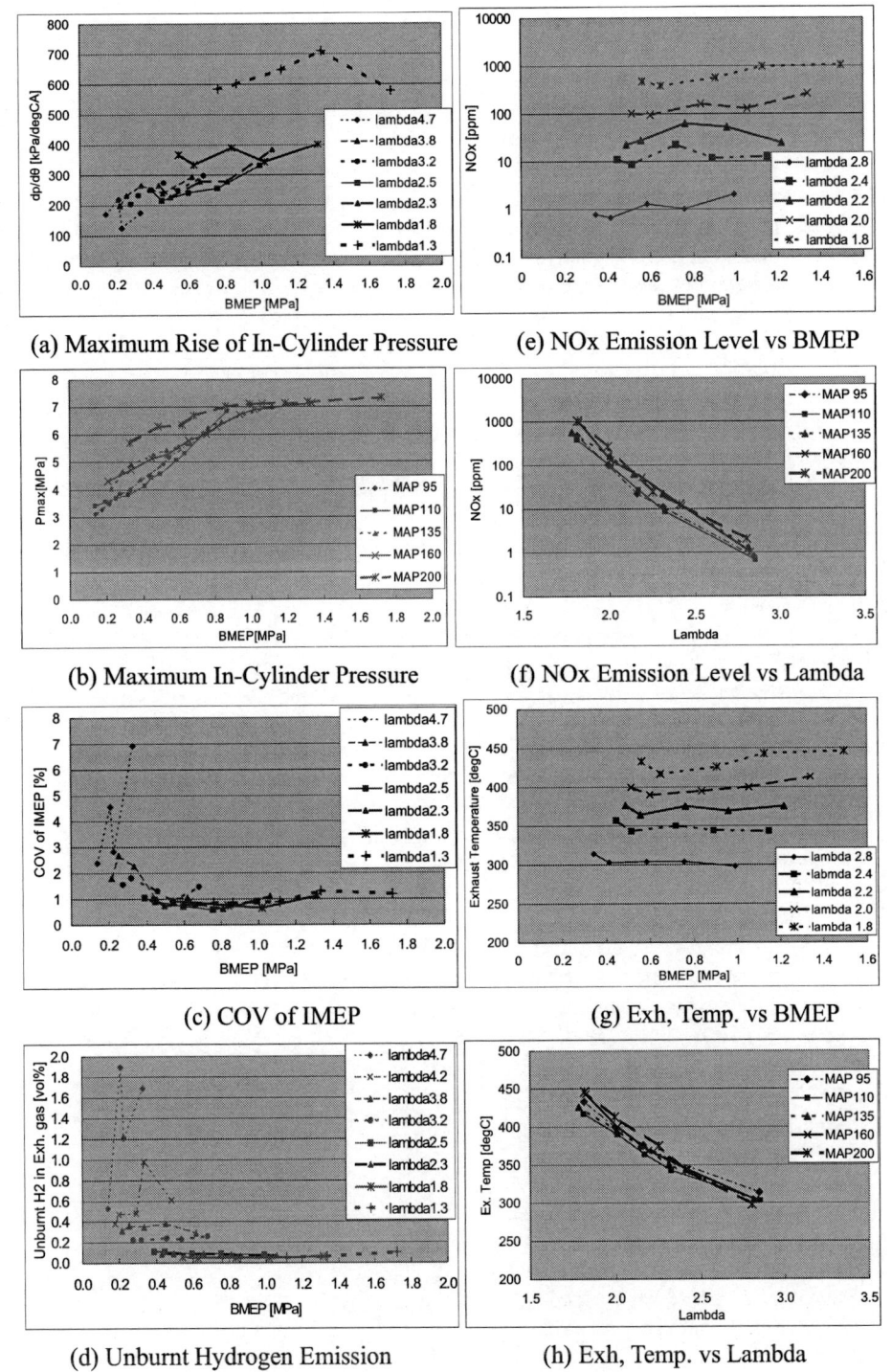

(c) COV of IMEP

(d) Unburnt Hydrogen Emission

(e) NOx Emission Level vs BMEP

(f) NOx Emission Level vs Lambda

(g) Exh, Temp. vs BMEP

(h) Exh, Temp. vs Lambda

Figure 5. Combustion and Emission Analysis

On the other hand, exhaust temperature was also dependent on lambda rather than intake manifold pressure or BMEP. Around lambda 2.4 exhaust temperature was below 350degC in all load conditions. High efficiency turbocharger, electrically assisted turbo system or a new effective boosting system will be required to obtain practicable torque output of a hydrogen engine.

3.3 CFD of In-cylinder Hydrogen Mixture
General physical property of hydrogen was considered in the calculation conditions of our initial RANS CFD analysis. Two cases of calculation were performed with different injection timing. One injection timing was early, and the other was late in the compression stroke. The CFD results are shown in Figure 6.

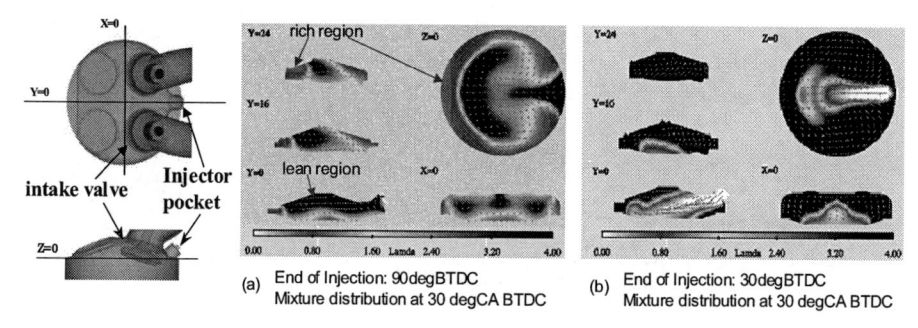

(a) End of Injection: 90degBTDC
Mixture distribution at 30 degCA BTDC

(b) End of Injection: 30degBTDC
Mixture distribution at 30 degCA BTDC

Figure 6. In-Cylinder CFD

Both results showed heterogeneous hydrogen distribution in the combustion chamber. The CFD results suggested difficulty in mixing with the air in the cylinder. We need to know hydrogen jet characteristics for its adequate application to approach an idea for improvement of hydrogen mixing with air in the cylinder and to improve CFD calculation accuracy.

3.4 Visualization of Hydrogen Jet Growth
In order to focus on hydrogen jet characteristics of mixing with air, which must affect combustion condition in the cylinder, a vessel chamber study was conducted with laser visualization technique[3]. Laser sheet Mie-scattering visualized whole jet shape and time elapsed distance between tip of growing jet and the nozzle of the injector as shown in Figure 7 together with that of calculated results.

Hydrogen jet and nitrogen jet included small amounts of water vapor remained in dry ambient gas charged after previously tested gas was dumped out of the vessel. Captured images of those jets showed whitened by laser sheet due to the vapor molecules as shown in

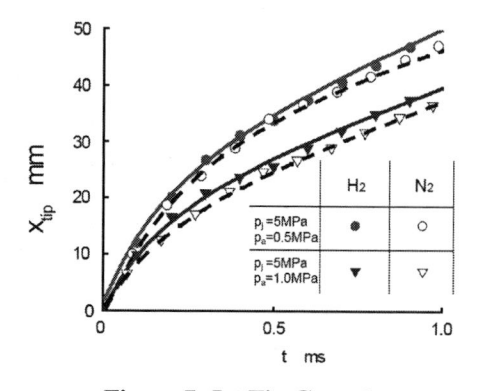

Figure 7. Jet Tip Growth

Figure 8. Because the injector nozzle had inclined angle of 7.5deg, images of the hydrogen jet showed biased direction. But the angle was stable and did not influence the quality of the visualization tests.

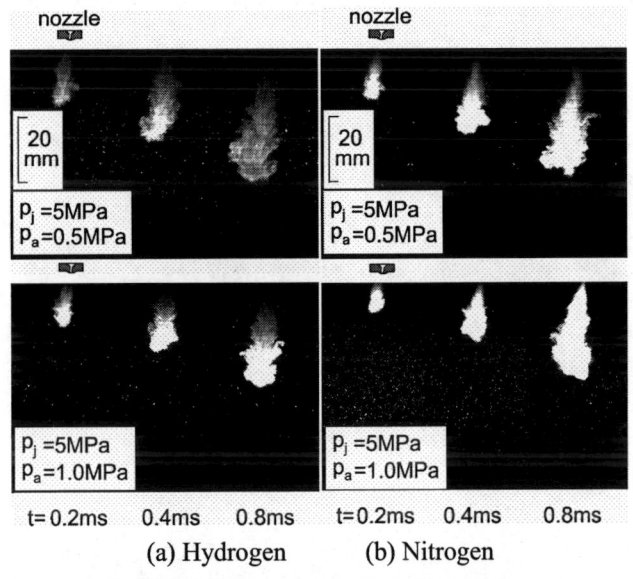

Figure 8. Jet Image by Mie-Scattering

The momentum theory was applied to the calculation of the distance of jet tip from the nozzle using following equations;

$$\frac{X_{tip}}{d} = \sqrt{\frac{1}{\tan\theta}\frac{t}{t_0}} \qquad (3\text{-}1)$$

$$t_0 = \sqrt{\frac{\rho_a}{\rho_j}\frac{d}{U_0}} \qquad (3\text{-}2)$$

where d is nozzle diameter, t is elapsed time from injection start, θ is half of diffusion angle of jet, ρ_a and ρ_j are density of the ambient gas and injected gaseous fluid, respectively, U_0 is sonic velocity of the injected gaseous fluid at the ambient temperature.[3] In the experiments, diffusion angle of jets in all test conditions was almost same, and then $\tan\theta$ can be assumed as constant. Ratio of normalized distance *Xtip/d* to the non-dimensional time *t/t₀* was equal without regard to any condition. Figure 7 shows good agreement between the visualization results and the calculations. Hydrogen jet much accelerated its penetration just after injection compared to nitrogen jet. After 0.3ms or later, however, tip of the hydrogen did not expend so much.

The whole shape of the hydrogen jet visualized by laser sheet was compared to that of nitrogen at 0.2ms, 0.4ms, 0.8ms of elapsed time as shown in Figure 8. Hydrogen jet showed larger inequality shape than nitrogen jet in the upstream side near the nozzle. It is considered that hydrogen jet obtained large amount of ambient gas entrainment into the jet in upstream region. Once hydrogen jet with low density entrained ambient gas with high density, the hydrogen jet decelerated accordingly. Hydrogen jet did not show large expansion, consequently, notwithstanding its very low density.

3.5 PIV results discussion on Entrainment

In order to focus further on mixing process of hydrogen jet and ambient gas, PIV was employed to demonstrate entrain behavior around hydrogen jet[3]. The high-speed camera was zoomed up to capture detail entrainment with better image resolution in the early period of jet growth. More than one zoomed-up images were combined to show whole grown jet in later timing. PIV results are shown in Figure 9.

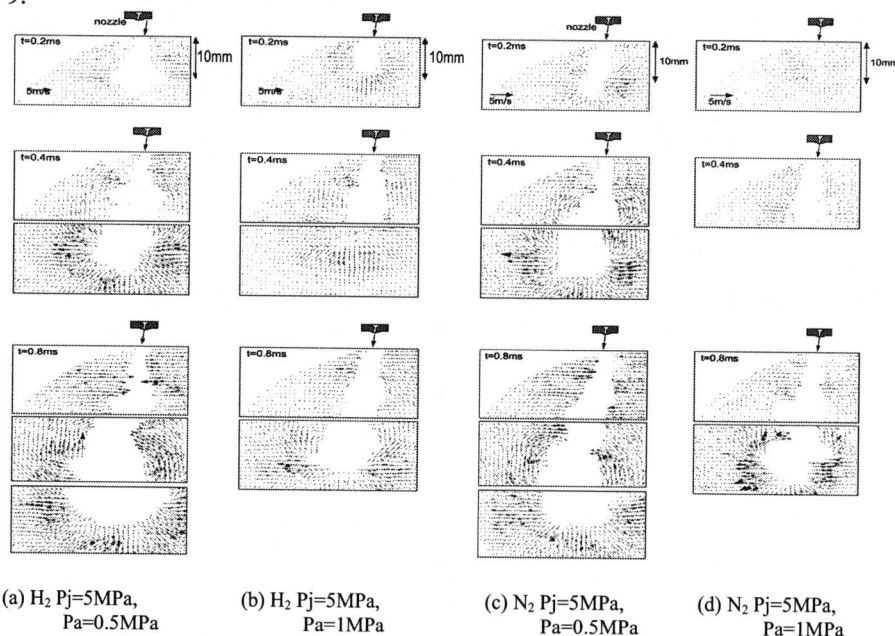

(a) H_2 Pj=5MPa, Pa=0.5MPa
(b) H_2 Pj=5MPa, Pa=1MPa
(c) N_2 Pj=5MPa, Pa=0.5MPa
(d) N_2 Pj=5MPa, Pa=1MPa

Figure 9. Entrainment Image by PIV

Entrainment of the hydrogen was more active than that of nitrogen in the upstream of the jet at 0.4ms or 0.8ms. Jet velocity variation downstream of the injector nozzle or jet deceleration of hydrogen was apparently larger than that of nitrogen. Such a large velocity variation generated intense turbulence and deformed jet outlook shape and encouraged entrainment in the upstream region. Around the tip of the hydrogen jet, however, did not show larger entrainment than that of nitrogen. It must be the same reason why the jet growth downward did not accelerate at 0.4ms or later. Hydrogen jet did not spontaneously expand along elapsed time. This nature of the hydrogen jet could be a reason behind unburned hydrogen emission and instability of combustion in lean and low load conditions. For enhancement of hydrogen mixing with air, hydrogen jet distribution should be key design subject. For example, multi-hole injectors would be one of the solutions for that with its benefit of flexibility with wide jet allocation.

4 CONCLUSION

(1) It has been found feasible in hydrogen-fuelled engines to achieve low NOx emission equivalent to the world lowest level. A fuel management system is preferable to control excess air ratio around lambda 2.5 to keep low NOx.
(2) It is possible for hydrogen-fuelled engines to have similar power output to gasoline-fuelled engines with same displacement if a future turbocharger can boost with 350degC exhaust gas or lower.
(3) Insufficient mixing of hydrogen with air in the cylinder would be a reason behind unburned hydrogen emission out of the engine. Hydrogen jet design of a injector to overcome its low density jet would be a key point to improve its mixture and engine combustion in low loads and lean operations.

5 ACKNOWLEDGMENT

The authors would like to thank all colleagues and students concerned in the experiment and computation works.

REFERENCES

(1) Inoue, T., et al.: "Experimental Study on Application of Hydrogen Gas Injection at High Pressure into a Small Displacement Spark-Ignition Engine", Proceedings of JSAE Spring Conference, Paper No.20055219, (2005) (in Japanese)
(2) Nakagawa, K., et al.: "The Effect of Boost Pressure on the Engine Performance by using a Small Hydrogen Fuelled Engine with Direct Injection in the Early Compression Stroke", Proceedings of JSAE Spring Conference, Paper No.20075324, (2007) (in Japanese)
(3) Kato, R, et al.: "Flow and Mixture-Formation Process of an Unsteady Hydrogen Jet", Proceedings of JSME Kansai Branch 82nd Conference, Paper No.113,, (2007) (in Japanese)

FUELS AND LUBRICANTS

Fuel dilution effects in a direct injection of natural gas engine

G.P. McTaggart-Cowan[1], S.N. Rogak[2], P.G. Hill[2], S.R. Munshi[3], W.K. Bushe[2]

[1] Wolfson School of Mechanical and Manufacturing Engineering, Loughborough University, UK
[2] Department of Mechanical Engineering, University of British Columbia, Canada
[3] Westport Power Inc., Canada

ABSTRACT

This work reports the effects of fuel dilution in a heavy-duty gaseous-fuelled direct-injection engine with pilot ignition. Diluting the natural gas with 20% and 40% nitrogen results in no significant change in the combustion duration, while reducing the combustion intensity and increasing stability. Emissions of NO_x, PM, HC, and CO are reduced with no effect on efficiency. The results indicate the benefits of increased in-cylinder turbulence and are of particular relevance when considering fuel composition variations with non-conventional gaseous fuels.

NOTATION

50% IHR	Mid-point of integrated heat release	HC	Hydrocarbons
°CA	Crank angle degree	HRR	Heat release rate
ATDC	After top-dead-center	IHR	Integrated heat release
CNG	Compressed natural gas	LNG	Liquefied natural gas
EGR	Exhaust gas recirculation	NO_x	Oxides of Nitrogen (NO, NO_2)
GID	Gaseous fuel ignition delay time	PID	Pilot ignition delay time
GikWhr	Gross indicated kilowatt-hour	PM	Fine particulate matter

1. INTRODUCTION

Heavy-duty engine manufacturers are developing advanced in-cylinder and post-exhaust aftertreatment devices to ensure that diesel engines meet stringent new emission standards. One technique to reduce engine-out emissions of local and global air pollutants is to replace the diesel fuel with natural gas. A barrier to using natural gas is that its composition can vary significantly between suppliers, seasons, and geographical sources. The addition of unconventional and bio-derived gases can

© IMechE 2007

have an even greater effect on fuel composition [1]. These constituents can include heavier hydrocarbons or inert diluents. The presence of diluents, such as molecular nitrogen (N_2), can significantly influence the combustion and emissions. Varying the fuel composition can also provide insight into the fundamental combustion factors that influence pollutant formation. This study investigates the impacts of diluting natural gas with varying levels of N_2 on the combustion and emissions of a heavy-duty, compression-ignition, direct-injection engine.

1.1 Combustion System

One technology for natural gas fuelling of heavy-duty engines, developed by Westport Power Inc., uses natural gas injected directly into the combustion chamber late in the compression stroke. This technology retains the performance and efficiency of an equivalently sized diesel engine [2]. As the pressure and temperature in the combustion chamber are not sufficient for natural gas to reliably auto-ignite, a small amount of diesel fuel is injected prior to the natural gas. The auto-ignition and combustion of this pilot fuel provides the ignition source for the gaseous fuel, which then burns in a predominantly non-premixed combustion event. This provides a heat-release rate that is similar to that for conventional diesel fuelling. CO_2 emissions are reduced by the fuel's lower carbon to energy ratio, while the lower adiabatic flame temperature significantly reduces oxides of nitrogen (NO_x) emissions and the lower sooting tendency of natural gas reduces fine particulate matter (PM) emissions. Test-cycle engine emissions have been certified at levels lower than the US EPA's 2007 requirements [3]. Further, as the natural gas is not premixed in the combustion chamber, emissions of unburned fuel are significantly lower than from many other natural gas fuelling technologies [4].

Using exhaust gas recirculation (EGR) can achieve further order-of-magnitude reductions in NO_x emissions; however, the combustion is degraded and emissions of unburned fuel, CO, and PM are excessive [5]. At high EGR levels, more than 90% of the PM originates from the natural gas [6]. These results apply for conventional natural gas; the effect of variations in the concentration of diluent in the gaseous fuel under these conditions has not been studied previously.

1.2 Fuel dilution

The effect of diluting a gaseous fuel with nitrogen has been investigated in various contexts. In automotive premixed-charge spark-ignition engines, NO_x emissions and pre-ignition knock can be reduced through the addition of N_2 to both gaseous and liquid fuels [7,8]. In more advanced premixed combustion systems, such as homogeneous charge compression ignition, fuel or charge dilution is a key technique for controlling the ignition and combustion processes [9]. For the same total fuel energy content in a premixed-charge system, diluting the fuel has essentially the same effect as diluting the oxidizer.

For non-premixed combustion, the effects of fuel dilution may vary substantially from those of oxidizer dilution. Oxidizer dilution with N_2 behaves very similarly to conventional EGR in reducing flame temperatures [10]. In industrial boilers (low pressure non-premixed turbulent combustion), fuel dilution reduces NO_x more effectively than does oxidizer dilution [11]. This is attributed to enhanced mixing rates and reduced residence time for the burned gases before mixing quenches the

NO$_x$ reactions; the reaction-zone chemistry is insensitive to the source of the diluent. Fuel dilution levels above 20% also result in no detectable soot formation. Similar results demonstrating the importance of mixing have been shown in fundamental transient jet studies where injecting an inert gas prior to the fuel substantially enhances the turbulent combustion rate [12]. In other fundamental non-premixed combustion studies, N$_2$ is added to methane to reduce fuel concentrations. Investigations of laminar co-flow non-premixed flames indicate that soot volume fractions are reduced proportionally with the reduction in methane concentration [13]. In a non-premixed opposed flow diffusion flame, fuel dilution with N$_2$ less than 80% (by volume) has no significant direct effect on the chemical kinetics [14].

These results suggest that the principal influence of N$_2$ addition is reduced fuel energy density. There is no evidence of direct effects on the reaction kinetics. For a non-premixed transient gaseous jet, such as in a direct-injection engine, a lower fuel energy density will require a longer injection duration to ensure that an equivalent amount of chemical energy is available for the combustion event. This will increase the total mass injected, significantly increasing the total kinetic energy transfer to the combustion chamber gases. Changing the density of the injected fuel will influence both the penetration and the mixing of the gaseous jet. These effects will likely influence the local fuel-oxidizer stoichiometry in the reaction zone, affecting both the premixed and non-premixed phases of the combustion process. Simultaneously, as N$_2$ is relatively inert, the chemical kinetics are not directly affected. This provides an insightful comparison of the relative importance of mixing and chemical kinetic limitations in the combustion process.

2. EXPERIMENTAL

The research facility used in this project is a Cummins ISX series heavy-duty diesel engine modified for single-cylinder operation and adapted to operate on direct injection of natural gas fuelling. The single-cylinder engine, described in Table 1, has performance and emissions similar to an equivalent natural-gas fuelled multi-cylinder ISX engine [15,16]. The diesel and natural gas injection processes are controlled electronically using a single multi-fuel injector. Combustion air is supplied from an industrial rotary-screw air compressor fitted with multi-stage water and oil separators. A back-pressure valve controls the exhaust system pressure, which drives exhaust gas through the EGR system. An intake pressure regulator and coolers on both the fresh air and EGR streams permit control of the charge temperature, pressure, and dilution level independent of the engine operating mode.

The engine facility is fully instrumented, with measurements of air and fuel flow (both diesel pilot and natural gas) as well as exhaust composition (CO, HC, NO$_x$, PM). The gaseous fuel flow measurement uses a Coriolis-force mass flow sensor, which provides the gaseous mass flow rate independent of the fuel composition. Fuel composition is determined using gas chromatography. PM emissions are measured using a micro-dilution system, where raw exhaust is diluted (at a volume ratio of 15:1) by clean, dry nitrogen. The total PM mass in the diluted stream is evaluated using a tapered element oscillating microbalance (validated against gravimetric filter measurements [15]).

Table 1: Engine and injector specifications

Engine	Cummins ISX single cylinder 4-stroke, 4-valve
Fuelling	Direct injection; diesel pilot, gaseous main fuel
Displacement (/cylinder)	2.5 L
Compression Ratio	17:1
Bore/Stroke/Connecting Rod Length	137/169/261 mm
Injector	Westport Power Inc. dual-fuel concentric needle
Injection control	Separate diesel and natural gas solenoids
Injector holes	7 pilot, 9 gas
Injection angle	18° below fire deck

The combustion process is monitored using a water-cooled in-cylinder pressure transducer and a ½° crank-angle encoder to identify piston location. This data is used to calculate the gross indicated work per cycle (work done during the compression and power strokes only) and the gross indicated power, which is used to normalise the fuel consumption and the emissions measurements. The heat-release rate (HRR) can be calculated from the pressure trace [17], and represents the rate of energy release from the combustion processes less wall heat transfer and crevice flow losses. The definition of the combustion timing and duration are based on the relative fraction of the total energy released between the start of combustion (identified from the HRR) and the timing of interest, presented as the integral of the heat release (IHR). Combustion timing is defined as the crank-angle location at which half of the heat has been released (50%IHR), while the burn duration refers to the total duration of combustion (from 10% IHR to 90% IHR).

2.1 Operating Conditions

The engine operating condition used in this work is representative of a high-load steady-state cruising mode for a heavy-duty engine. Using an EGR level of 30% maintains low NO_x emissions without degrading the combustion event. Details of the operating condition and fuel blends used in this work are shown in Table 2.

Table 2: Engine operating conditions and fuel blends

Speed (RPM)	1200		
Gross Indicated Power (kW) [% load]	35 [75%]		
Gross Ind. Mean Effective Pressure (bar)	13.5		
Gaseous Fuel Pressure (MPa)	21		
EGR (mass %)	30		
Intake Oxygen Mass Fraction	0.19		
Combustion Timings (50% IHR)	5,10,15°ATDC		
Fuel - %N_2 by volume	0	20	40
- %N_2 by mass	0	30.5	53.8
Fuel energy density (MJ/kg)	45	31.3	20.8
Fuel density at STP (kg/m³)	0.71	0.82	0.93

Varying the timing of the combustion provided a range of combustion conditions at constant charge composition and overall equivalence ratio. The diesel injection timing was used to control the combustion timing; the gaseous fuel injection started a fixed 1.0 ms after the end of the diesel injection. The mid-point of the combustion event was varied between 5° crank angle (CA) after top-dead-centre (ATDC) and 15°CA ATDC. (Later timings generate lower NO_x emissions but reduce efficiency and increase combustion instability [5,17]). The injection timing was adjusted for the different fuel blends to hold the 50%IHR at the specified value. For all timings, the engine's power output was held constant by varying the mass flow-rate of the gaseous fuel. The pilot quantity was fixed at 5% of the total fuel on an energy basis.

3. RESULTS

Each test condition was repeated three times to maximize accuracy and minimize potential bias. The presented emissions and fuel consumption are averages of these values. The in-cylinder heat-release rates, and corresponding stability and duration results, are based on 45 consecutive cycles collected at a single test point, selected to be representative of the combustion performance for the other replications.

3.1 Combustion process

To maintain the combustion timing with the diluted fuel, the injection process needs to begin sooner. For the 40% N_2 case, the pilot injection starts approximately 3°CA earlier. At the earliest combustion timing, the slightly lower combustion chamber temperature and pressure with the diluted fuel results in a small increase in the pilot-fuel ignition delay time (PID), shown in Figure 1. Conversely, the ignition delay time of the gaseous fuel, also shown in Figure 1, is reduced, especially between the natural gas and 20%N_2 cases. This is most likely a result of changes in the fuel-air stoichiometry in the gas jet as it approaches the diesel pilot flame. The higher density (and hence higher momentum and greater penetration rate) of the gaseous jet may also bring the ignitable fuel-air cloud closer to the pilot flame at an earlier time. However, the pressure trace provides insufficient information to reliably identify the principal causes of the shorter gaseous fuel ignition delay (GID).

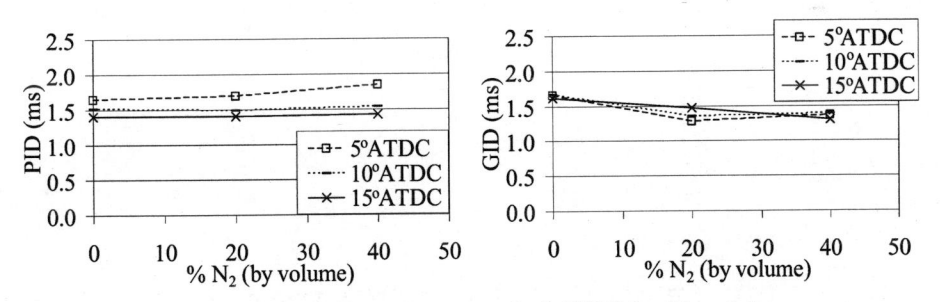

Fig. 1: Pilot (PID) and gaseous fuel (GID) ignition delay

Once the gaseous fuel ignites, N_2 dilution reduces the intensity of the initial, partially-premixed, combustion phase. This is a result of less methane having mixed to a combustible level prior to ignition, due to both the shorter ignition delay time and the lower concentration of methane in the diluted fuel. Once this phase is completed, the combustion continues, limited primarily by the rate at which fuel and oxidizer are mixing. Both these effects are demonstrated in the HRR plots, shown in

Figure 2 for the early and late combustion timings. Without nitrogen in the fuel, and even at the 20% N$_2$ case, the partially-premixed phase dominates the combustion event. With the diluted fuel, the heat-release curve is flattened, indicating a more mixing-controlled combustion. These mixing limitations are primarily a result of the reduced chemical energy injection rate due to the fuel's lower energy density. To provide the same total chemical energy for the combustion, the injection is extended; for 40% N$_2$, the injection duration doubles and it ends 10°CA later.

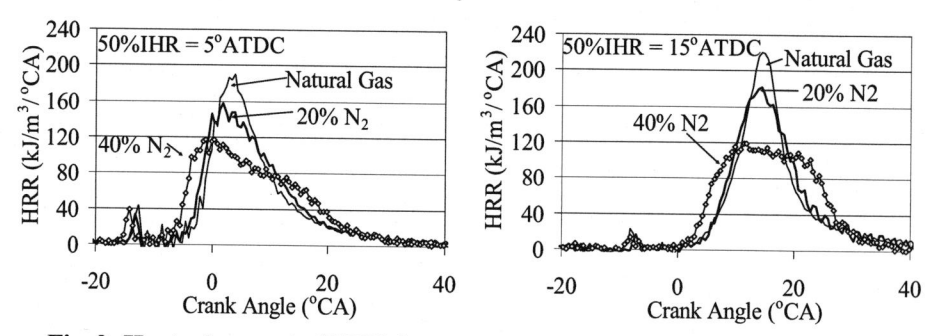

Fig. 2: Heat release rate (HRR) for all three fuel compositions, for early and late combustion timings

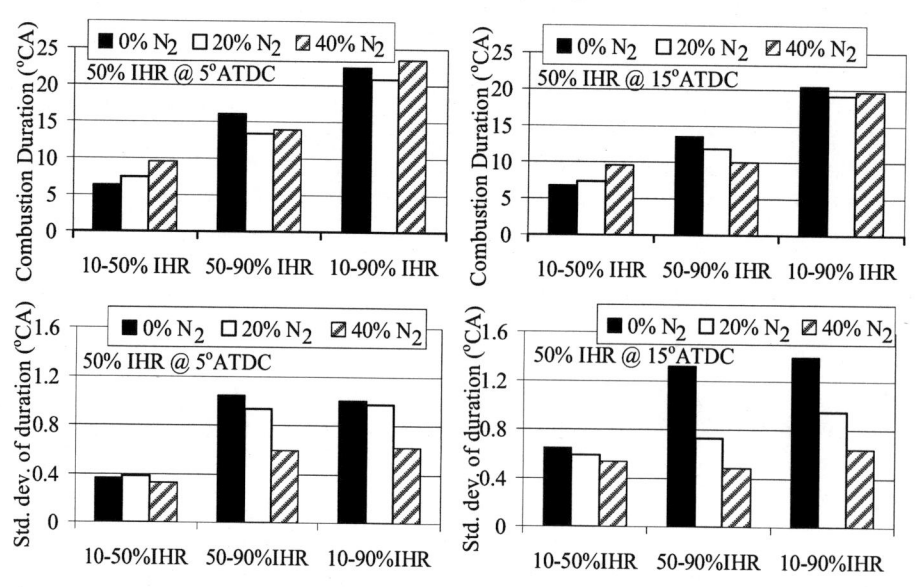

Fig. 3: Early (10%-50%IHR), late (50%-90% IHR) and total (10%-90% IHR) combustion duration and stability, for early and late combustion timings

Although the injection duration is longer and the peak heat-release rate is lower with N$_2$ in the fuel, the overall combustion duration is not affected, as shown in Figure 3. The reason is demonstrated by comparing the early and late combustion durations (10%-50%IHR and 50%-90%IHR, respectively). Fuel dilution increases the early combustion duration, due to the smaller quantity of fuel available to burn in the partially-premixed phase, but reduces the late combustion duration. This reduction is a result of the higher kinetic energy transferred to the combustion chamber resulting in more intense late-cycle turbulence. This leads to more rapid mixing of the fuel injected late in the injection process, resulting in a higher heat-release rate late in the cycle (as indicated in the HRR, Figure 2). These results demonstrate the importance of turbulent mixing, especially late in the combustion phase.

N_2 dilution also reduces the variability of the combustion event, as shown in Figure 3 by the cycle-to-cycle standard deviation of the combustion duration for the overall, early and late combustion phases (10%-90%, 10%-50%, and 50%-90% IHR, respectively). The reduction in variability is more significant at the later combustion phases; variability in the early combustion is only slightly reduced. This provides further support for the supposition that the N_2 addition is significantly enhancing the late-cycle mixing, resulting in more rapid and repeatable combustion. This improvement in combustion stability leads to a small (~3%) but consistent increase in gross indicated fuel conversion efficiency with fuel dilution.

3.2 Emissions

Dilution of the gaseous fuel with nitrogen has a substantial effect on both gaseous and PM emissions. These impacts are attributable, in general, to the effects of the N_2 on the fuel-air stoichiometry and on the combustion process. Previous studies suggest that N_2 in the fuel plays little direct role in the reaction kinetics [11].

The effect of N_2 dilution on the emissions of NO_x, CO, hydrocarbons (HC) and PM are shown in Figure 4. At all the timings, the NO_x emissions increase slightly with 20% N_2 dilution, but drop significantly at 40% N_2. These reductions are most likely a result of the lower intensity combustion leading to lower combustion temperatures. However, the reductions are not nearly as significant as suggested by the reduction in the adiabatic flame temperature under stoichiometric conditions with N_2 dilution.

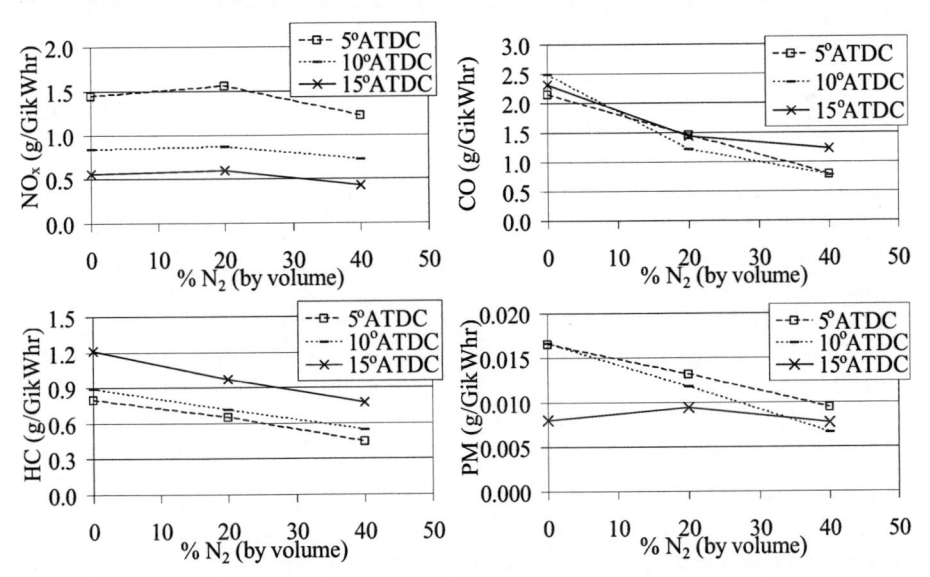

Fig. 4: Emissions of NO_x, HC, CO and PM as functions of fuel dilution and combustion timing

According to previous results, the reduction in flame temperature with 40% N_2 dilution (~80K) should reduce NO_x emissions by approximately 40% [5,15], twice the observed reduction. Changes in the stoichiometry in the reaction zone may be leading to variations in actual combustion temperature that are not predicted using the calculated adiabatic (stoichiometric) flame temperature. Changes in mixing rates could also result in longer residence times for the burned gases at high temperature, allowing the thermal NO reactions to proceed closer to equilibrium.

Hydrocarbon emissions, which are more than 95% unburned methane, decrease linearly with N_2 dilution. These reductions are consistent in magnitude, at 160 mg/kWhr per 20% increase in N_2, for all combustion timings. This reduction is equivalent to 20% of the unburned fuel emitted at the lowest emissions case. This suggests that the HC emissions are related to a fixed volume of unburned fuel. This may be a result of the fuel retained in the injector sac and nozzle passages at the end of the injection process, which then vents into the combustion chamber as the cylinder pressure falls during the expansion stroke. It may also be due to early-injected fuel which mixes beyond the combustible limit before ignition occurs.

The CO emissions are also significantly reduced with the addition of N_2 to the fuel at most timings. The fact that the combustion is significantly more stable with N_2 is most likely one of the principal causes of this reduction, since CO emissions are often a result of incomplete combustion, especially late in the combustion cycle due to bulk quenching of the reactants. Furthermore, these incomplete-combustion processes are also often associated with emissions of particulate matter.

PM emissions are substantially reduced by N_2 dilution at the 5° and 10°ATDC timings, as shown in Figure 4. At the 15°ATDC timing, N_2 dilution has no significant effect on the total PM mass emissions. This may be a result of the very low emissions levels at this timing. The general reduction in PM agrees with previous studies of low-pressure non-premixed combustion systems which found that fuel dilution prevented in-flame PM formation [11]. Whether similar effects are occurring in the current high-pressure combustion system, or whether late-cycle oxidation is being enhanced, is a subject of ongoing research.

4. DISCUSSION

One of the main impacts of this study is in improving fundamental understanding of pilot ignited direct injected natural gas combustion systems. Increasing the mass of the gaseous jet significantly increases late-cycle turbulence without significantly affecting the chemical kinetics. The results clearly demonstrate that, under the conditions tested, increasing mixing and reducing the peak combustion intensity significantly benefit engine-out emissions. These positive results suggest that the technique also has potential in commercial applications.

For natural gas in transportation applications, careful monitoring of fuel composition may be needed to minimize emissions while maintaining engine performance. This study indicates that diluents in the fuel can, in fact, be beneficial. The concentration of these diluents could even be used as a tool to counteract higher levels of other more polluting species, allowing more complete usage of the fuel stock with less processing requirements. Provision of such fuels for either stationary or mobile (on- and off-road) applications appears to offer the potential to reduce emissions without impairing engine performance.

On-vehicle storage of gaseous fuels is an ongoing challenge. Both liquefied (LNG) and compressed gaseous (CNG) storage systems are in use in heavy-duty applications. LNG, while more energy intensive and requiring more complicated fuelling systems, provides higher volumetric energy density and easy purification. Fuel dilution in an LNG system would most likely require separate onboard storage. This would allow control of the fuel composition, but would substantially increase fuel system complexity. The use of diluents is much more feasible in a CNG system. The main drawbacks are reduced storage energy density and higher compression work requirements. Adding 40% N_2 to CNG would nearly double the storage volume and compressor work requirements; however, the power required to compress the fuel remains below 2% of its energy content [15]. This investment may be justified by lower emissions and reduced exhaust aftertreatment requirements.

5. CONCLUSIONS

The principal impacts of diluting a gaseous fuel with nitrogen are:
1. A reduction in the initial combustion intensity and the peak heat release rate due to lower chemical energy in the partially premixed charge following ignition. The late-phase combustion process is enhanced through more rapid mixing and combustion, primarily attributed to the higher gaseous jet kinetic energy.
2. A reduction in combustion instability. This is due primarily to enhanced late-cycle combustion and a more consistent premixed combustion phase.
3. A substantial reduction in total PM mass and CO emissions at most timings, due to more complete late-cycle burn up as well as reduced in-flame PM formation.
4. A reduction in unburned fuel emissions by a constant amount for a given quantity of fuel. This suggests that sources of HC emissions at these operating conditions are volume-related, including fuel retained in the injector nozzle and gas plenum or overleaning of gaseous fuel injected early in the injection process.
5. A slight reduction in emissions of NO_x. This finding suggests that either the combustion temperature is not being reduced as much as anticipated, or that the residence time in the post-flame gases is greater, allowing the NO-forming chemical reactions to more closely approach equilibrium.

6. ACKNOWLEDGEMENTS

Westport Power Inc. and the Natural Sciences and Engineering Research Council of Canada provided support for this project. The authors thank the staff at UBC, and in particular H. Jones and R. Parry, for their assistance in conducting this work.

7. REFERENCE LIST

[1] Richards, G.A., M.M. McMillian, R.S. Gemmen, *et al*. Issues for Low-Emission, Fuel-Flexible Power Systems. Progress in Energy and Combustion Science. **27**. 2001. Pp. 141-169.

[2] Harrington, J. S. Munshi, C. Nedelcu, *et al*. Direct Injection of Natural Gas in a Heavy-Duty Diesel Engine. SAE Technical Paper 2002-01-1630. 2002.

[3] California Air Resources Board. Executive Order A-343-0003. Available from: www.arb.ca.gov/msprog/onroad/cert/mdehdehdv/2006/westport_hhdd_a343000 3_14d9_1d2-0d02_cngplusdiesel.pdf. Accessed 27-03-2007.

[4] McTaggart-Cowan, G.P., C.C.O. Reynolds and W.K. Bushe. Natural Gas Fuelling for Heavy-Duty On-Road Use: Current Trends and Future Direction. International Journal of Environmental Studies. **63**(4). 2006. Pp. 421-440.

[5] Hill, P.G. and G.P. McTaggart-Cowan. Nitrogen Oxide Production in a Diesel Engine Fuelled with Natural Gas. SAE Technical Paper 2005-01-1727. 2005.

[6] Jones, H.L., G.P. McTaggart-Cowan, S.N. Rogak, *et al*. Source Apportionment of Particulate matter from a Direct Injection Pilot-Ignited Natural Gas Fuelled Heavy Duty DI Engine. SAE Technical Paper 2005-01-2149. 2005.

[7] Nellen, C. and K. Boulouchos. Natural Gas Engines for Cogeneration: Highest Efficiency and Near-Zero Emissions through Turbocharging, EGR and 3-Way Catalytic Converter. SAE Technical Paper 2000-01-2825. 2000.

[8] Crookes, R.J. Comparative Bio-Fuel Performance in Internal Combustion Engines. Biomass & Bioenergy. **30**. 2006. Pp. 461-468.

[9] Zhao, H., Z. Peng and N. Ladommatos. Understanding of Controlled Autoignition Combustion in a Four-Stroke Gasoline Engine. Proc. Instn. Mech. Engrs. Part D. **215**. 2001. Pp. 1297-1309.

[10] Ladommatos, N., S. Abdelahlim and H. Zhao. The Effects of Exhaust Gas Recirculation on Diesel Combustion and Emissions. International Journal of Engine Research. **1**(1). 2000. Pp. 107-126.

[11] Feese, J.J. and S.R. Turns. Nitric Oxide Emissions from Laminar Diffusion Flames: Effects of Air-Side versus Fuel-Side Diluent Addition. Combustion and Flame. **113**. 1998. Pp. 66-78.

[12] Kaneko, T., T. Fujii, Y. Matsuda and T. Chikahisa. NOx Reduction in Diesel Combustion by Enhanced Mixing of Spray Tip Region. JSME International Journal, Series B. **48**(4). 2005. Pp. 665-670.

[13] Gulder, O.L. Effects of Oxygen on Soot Formation in Methane, Propane, and n-Butane Diffusion Flames. Combustion and Flame. **101**. 1995. Pp. 302-310.

[14] Fotache, C.G., T.G. Kreutz and C.K. Law. Ignition of Counterflowing Methane versus Heater Air under Reduced and Elevated Pressures. Combustion and Flame. **108**. 1997. Pp. 442-470.

[15] McTaggart-Cowan, G.P., W.K. Bushe, P.G. Hill and S.R. Munshi. A supercharged heavy-duty diesel single-cylinder research engine for high-pressure direct injection of natural gas. International Journal of Engine Research. **4**(4). 2004. Pp. 315-330.

[16] McTaggart-Cowan, G. Pollutant Formation in a Gaseous-Fuelled, Direct Injection Engine. PhD. Thesis. University of British Columbia. 2006.

[17] Heywood, J.B. Internal Combustion Engine Fundamentals. McGraw-Hill. 1988.

A study of alcohol blended fuels in a new optical spark-ignition engine

J.S. Malcolm, P.G. Aleiferis
Department of Mechanical Engineering, University College London, UK

A.R. Todd, A. Cairns, A. Hume, H. Blaxill
MAHLE Powertrain Ltd., UK

H. Hoffman, J. Rueckauf
MAHLE International GmbH, Germany

ABSTRACT

A new single-cylinder optical spark-ignition engine has been designed, developed and employed to evaluate combustion of blends of gasoline, *iso*-octane and a variety of alcohols under part-load engine operation at 1500 RPM with port fuel injection. In particular, six fuels were tested; a pump-grade gasoline and a commercial E85, as well as *iso*-octane and splash-blended mixtures of *iso*-octane with 25% ethanol, 85% ethanol and 25% butanol. The latter alcohol is a potential second generation biofuel, so far subject to little detailed research. Differences in combustion between the tested fuels were studied using high-speed crank-angle resolved natural light flame imaging in conjunction with in-cylinder pressure analysis over batches of 100 cycles. The flame images were processed to infer the evolution of an equivalent flame radius during the early stages of combustion. The results demonstrated the effect of alcohol addition to *iso*-octane and benchmark comparisons with commercial grades of gasoline and E85.

1 INTRODUCTION

Evolving emissions legislation and concerns for diminishing fuel reserves continue to prompt automotive engine manufacturers to seek alternative modes of engine operation. In recent years European CO_2 emissions targets have been met through increased Diesel sales. However, the distillation of crude oil results in large quanitities of both Diesel and gasoline fuel (1). In order to meet future global emissions goals, in the short term it will be necessary to improve the fuel consumption of the gasoline engine and in the longer term uncover alternative sustainable sources of fuel. Consequently, in recent years there has been increasing interest in the use of alcohol-based fuels in gasoline-like engine applications, as such fuels have the potential to be employed in a near CO_2 neutral manner via efficient conversion of biomass. Current first generation biofuels for Spark Ignition

© IMechE 2007

(SI) engines have become generally available in the form of gasoline-ethanol blends; where present quality standards allow fuel manufacturers to include up to ~5% bioethanol within the existing commercial gasoline pool without significant implication on vehicle durability or performance. However, primarily due to high anti-knock rating, higher concentrations of ethanol are desirable and are well known to present the opportunity for improved SI engine thermal efficiency and increased performance (2–6). One perceived thermodynamic disadvantage of ethanol is its low heating value (~2/3 that of gasoline) which may present difficulties with consumer acceptance of higher ethanol content fuels without appropriate economic adjustments. However, at part-load such penalties of ethanol have been suggested to be avoided by the use of advanced engine operating strategies such as aggressive downsizing (7) or controlled autoignition (8).

Although there is some background work on engine performance and emissions with lower alcohols like ethanol and methanol (9–17), there is very little published work on optical studies of combustion of alcohols in SI engines, thus there is very little fundamental understanding as to how alcohol addition to gasoline fuels affects in-cylinder mixture formation and combustion. Recently, butanol has attracted a lot of attention because it exhibits some similar properties to those of gasoline and it is thought to make a realistic option as an additive to gasoline which may be used to make up high proportions of alcohol-based fuels for SI engines. Butanol has the added benefit that it is not as aggressive to standard automotive parts as ethanol is and thus can be used with fewer changes to existing hardware designs. However, there is very little literature on butanol use in SI engines (18–22) and, in particular, no comprehensive optical work. Additionally, although there is a body of literature on laminar and turbulent burning velocities for typical SI engine reference fuels like *iso*-octane and *n*-heptane, there are limited data on alcohol flame speeds (23–27). Even these are mainly focused on methanol's and ethanol's laminar burning at atmospheric conditions and, in fact, it seems that there is a complete lack of published detailed data on butanol. The aim of the current study is to shed more light on the area of how alcohols perform inside the cylinders of SI engines and compare them to today's leading fossil technologies. Specifically, this paper reports on some preliminary work using port injection of different fuel blends. A variety of mixtures of gasoline, *iso*-octane, ethanol and butanol were tested on a new optical engine.

2 EXPERIMENTAL SETUP

2.1 Optical research engine

The research engine used for this work was recently designed and developed by MAHLE to provide a low-cost optical single-cylinder assembly. The conventional engine bore was replaced with a quartz liner covering the full length of the stroke. The gable ends of the combustion chamber were removed and the liner was contoured to fit the pent-roof and allow optical access into the chamber (Figure 1).

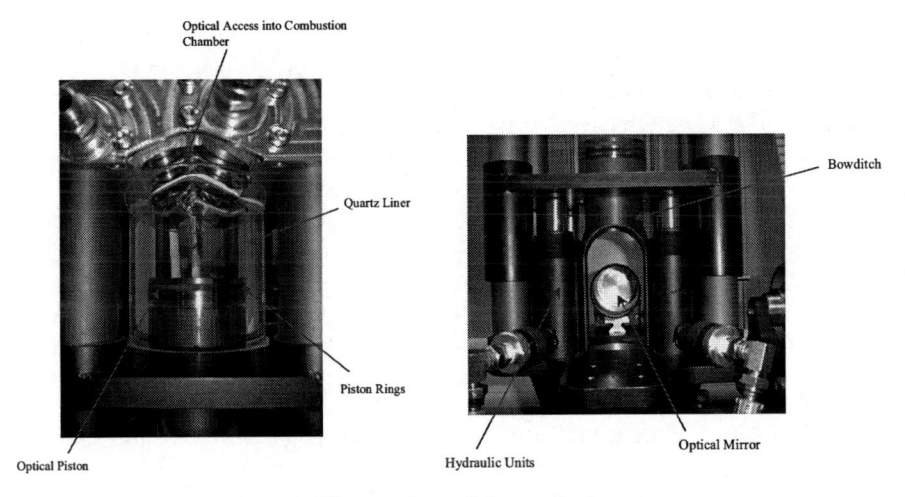

Figure 1. Illustration of the optical access.

The standard piston was also replaced with an optical upper piston assembly including a sapphire window that allowed ~65% of the bore diameter to be visualised from below. The piston rings were produced from Torlon to help ensure durability of the optical liner. A 45° mirror is used in combination with a Bowditch piston to allow optical access through the piston crown. The design configuration of the assembly permits testing with different size bores, up to 95 mm. The crankcase was a Lister-Petter Diesel which required little modification but imposed a maximum permissible engine speed of 2000 Revolutions Per Minute (RPM). The cylinder head fitted to the engine at the time of this work was a modified four-cylinder production component with four valves per cylinder. The combustion chamber was originally based on the side direct fuel injection principle, with a so-called tumble flap used to allow some control over the tumbling bulk in-cylinder motion. For the purposes of the current work, the head was modified to allow port fuel injection and the tumble flap was removed. The specific details of the engine are provided in Table 1. It should be noted here that in this paper 0° Crank Angle (CA) corresponds to intake Top Dead Centre (TDC) and crank angle timings will be mainly presented with respect to that as °CA After intake TDC (ATDC).

Table 1. Engine characteristics.

Bore [mm] × Stroke [mm]	82.5 × 88.9
Compression Ratio	9.8:1
Intake Valve Timing	Opens 711° CA ATDC, Closes 232° CA ATDC
Exhaust Valve Timing	Opens 482° CA ATDC, Closes 718° CA ATDC
Spark Plug	NGK triple-electrode

Operating conditions
For the results reported here the engine was operated in port fuel injection configuration using injection pressure of 3.6 bar. Both the engine head and cylinder

liner were heated to 80 °C to represent typical warm engine conditions. The engine was operated at 1500 RPM with the Bowditch piston extension and the 45° mirror for optical access to in-cylinder phenomena. The throttle was set to give a plenum pressure of 0.5 bar for typical part-load operation. The engine control system was supplied by MBE Powercontrol Technologies and was specially modified to provide flexible trigger interfacing for the current application.

2.2 Data acquisition

The data acquisition system has the capability of measuring eight channels at 0.2° CA resolution with the most important of these being the pressures, both in-cylinder and plenum, as well as the crank angle and cycle markers. From these measurements it is possible to perform statistical analysis of the Indicated Mean Effective Pressure (IMEP), the amplitude and timing of peak in-cylinder pressure, *etc.*, for a series of consecutive cycles. Heat release analysis can also be applied on the pressure traces for identification of the timing of various percentages of Mass Fraction Burned (MFB) in each cycle and further statistical studies. An engine timing unit (AVL 427 ETU) is used for synchronisation between pressure recording and all other scientific equipment, including triggering of optical instrumentation such as digital cameras.

2.3 Fuels

Six different fuels were selected for this study. The first of these fuels was standard pump-grade gasoline (95 RON). This was used as the baseline 'control' fuel. With gasoline being a multi-component fuel, and in order to get a better understanding of the combustion process, single-component pure *iso*-octane was also employed. *Iso*-octane was used both in its pure form and in mixtures with various alcohols in differing quantities. In particular, mixtures of 25% ethanol and 75% *iso*-octane, 85% ethanol and 15% *iso*-octane, as well as 25% butanol and 75% *iso*-octane, were splash blended using pure components from chemical suppliers. For comparison with the multi-component gasoline, a commercial grade of E85 (85% ethanol, 15% gasoline) was also employed. This was not splash-blended but was supplied by a fuel company as an already-prepared fuel. The ignition timing was first mapped to identify the point of Minimum spark advance for Best Torque (MBT) for each fuel blend. Two injection strategies were tested: one against open intake valves at 35° CA ATDC and a second one against closed intake valves at 655° CA ATDC. In this paper we will only present results for the closed-valve injection strategy. For all fuels it was decided that data should be acquired at one fixed ignition timing, namely the MBT of gasoline, and at the MBT point of each fuel. Fixing the ignition timing allowed for direct comparisons to be carried out between the combustion processes of all fuels with nominally same in-cylinder flow fields and temperatures/pressures at the start of combustion. The experiments at the MBT of each fuel were also necessary to identify differences between the fuels at optimised combustion phasing. Both stoichiometric and lean operation was observed. The lean point was set at an equivalence ratio of $\Phi=0.83$ (*i.e.* air excess ratio of $\lambda=1.2$) which was close to the lowest misfire limit for all blends. The fuels and ignition timings employed are summarised in Table 2.

Table 2. Test matrix of fuels and ignition timings.

Fuel	MBT (Φ=1.0, λ=1.0)	MBT (Φ=0.83, λ=1.2)
Gasoline	32	40
Iso-Octane (I100)	36	44
Butanol 25%, *Iso*-Octane 75% (B25 I75)	34	44
Ethanol 25%, *Iso*-Octane 75% (E25 I75)	30	36
Ethanol 85%, *Iso*-Octane 15% (E85 I15)	28	38
Ethanol 85%, Gasoline 15% (E85)	24	38

2.4 Image acquisition

The imaging system was based on a Photron APX-RS CMOS camera with a Nikon 60 mm lens. Flame growth was recorded on 640×480 pixel frames over 100 cycles for each test case. At stoichiometric conditions, images were acquired at a frame rate of 9 kHz, corresponding to one image per 1° CA at 1500 RPM. For lean conditions, the frame rate was set to 4.5 kHz and the flame was captured with an integration time of 2° CA to compensate for the low levels of flame luminosity at those conditions. A typical imaging sequence of flame growth in a single cycle is shown in Figure 2 for a blend of 25% ethanol and 75% *iso*-octane (E25 I75) at stoichiometry. The timings shown correspond to Crank Angle degrees (° CA) After Ignition Timing (AIT). Image processing was carried out to calculate an 'equivalent' mean flame radius. This was achieved by first averaging the flame images at the same crank angle timing over the total set of 100 cycles (*i.e.* ensemble averaging) and then thresholding the mean image to calculate the 'projected' enflamed area at each crank angle. This was used to calculate an equivalent flame radius from the area of an equivalent circle.

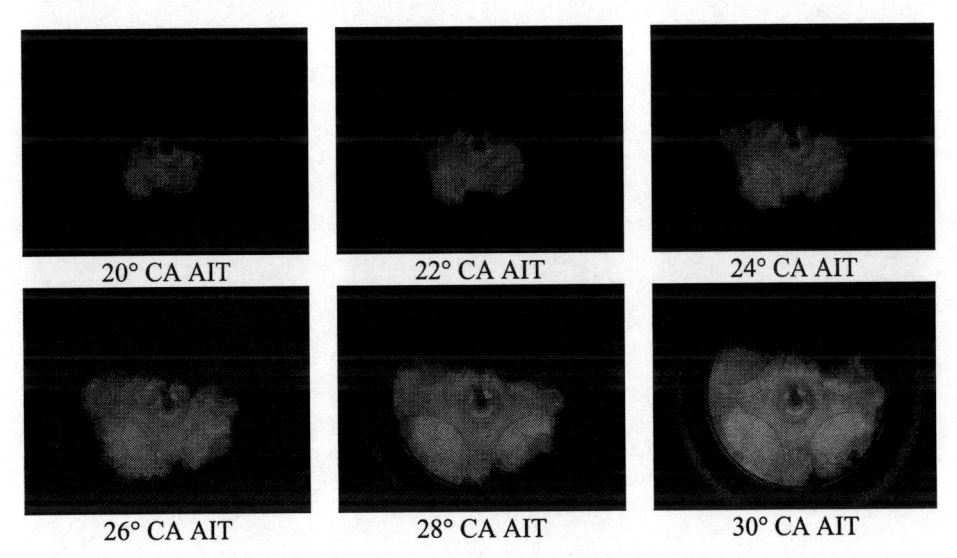

Figure 2. Flame Growth for 25% Ethanol and 75% *Iso*-Octane Fuel Blend.

3 RESULTS AND DISCUSSION

3.1 Pressure analysis

Figure 3 shows the mean pressure traces for all conditions. At stoichiometric fuelling with ignition timing fixed at the MBT of gasoline (328° CA ATDC), it can be observed in Figure 3(a) that the fuels split into two distinguishable groups. The first one producing peak pressures of the order 21–23 bar consists of gasoline, E85 and the 85% ethanol 15% *iso*-octane blend (E85 I15). The second one producing peak pressures in the range 19–20 bar consists of *iso*-octane and its lower by volume alcohol mixtures (E25 I75, B25 I75). These two groups are also visible in the MFB traces in Figure 4(a), where *iso*-octane, E25 I75 and B25 I75 overlay each other. As the ignition timing was fixed for this test, the observed differences can be associated with the burning characteristics of the fuels when these are ignited at the same pressure/temperature conditions. For reference, it may be pointed out that ethanol has a higher laminar burning velocity at stoichiometry than *iso*-octane (26) and this seems to be in agreement with the observations in Figure 3(a). In the absence of burning velocity data for butanol, no direct conclusion can be drawn, but it is worth pointing out that the 25% addition of ethanol to *iso*-octane (E25 I75), as well as 25% addition of butanol to *iso*-octane (B25 I75), does not seem to affect significantly the burning characteristics of pure *iso*-octane. The addition of these alcohols, however, does decrease the coefficient of variation of IMEP as shown in Table 3.

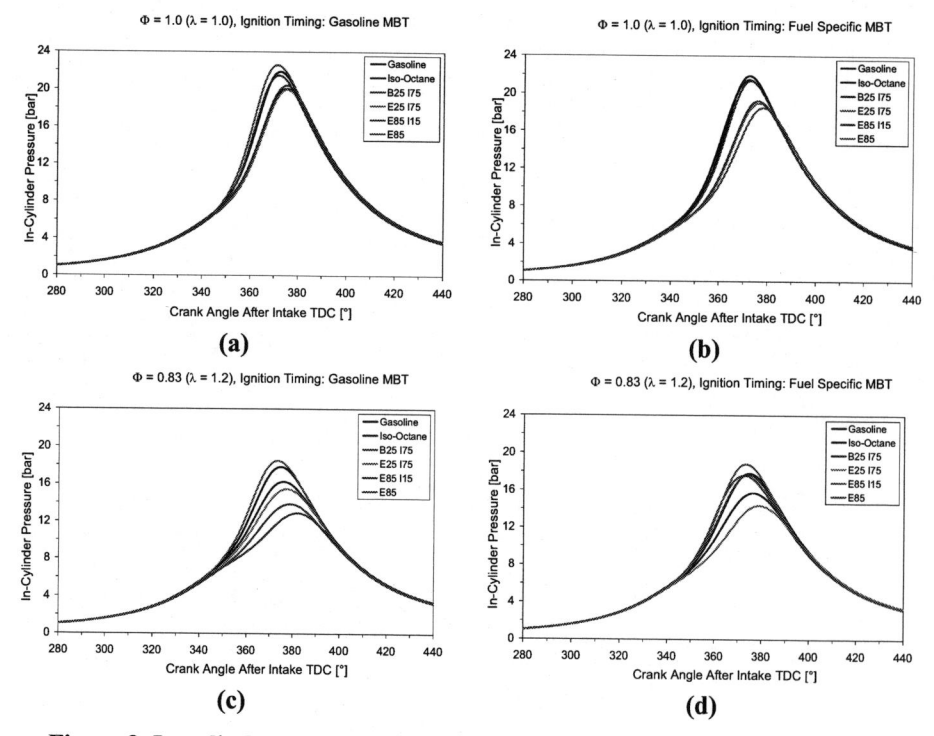

Figure 3. In-cylinder pressure for different fuel blends at stoichiometric and lean conditions with different ignition timings.

Table 3. Coefficients of variation of IMEP (COV$_{IMEP}$).

Fuel (Ignition Timing: Gasoline MBT)	COV$_{IMEP}$ [%] (Φ=1.0, λ=1.0)	COV$_{IMEP}$ [%] (Φ=0.83, λ=1.2)
Gasoline	0.89	1.35
Iso-Octane (I100)	0.81	1.95
Butanol 25%, Iso-Octane 75% (B25 I75)	0.69	2.67
Ethanol 25%, Iso-Octane 75% (E25 I75)	0.66	2.19
Ethanol 85%, Iso-Octane 15% (E85 I15)	1.14	3.42
Ethanol 85%, Gasoline 15% (E85)	1.05	1.52
Fuel (Ignition Timing: Fuel Specific MBT)	COV$_{IMEP}$ [%] (Φ=1.0, λ=1.0)	COV$_{IMEP}$ [%] (Φ=0.83, λ=1.2)
Gasoline	0.89	1.35
Iso-Octane (I100)	0.86	2.32
Butanol 25%, Iso-Octane 75% (B25 I75)	0.76	2.23
Ethanol 25%, Iso-Octane 75% (E25 I75)	0.64	2.64
Ethanol 85%, Iso-Octane 15% (E85 I15)	0.80	0.98
Ethanol 85%, Gasoline 15% (E85)	1.25	1.15

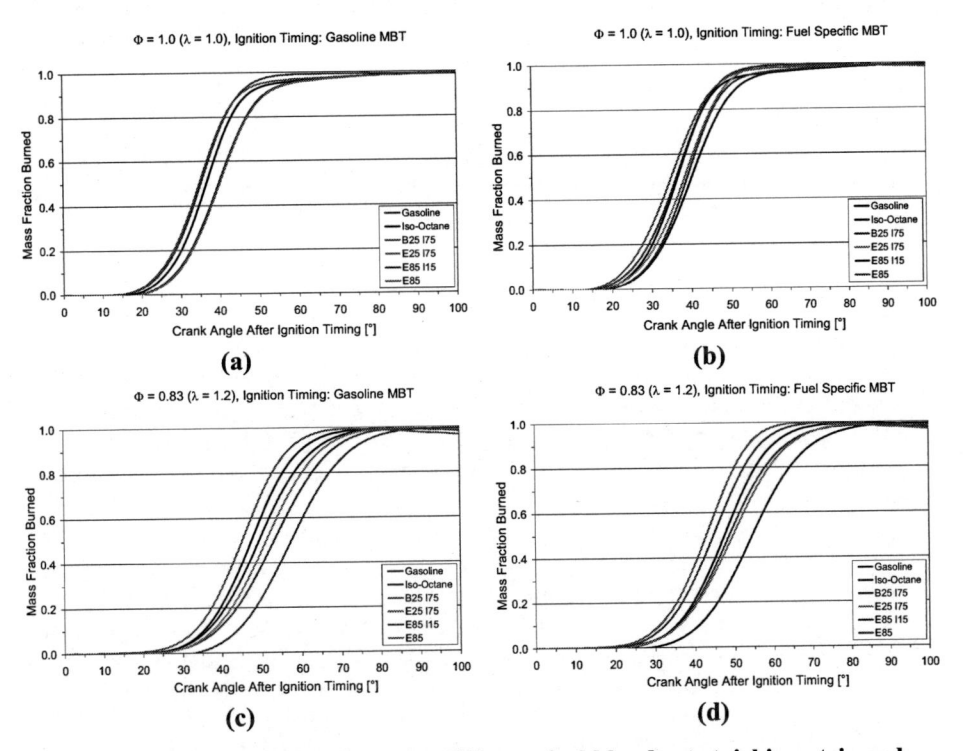

Figure 4. Mass fraction burned for different fuel blends at stoichiometric and lean conditions with different ignition timings.

From the stoichiometric tests that used the MBT ignition timings of each fuel (Table 2), it seems that there is a reversal of the previous trends in Figure 3(b), as far as the *iso*-octane and the alcohol mixtures are concerned. The three fuels to give the highest peak pressure are gasoline, *iso*-octane and B25 I75, whilst the three that give the lower peak pressures are the ethanol fuels (E25 I75, E85 I15 and E85). It is interesting to highlight that the butanol blend follows the trend of *iso*-octane and gasoline, whilst the blend with the same percentage of ethanol does not. When looking at the corresponding MFB curves in Figure 4(b), it is important to keep in mind that the conditions at ignition timing are different in terms of pressure and temperature. Data on laminar burning velocities of pure *iso*-octane and pure ethanol in a stoichiometric environment, suggest that an increase of 100 K (similar to the differences in in-cylinder temperatures experienced over the ignition timings of this test) can increase the laminar burning velocity by as much as 50–80% (26). The pressure difference also has some impact (a larger pressure decreases the laminar burning velocity) but this is of less significance and in practice expected to be of the order 10% when the pressure approximately doubles (26), *e.g.* as the in-cylinder pressure increases from 2.5 bar to 5 bar when the ignition timing is adjusted from about 40° CA to 20°CA before compression TDC.

For lean conditions and with fixed ignition timing at the MBT of gasoline (320° CA ATDC, Table 2) in Figure 3(c), the commercial E85 is still producing a higher peak pressure (and IMEP) than the rest of the fuels, but the 85% ethanol splash blend with 15% *iso*-octane produces the lowest peak pressure. The origin of that difference for the nominally identical ethanol content fuels is not clear at this stage but may be attributed to the multi-component nature of the gasoline constituents of the E85. The addition of alcohols to *iso*-octane seems to bring the peak pressure down, with butanol having a larger effect than the blend of ethanol with nominally same *iso*-octane content. It should be pointed out that the ignition timing for the butanol blend B25 I75 had to be advanced to achieve MBT conditions, whilst the ethanol blend E25 I75 had to be retarded and this may act to explain the higher peak pressure of the ethanol blend in comparison to that of the butanol blend when the ignition timing was fixed in Figure 3(c). The lower mean peak pressure recorded for the higher percentage ethanol splash blend may be explained by the lower laminar burning velocity of ethanol (in comparison to that of *iso*-octane) at lean conditions, especially at an equivalence ratio of the order $\Phi=0.8$ (26). This is upheld by the MFB traces in Figure 4(c). The low peak pressure of E85 I15 is also consistent with the high coefficient of variation of IMEP shown in Table 3, which may indicate that this fuel is closer to its misfire limit than the rest.

For the lean test condition using the fuel specific MBT ignition timing in Figure 3(d), the E85 was found to continue giving high peak pressures, whilst the butanol blend B25 I75 and the E85 I15 blend were seen to give similar in-cylinder pressures to the gasoline fuel. Again, it is interesting to note that the 25% butanol addition to *iso*-octane seems to bring the B25 I75 mixture close to the burning characteristics of gasoline. This is also visible in Figure 4(d), where the increase in butanol content over the pure *iso*-octane produces similar MFB traces to that of gasoline and is much faster than that of pure *iso*-octane for optimised combustion phasing.

3.2 Image analysis

Figure 5 shows the flame radii as calculated from the images for each test case. It should be pointed out that the maximum flame radius shown corresponds to the edge of the optical crown which was 53 mm in diameter. To provide a link between Figure 5 and the MFB traces shown earlier in Figure 4, the flame radius was used to estimate the burned volume of the gas on the basis of an equivalent sphere. According to (28), the volume fraction burned y_b was then related to the mass fraction burned x_b via the ratio of the unburned to the burned gas densities, ρ_u and ρ_b respectively: $x_b = \left[1 + \dfrac{\rho_u}{\rho_b}\left(\dfrac{1}{y_b} - 1\right)\right]^{-1}$. For a flame radius of 15 mm, corresponding on average to 20° CA AIT for stoichiometric burning and to 10° CA later for lean burning, the mass fraction burned was calculated to be of the order 5–10% for a density ratio ρ_u/ρ_b of the order 3–5. This is in agreement with Figure 4 and it highlights that flame radii in Figure 5 demonstrate phenomena very early in the cycle that the pressure analysis cannot capture reliably on a cycle-by-cycle basis.

In Figure 5(a), for stoichiometric conditions and fixed ignition timing of 328° CA ATDC (MBT of gasoline, Table 2), the fastest fuel with a flame radius of 15 mm at 18° CA AIT is E85, and the slowest to the same radius is the 25% ethanol 75% *iso*-octane blend (E25 I75) at 25° CA AIT. This corresponds reasonably well with the MFB traces in Figure 4(a). Gasoline is in the mid range of the curves in both flame radius and MFB graphs, 5(a) and 4(a), respectively. Pure *iso*-octane and the butanol blend B25 I75 are also in the middle of the range and they appear to have larger flame growth than that of gasoline, which is not the case in the MFB data. Although they appear to be faster, the rate of growth of flame radius is slower than that of gasoline which may help account for the higher pressure achieved by gasoline in Figure 3(a). Although the *iso*-octane and the E25 I75 blend overlay in both in-cylinder pressure and MFB graphs, the ethanol blend appears to start later but has a faster rate of growth after combustion has begun in Figure 5(a). At stoichiometric conditions with ignition timings optimised for each fuel independently in Figure 5(d), the butanol blend B25 I75 is faster at early growth than the rest of the fuels, whilst E85 exhibits a relatively slow growth in comparison to observations that apply to the MFB traces in Figure 4(b). The origin of this is not currently clear and forms part of our work in progress, but by examining the raw data, it was found that the flames of the butanol blend had greater luminosity than that of the E85 flames and this probably accounts for the trends in Figure 5(b). *Iso*-octane and gasoline follow trends in Figure 5(b) that seem to agree with the MFB data in Figure 4(b).

At lean operating conditions and for all the fuels ignited at the MBT of gasoline (340° CA ATDC, Table 2), the flame radii in Figure 5(c) show similar trends to the MFB curves in Figure 4(c), with the commercial E85 being the fastest and the splash-blended E85 I15 being the slowest. Additionally, it seems that pure *iso*-octane is marginally faster than the lower alcohol blends which is in agreement with earlier comments about the differences in laminar burning velocity. Gasoline appears the be slow in starting but has the highest rate of growth once combustion has begun. The order of flame growth at fuel specific MBT ignition timings for lean operation in Figure 5(d) shows good agreement with the MFB data in Figure 4(d). Specifically, commercial E85 is the fastest, *iso*-octane is the slowest, and the lower percentage alcohol blends perform very close to gasoline after gasoline's slow start.

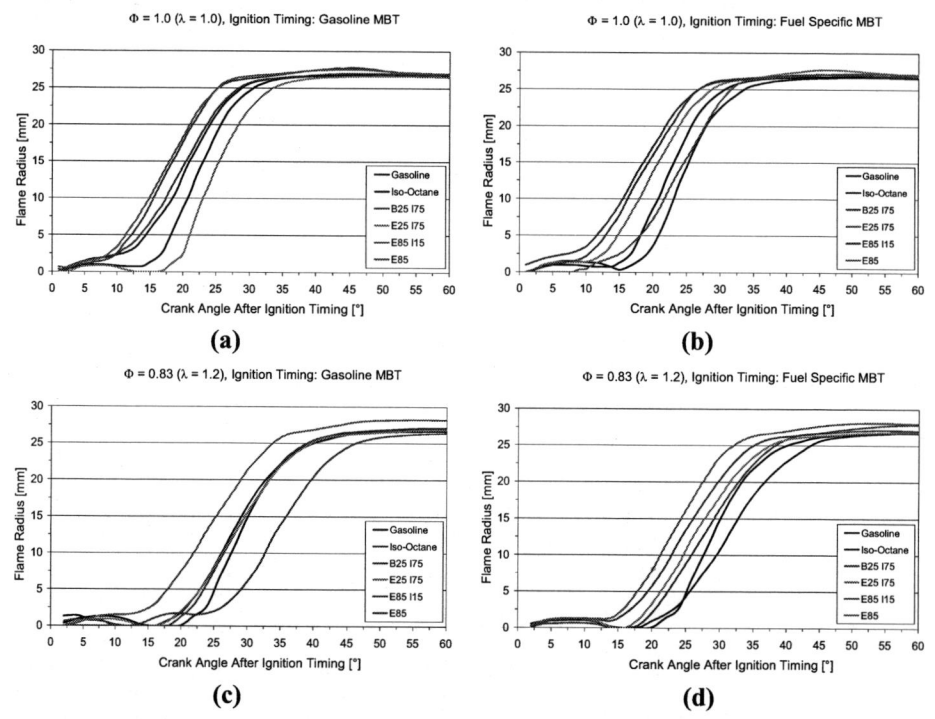

Figure 5. Evolution of flame radius for different fuel blends at stoichiometric and lean conditions with different ignition timings.

4 CONCLUSIONS

The combustion of blends of gasoline, *iso*-octane and a variety of alcohols under part-load engine operation at 1500 RPM was studied with port fuel injection in a newly designed optical spark-ignition engine. In particular, six fuels were tested at stoichiometry and at a lean equivalence ratio of $\Phi=0.83$ ($\lambda=1.2$); a pump-grade gasoline and a commercial E85, as well as *iso*-octane (I100) and splash-blended mixtures of *iso*-octane with 25% ethanol (E25 I75), 85% ethanol (E85 I15) and 25% butanol (B25 I75). The tests were carried out at fixed ignition timing, namely the MBT of gasoline, to examine the combustion of these fuels with nominally same in-cylinder pressure, temperature and flow environments. Additionally, the ignition timing was optimized independently for each fuel to achieve MBT. Differences between the fuels were observed using high-speed crank-angle resolved natural light flame imaging and in-cylinder pressure analysis over batches of 100 cycles. The pressure data were used to calculate mass fraction burned traces for each fuel and the flame images were processed to infer the evolution of an equivalent flame radius during the early stages of combustion. The conclusions are summarised as follows:

- The commercial E85 was found to produce higher peak pressures, greater IMEP, faster mass fraction burned traces and faster flame radius growth than the rest the fuels for most test cases, irrespective to changes in ignition timing. Additionally, E85 was found to be relatively insensitive to changes in air-to-fuel ratio from stoichiometric ($\Phi=1.0$ or $\lambda=1.0$) to lean ($\Phi=0.83$ or $\lambda=1.2$).
- The splash-blended E85 I15 fuel was found to follow similar patterns to those expected from laminar burning velocity correlations. In particular, for fixed ignition timing at stoichiometric conditions it was faster than pure *iso*-octane, but at lean conditions it was slower.
- Addition of 25% ethanol or 25% butanol to *iso*-octane did not affect appreciably the combustion characteristics of *iso*-octane at fixed ignition timing as observed through thermodynamic analysis of the pressure traces for stoichiometric conditions. For lean conditions though, the effect was to marginally slow down the combustion process.
- Addition of 25% butanol to *iso*-octane was found to burn faster than pure *iso*-octane for optimised ignition and to exhibit behaviour more akin to gasoline.

REFERENCE LIST

(1) Hurden, J., DeBoer, C. and Cairns, A., "Towards the Optimum Use of Crude Oil in Passenger Cars", Global Powertrain Congress, 2004.
(2) Brinkman, N.D., "Ethanol Fuel – A Single Cylinder Engine Study of Efficiency and Exhaust Emissions", SAE 810345, 1981.
(3) Nakata, K., Utsumi, S., Ota, A., Kawatake, K., Kawai, T. and Tsunooka, T., "The Effect of Ethanol on a Spark Ignition Engine", SAE 2006-01-3380, 2006.
(4) Fachetti, A. and Kremer, F.G., "Alcohol as Automotive Fuel – Brazilian Experience", SAE 2000-01-1965, 2000.
(5) Turner, J.W.G., Pearson, R.J., Holland, B. and Peck, A., "Development of a High Performance Sports Car for Operation on a High Alcohol-Blend Fuel", JSAE 20075038, 2007.
(6) Cowart, J.S., Caton, P.A. and Hamilton, L.J., "An Experimental and Modelling Investigation into the Comparative Knock and Performance Characteristics of E85, Gasohol [E10] and Regular Gasoline [87 (R+M/2)]", SAE 2007-01-0473, 2007.
(7) Kapus, P.E., Fuerhapter, A., Fuchs, H. and Fraidl, G.K., "Ethanol Direct Injection on Turbocharged SI Engines – Potentials and Challenges", SAE Paper 2007-01-1408, 2007.
(8) Zhang, Y., He, B.-Q., Xie, H. and Zhao, H., "The Combustion and Emission Charactersistics on a Port Fuel Injection HCCI Engine", SAE 2006-01-0631, 2006.
(9) Gautam, M. and Martin D.W., "Combustion Characteristics of Higher Alcohol/Gasoline Blends", Proc. IMechE, Part A, Vol. 214, pp. 497–511, 2000.
(10) Gautam, M., Martin, D.W. and Carder, D., "Emissions Characteristics of Higher Alcohol/Gasoline Blends", Proc. IMechE, Part A, Vol. 214, pp. 165–182, 2000.

(11) Dai, W., Cheemalamarri, S., Curtis, E.W., Boussarsar, R. and Morton, R.K., "Engine Cycle Simulation of Ethanol and Gasoline Blends", SAE 2003-01-3093, 2003.
(12) Martinez, F.A. and Ganji, A.R., "Performance and Exhaust Emissions of a Single-Cylinder Utility Engine Using Ethanol Fuel", SAE 2006-32-0078, 2006.
(13) Sanquist, H., Karlsson, M. and Denbratt, I., "Influence of Ethanol Content in Gasoline on Speciated Emissions from a Direct Injection Stratified Charge SI Engine", SAE 2001-01-1206, 2001.
(14) Guerrieri, D.A., Caffrey, P.J. and Rao, V., "Investigation into the Vehicle Exhaust Emissions of High Percentage Ethanol Blends", SAE 950777, 1995.
(15) Topgül, T., Yücesu, H.S., Cinar, C. and Koca, A., "The Effects of Ethanol-Unleaded Gasoline Blends and Ignition Timing on Engine Performance and Exhaust Emissions", Renewable Energy, Vol. 31, pp. 2534–2542, 2006.
(16) Yücesu, H.S., Topgül, T., Cinar, C. and Okur, M., "Effect of Ethanol-Gasoline Blends on Engine Performance and Emissions in Different Compression Ratios", Applied Thermal Enginering., Vol. 26, pp. 2272–2278, 2006.
(17) Davis, G.W. and Heil, E.T., "The Development and Performance of a High Blend Ethanol Fueled Vehicle", SAE 2000-01-1602, 2000.
(18) Popuri, S.S.S. and Bata, R.M., "A Performance Study of *Iso*-Butanol, Methanol- and Ethanol-Gasoline Blends Using a Single Cylinder Engine", SAE 932953, 1993.
(19) Bata, R.M. and Elrod, A.C., "Butanol as a Blending Agent with Gasoline for IC Engines", SAE 890434, 1989.
(20) Alasfour, F.N., "Butanol – A Single Cylinder Study: Engine Performance", International Journal of Energy Research, Vol. 21, pp. 21–30, 1997.
(21) Alasfour, F.N., "NO_X Emission from a Spark-Ignition Engine Using 30% *Iso*-Butanol-Gasoline Blend: Part 1: Preheating Inlet Air", Applied Thermal Engineering, Vol. 18, pp. 245–256, 1998.
(22) Alasfour, F.N., "NO_X Emission from a Spark-Ignition Engine Using 30% *Iso*-Butanol-Gasoline Blend: Part 2: Ignition Timing", Applied Thermal Engineering, Vol. 18, pp. 609–618, 1998.
(23) Metghalchi, M. and Keck, J.C., "Burning Velocities of Mixture of Air with Methanol, *Iso*-Octane and Indolene at High Pressure and Temperature", Combustion and Flame, Vol. 48, pp. 191–210, 1982.
(24) Ryan, T.W. and Lestz, S.S., "The Laminar Burning Velocity of *Iso*-Octane, *n*-Heptane, Methanol, Methane, and Propane at Elevated Temperature and Pressures in the Presence of Diluent", SAE 800103, 1980.
(25) Takashi, H. and Kimitoshi, T., "Laminar Flame Speeds of Ethanol, *n*-Heptane, *Iso*-Octane Air Mixtures", FISITA 2006, Yokohama, Japan, Paper SC40.
(26) Gülder, Ö.L., "Correlations of Laminar Combustion Data for Alternative SI Engine Fuels", SAE 841000, 1984.
(27) Liao, S.Y., Jiang, D.M., Huang, Z.H., Zeng, K. and Cheng, Q., "Determination of the Laminar Burning Velocities for Mixtures of Ethanol and Air at Elevated Temperatures", Applied Thermal Engineering, Vol. 27, pp. 374–380, 2007.
(28) Beretta, G.P., Rashidi, M. and Keck, J.C., "Turbulent Flame Propagation and Combustion in Spark-Ignition Engines", Combustion and Flame, Vol. 52, pp. 217–245, 1983.

Influence of gasoline engine lubricant on tribological performance, fuel economy and emissions

Peter M. Lee, Martin Priest
School of Mechanical Engineering, University of Leeds, UK

ABSTRACT

The requirement for increased performance, improved fuel economy and reduced emissions is constantly sustaining the demand for research into combustion, fuels and lubricants. Due to the nature of the operation of an engine and the current market climate the lubricant not only has to respond to these requirements, but also to changes in engine design, fuelling methods and fuel types, increased power densities and developments in emissions formation and after-treatment. This paper will describe advances made at the authors' institution to elucidate the influence of gasoline engine lubricant on tribological performance, fuel economy and emissions, giving examples of work undertaken and then look to future possible lubricant demands.

1. INTRODUCTION

The lubricant influences the tribological performance, fuel economy and the emissions of the internal combustion engine. The interaction between the lubricant, the piston ring pack and its environment plays the most significant role in the control of oil consumption, affecting emissions, fuel economy and fuel economy retention through friction, and overall engine performance, such as wear. It is an artefact of the design of the ring pack and the need for there to be some lubrication between rings and liners that a small amount of oil passes the top ring and enters the combustion chamber. This results in oil consumption as it is lost to the exhaust where it adds to the emissions. This consumed oil also reduces the effectiveness of the gasoline after-treatment catalysts by forming a film of phosphorus over the catalytic surface. For this reason oil manufacturers are being legislated to reduce the quantity of sulphur and phosphorus contained in engine oil (from 0.12% in 1994 through 0.10% in 2001 for GF-3 to 0.08% in 2004 for GF-4). Some of the thinnest oil films in an engine are found between the top ring and the liner, at times operating in boundary lubrication conditions, as shown in Figure 1, hence permitting contact between these components, and it is partly for this reason that the piston assembly accounts for 40-60% of total engine friction.

© *University of Leeds 2007*

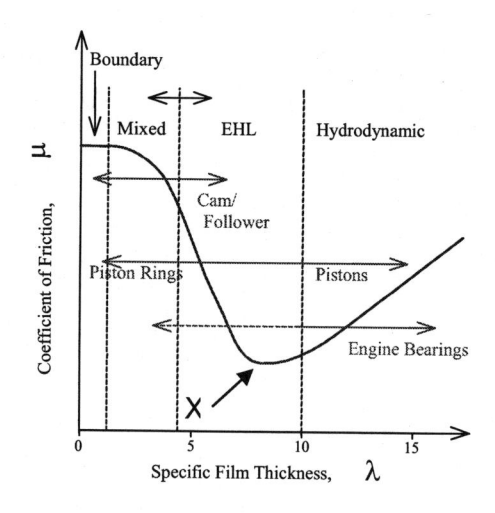

Figure 1 – Stribeck diagram showing engine components and lubricant regimes

With the continued drive towards fuel economy improvements from lubricants the average lubricant viscosity is being reduced in order to attain operation at the lowest coefficient of friction, point X in Figure 1. This reduces friction in the engine due to the reduction of shear of the lubricant where a lubricant film is present. However lower viscosity lubricants are inherently more volatile which may lead to them releasing more hydrocarbons and constituents into the exhaust stream. In the development of the new GF-5 specification this is being considered by additive companies developing low volatility phosphorus additives to prevent this entering the exhaust stream and damaging the catalytic converters. It is not possible to simply remove the phosphorus as it is an efficient tried and tested anti-wear and anti-oxidant additive. The use of reduced viscosities for thinner oil films also increases the chance of component contact and hence wear at piston top dead centre ring / liner interface.

Due to the mature design of the ring pack the vast majority of the oil in the ring pack is however not lost to the combustion chamber but flows back down to the sump and is replenished by 'fresh' oil from the sump. When the oil is in the ring pack it experiences high temperatures, above 200°C, and mixes with the blow-by gas and un-burnt oxygen in this region. It is due to this that the oil degrades, primarily by oxidation, and after sufficient degradation the viscosity increases resulting in a reduction in fuel economy retention. Degraded oil also results in reduced overall engine protection. It is this degradation that provides the impetus for the oil to be replaced and hence it is crucial to elucidate in order to prolong the useful life of the oil. It is the behaviour of the oil, the reaction processes that occur and the flow mechanisms of the oil through the ring pack that need to be understood in order to aid the development of oils that will meet these ever increasing and conflicting demands.

2. WORK UNDERTAKEN

At Leeds much work has been undertaken to understand the tribological behaviour of oils and degraded oils and their effect on engine friction, wear, flow mechanisms in the engine and film formation. A number of specialist techniques have been developed for use with a single cylinder Ricardo Hydra gasoline research engine. This engine is based on a General Motors 2.0 litre, 4 cylinder production engine (c. 1988) of 86mm bore, 86mm stroke and indirect unleaded gasoline injection. A selection of these techniques and results are discussed below.

2.1 Lubricant degradation in the Top piston Ring Zone (TRZ) of the piston assembly

Degradation of the lubricant in the piston ring zone of the internal combustion engine has by far the largest influence on the overall performance of the engine and the useful life of the lubricating oil. As the oil degrades in this region acids are formed and the antioxidants in the oil are consumed. Slowly the viscosity of the oil increases and this results in increased friction and reduced fuel economy. It is from this region that the lubricating oil enters the combustion chamber and is burnt, adding to the emissions. It is therefore imperative that an understanding of the tribology in this region is investigated. To enable this oil has been sampled from the rear of the top piston ring during the fired operation of the engine. This oil is extracted by drilling a hole through the back of the piston ring groove to the inside of the piston and a pipe taken from here through a protective stainless steel sheath to the bottom of the crankshaft. At this point it is constrained only to prevent contact with the crankshaft as it is routed to the exterior of the engine and into collection vials. The pressure difference between the combustion chamber and the atmospheric pressure outside the engine forces the oil, ideally as a mist contained within the gas, along the pipe, Figure 1. This gas/mist stream is directed onto the side of the collection vial where the oil drops out of the gas stream and collects in the vial and the gas extracted. This method was first proposed by Saville et al [1] and used on diesel engines. Much development work has been undertaken resulting in the reliable system that now operates on this petrol engine and a number of studies have been reported [2-5].

The level of degradation of the lubricant was quantitatively determined by using the fourier transform infrared (FT-IR) absorption technique to measure the concentration of carbonyl-containing species present at wave numbers 1690-1750 cm^{-1} as described by Coates and Setti [6]. These carbonyl groups contain ketones, carboxcylic acids, esters and lactones. This work has clearly shown that degradation occurs in the TRZ of the engine and that little degradation occurs in the sump oil, as shown in Figure 2.

Additional work into the chemical mechanisms of the degradation of the oil was undertaken at the Chemistry Department, University of York. Readers interested are directed to papers from that institution [3, 7-9]. Some results for carbonyl concentration are presented in Figures 3 and 6, and described in section 2.3.

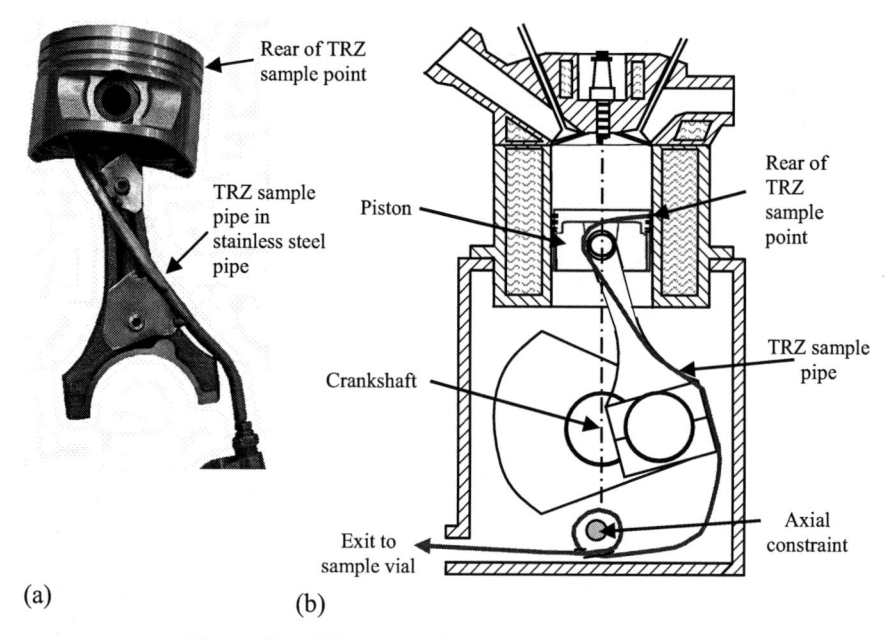

Figure 2 – **TRZ piston (a) and schematic of engine (b) with sample pipe, after [10]**

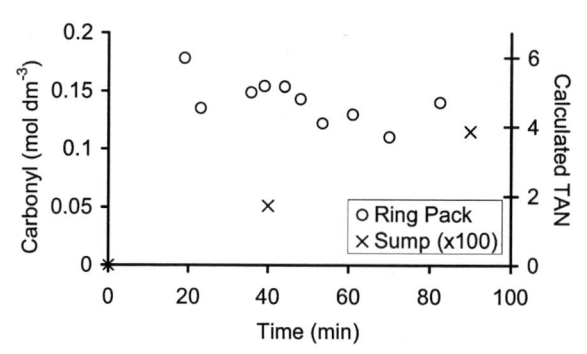

Figure 3 – **TRZ oil carbonyl levels and sump carbonyl levels (x100), after [3]**

2.2 Top Ring Zone (TRZ) oil characterisation

In order to fully understand the tribological effect of the oil in the TRZ of an engine it is necessary to characterise this oil. A comprehensive sampling programme was undertaken to collect 25ml of TRZ oil at 1500 r/min and three different engine loads using the TRZ sampling method described in the previous section. Fresh oil and used sump oil samples were also collected as listed in Table 1. Each 25ml of oil took 40hrs of engine running to complete and were subjected to a number of tribological, chemical and rheological tests that have aimed to characterise the oils. In order to do this work with such small quantities some standard tests had to be re-designed including the Mini Traction Machine (MTM™), (PCS Instruments, London, UK) and the TE77™ reciprocating tribometer, (Phoenix Tribology Limited,

Newbury, UK). The reader is referred to Lee et al [4, 5] for a more complete explanation of this work which is beyond the scope of this paper. The most notable result from this research was that the lubricant in the ring pack is significantly different from that in the sump and hence drawing conclusions from sump samples is misleading. Of significance is the level of volatiles, Figure 4, formed by quenching of the combustion process on the liner wall, found in these ring pack samples. These volatiles are representative of soot pre-cursers as well as the heavier mass fuel elements. It is possible for these volatiles to enter the gas state and be drawn into the exhaust gas, having a detrimental effect upon emissions. With consideration to future problems posed to lubricant formulation, initial work has suggested that bio-fuels are likely to have increased absorption levels into the TRZ oil and, ultimately, the sump. This will change the sump oil formulation from the optimum designed and may effect the frictional characteristics of the oil throughout the engine resulting in reduced fuel economy.

Table 1 – table of oil samples, after [4]

Oil sample name	Oil sample conditions
33% load TRZ*	33% maximum engine load at 1500 r/min
50% load TRZ	50% maximum engine load at 1500 r/min
75% load TRZ	75% maximum engine load at 1500 r/min
First 15 min	First 15 min from cold start at 1500 r/min
Fresh oil	Fresh Shell Helix Super 15W-40
40 h sump oil	Used sump oil after 40 h at 50% maximum engine load 1500 r/min

*TRZ = top ring zone

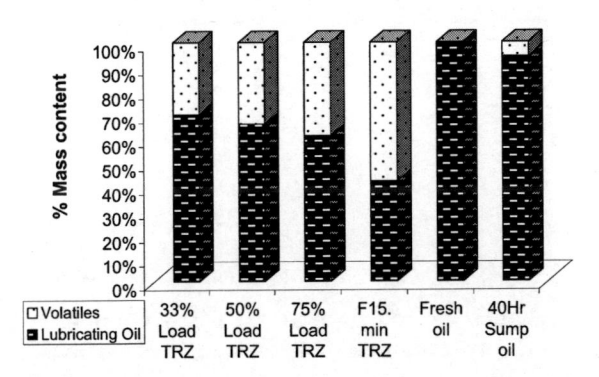

Figure 4 – Volatile content of oil samples measured by GC, after [4]

Figure 5 shows the friction results obtained from this extracted oil when tested in a Plint TE-77 reciprocating tribometer. The test parameters were set to simulate the ring and liner contact conditions in the most severe conditions just after combustion and the material used was that of the engine liner for both the pin and the plate. It is clear from the figure that there are distinct differences between all the samples, but most notably between the sump oil and fresh oil samples and the four top ring zone samples which is the oil that lubricates the ring/liner interface.

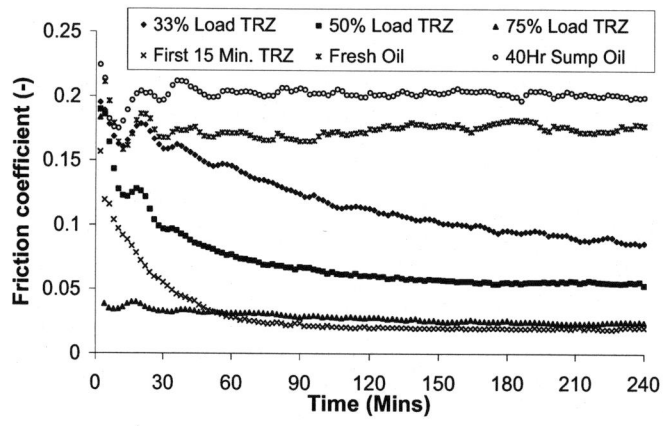

Figure 5 – Plint TE77 reciprocating tribometer flat cast iron pin on cast iron plate, 2.6MPa, 5mm stroke, 25Hz, 100°C, after [4]

2.3 Flow rates through and residence time in the ring pack

As oil primarily degrades in the ring pack it is important to know how much oil is present in the ring pack and the period of time for which that quantity of oil is exposed to the degradation process for any given engine condition. This knowledge will give indications for oil drain intervals and, combined with frictional analysis, some projected indication of the fuel economy retention times for oils. In order to understand this the flow of the oil through the ring pack and the residence time of the oil must be found.

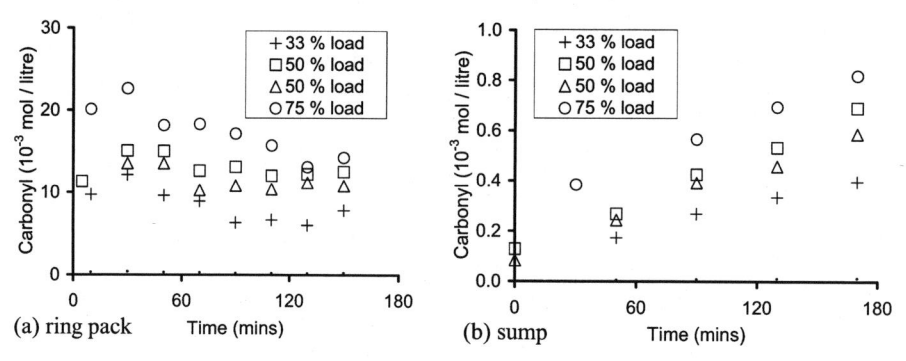

Figure 6 – Degradation levels measured by carbonyl in the ring pack (a) and sump (b) with respect to time at 1500rpm and loads as shown, after [11]

By taking both TRZ samples and sump samples, and measuring the carbonyl concentration of both sets of samples it is possible to calculate the flow rates of degraded oil from the ring pack to the sump. This is because if the sump is maintained at 70°C, a temperature at which negligible degradation occurs, the degradation of the sump oil is caused by the degraded oil returning from the ring pack. This results in a steady increase in degradation in the sump over time. Figure 6 shows the carbonyl species in the sump increasing (b) compared to the fairly constant level of degraded sample taken from the ring pack (a) with respect to time.

The results presented are for the same speed, 1500 r/min, and three different loads as indicated on the graph. The 50% load experiment was undertaken twice to investigate repeatability. The carbonyl content does not start at zero for the sump oil at the start of the test. This is because, despite flushing the engine, residual carbonyl from previous tests was present in the sump oil.

With the addition of a second sump containing a hexadecane marker, it is possible to change the sumps supplying the engine at the same time as samples start to be taken for analysis from the TRZ. This allows the increase in the hexadecane marker present in the TRZ to be measured as the oil in the first sump is replaced in the ring pack by the marked oil in the second sump. It is then possible to calculate the residence time of the oil in the ring pack based upon the time taken to replace the sump 1 oil in the ring pack with sump 2 marked oil. Work has been undertaken to study the effect of load and speed on this phenomena and it has been shown that degradation levels increase with engine load, due to increased TRZ temperatures. In order to undertake this work both sumps were maintained at the same temperature (70°C) and both sump oils were pumped through a switch valve designed to be fixed on the main gallery inlet of the engine. This ensured thermal shock to the engine was prevented, no air was entrained in the oil at switch over and that the changed oil was supplied to the engine as close as was practically possible. Table 2 presents the residence times obtained during this work and Figure 7 shows the flow rates. Again the length of this paper precludes a full explanation and readers are directed to Lee et al [12]

Table 2 – Residence times in the Piston Assembly with respect to engine speed and load, after [12]

		Speed (rpm)		
		1000	1500	2000
Load (%)	33		6.0	
	50	9.0	5.5	7.5
	75		4.0	

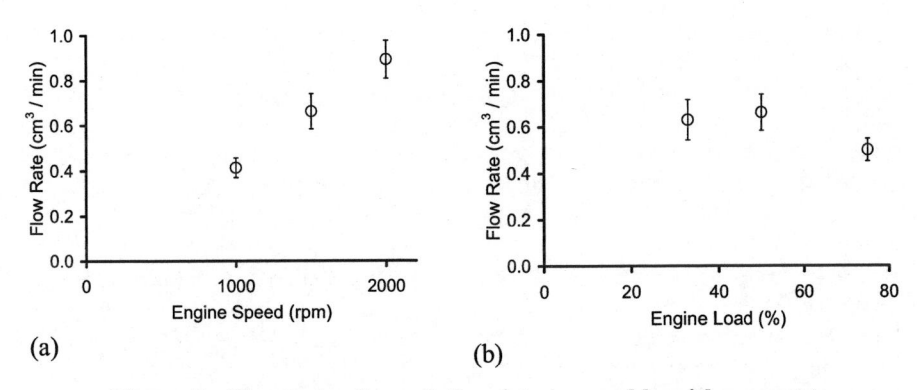

Figure 7 – Flow rates through the piston assembly with respect to (a) engine speed and (b) engine load, after [11]

Initial research [12], Figure 8, suggests that a higher viscosity lubricant will flow through the ring pack more slowly than a lower viscosity lubricant. Therefore as lubricants degrade and hence become more viscous they may remain in the ring pack, exposed to the degradation process, for longer. Although the speed at which the oil in the ring pack degrades will remain the same it will become highly degraded due to being present for longer and so it is likely to result in higher levels of sludge, varnish and then deposit formation before the lubricant is due to flow back to the sump. This will have additional detrimental effects on fuel economy over and above the increased viscosity of the lubricant as well as affecting the action of the rings and hence ultimately affecting emissions as they will no longer seal against the bore correctly.

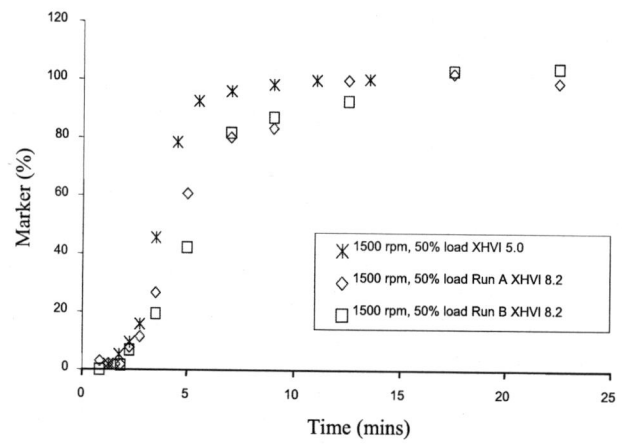

Figure 8 – Effect of lubricant viscosity on residence time, from [12]

2.4 Engine Friction Measurement

The vast majority of the frictional losses in an engine are attributable to the three main tribological sub-systems of an engine; the piston assembly, engine crankshaft bearings and the valve train. Experimental methods have been developed for the Hydra engine at Leeds that allow the losses in all three tribological components to be measured while the engine is fired. This involves the use of specially designed torque transducers on the drive pulleys of the valve train, the indicated mean effective pressure technique to obtain the piston assembly friction and the calculation of crankshaft bearing losses by subtraction. This method has confirmed that between 40 and 60% of the total engine friction is due to the piston assembly and has shown the effect of speed and lubricant viscosity on power-loss in the piston assembly, Figure 9. This work has been reported by Mufti et al [13, 14].

It can be seen from Figure 9 that the lubricant viscosity and additive package directly effect the power loss experienced in the piston assembly. Lower viscosity lubricants at lower temperatures result in lower power losses. However these lubricants have higher power losses at higher temperatures as they move from the lubricated to boundary or mixed lubrication regime earlier than higher viscosity lubricants. This has a negative impact on fuel economy due to the resulting higher friction.

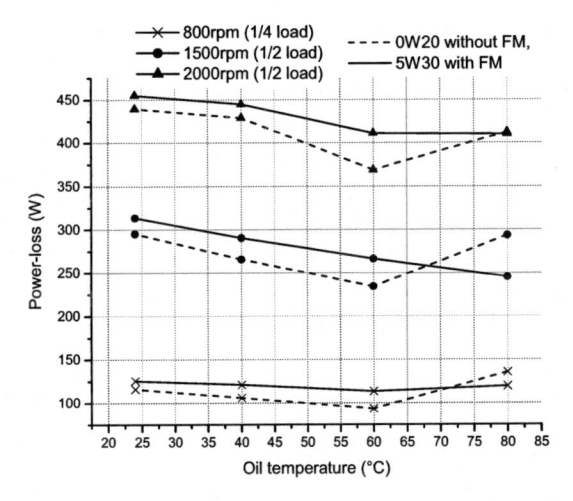

Figure 9 – Comparison of piston assembly friction power loss for SAE 0W20 without FM (friction modifier) and SAE 5W30 with FM lubricants, after [14]

This method has subsequently been used to validate the bench top work undertaken as described in section 2.5 and would be an invaluable tool in the investigation of the effect of lubricant degradation on the major tribological components of the engine.

2.5 Fuel administered friction modifiers

Although the use of lower viscosity lubricants to obtain improved fuel economy works for most of the engine it is not successful for all parts of the piston assembly. Where there are lubricant films, for example between the piston skirt and liner and mid stroke between the rings and the liner it does reduce frictional losses due to the lubricant shearing more efficiently. However when the piston is at top dead centre and bottom dead centre of the stroke, where boundary lubrication occurs, the thinner films result in less protection and higher friction is experienced due to contact between the rings and the liner. This is also undesirable for liner and ring wear. One possible method of resolving this is to use the effect of fuel dilution in the top ring zone to carry surface friction modifiers to the ring pack via the fuel as it is injected into the engine. Initial investigation using a bench top tribometer and model friction modifiers has been undertaken at the institution of the authors. A TE77™ was set up with a ring on liner section and two oil supplies. One oil contained a low viscosity oil to replicate the viscosity in an engine due to fuel dilution, the other sump contained the same oil, but with the chosen friction modifier at 2% w/w. The TE77 was run for an hour before the sump was changed to the friction modifier containing oil for an hour and then changed back to the original sump supply. Results showed that an immediate reduction in friction occurred and that this tailed off close to the original value when the oil was changed back to the original supply. This work was undertaken with three different additives as shown in Figure 10.

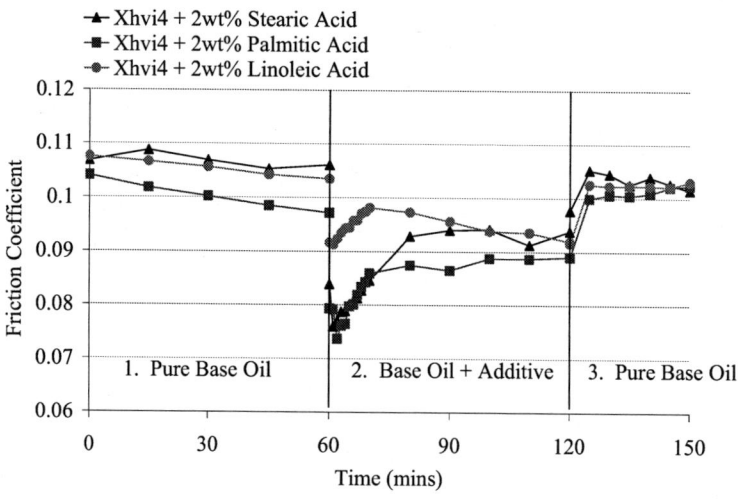

Figure 10 – TE77 piston assembly friction coefficients for non additivised oil and three different additivised oils.

Following on from this screener test chosen additives were run in the Hydra Engine under fired conditions with the additives placed in the fuel as oppose to the oil. When the fuel supply was changed a drop in friction was observed. This change was less than on the TE77 as the TE77 test operated in boundary condition for the majority of the time, where as the ring pack in the engine operated in boundary only at and near TDC and BDC. This work is explained in more detail in Smith et al. [15].

3. FUTURE OUTLOOK

There is no indication that the ever increasing and conflicting demands placed upon gasoline lubricants will abate in the foreseeable future. In the medium term OEMs are pushing towards fill for life oils and are increasing power densities using modern technologies such as gasoline direct-injection turbo charging, second stage turbo charging, improved air handling and EGR technologies, advanced combustion chamber design and variable valve timing combined with variable valve lift volume ('valvematic'). The new ILSAC GF-5 oil specification is scheduled for mid 2009 (with factory fill from 2010). Although specifications are yet to be finalised OEMs have identified major areas where they would like to see improvements. These include fuel economy and fuel economy retention throughout the oil drain interval, improved emission-control systems protection and increased sludge protection, deposit and oxidation control. ILSAC has identified low-temperature viscosity, high- and low-temperature corrosion, turbocharger protection and filter clogging protection as additional requirements. Aeration control is a renewed concern due to modern engines using oil as a hydraulic fluid in cam phaser devices, variable valve actuators, timing chain tensioners and hydraulic lash adjusters that allow for variable valve timing. Over the need to meet these increased requirements both OEMs and engine oil manufacturers have agreed that GF-5 motor oil should be backward compatible to prevent damage due to misapplication. As such GF-5 motor oils will be the most technologically advanced motor oil produced to date. Beyond this ILSAC GF-6 is planned. So what may be expected of lubricants beyond 2009? The European and US drive towards Bio-fuels will have an effect on lubricant /fuel compatibility with new chemical degradation processes occurring in the ring pack. Additive packages will most likely evolve still further to meet the three basic demands of improved fuel economy, better emission system compatibility and greater lubricant stability. Ever improving engine technology, downsizing and the various modes of hybridisation and new materials will result in further, as yet unknown, demands on the engine lubricant. Additionally a cradle-to-grave approach is likely, bringing consideration of biodegradability and recyclability. Currently engine lubricants are essentially additive packs mixed into base oil, which contains biological components. It is however highly likely that in order to meet the high anti-oxidant GF-6 specification the oil will have to be fully synthetic.

4 CONCLUSION

The engine lubricant has a direct effect upon overall engine performance, in particular initial fuel economy, fuel economy retention, exhaust emissions and emission aftertreatment devices. There is also interaction with between the lubricant and the fuel in the top ring zone region which has implications on fuel economy and lubricant degradation rates. Before the lubricant can play its full part in responding to the new technologies, maintaining fuel economy throughout extended oil drain intervals and further reducing emissions continued research is required to elucidate the fuel/lubricant interactions and the effect of lubricant degradation. With the continued increasing demand on the engine lubricant it is essential that it be considered as an integrated component in the same way as fuels, exhaust system controls and combustion systems are considered in order to achieve optimal system performance. Yet, to date, there is no evidence of this. Indeed, lubricant design is expected to respond to the often conflicting requirements placed on it by legislators, formulation restrictions, modern technologies and OEMs.

ACKNOWLEDGEMENTS

The research discussed in this paper has been funded by a range of sponsors including the EPSRC (Engineering and Physical Research Council, UK), Federal Mogul, Ford, Infineum, Jaguar and Shell. Many thanks go to co-workers at Leeds, in industry and in the Department of Chemistry at the University of York, UK.

REFERENCE LIST

1. Saville, S.B., et al., *A Study of Lubricant Condition in the Piston RIng Zone of Single-Cylinder Diesel Engines Under Typical Operating Conditions.* SAE 881586, 1988.
2. Gamble, R., *The Influence of Lubricant Degradation on Piston Assembly Tribology*, in *School of Mechanical Engineering*. 2002, The University of Leeds: Leeds.
3. Stark, M.S., et al., *Measurement of Lubricant Flow in a Gasoline Engine.* Tribology Letters, 2005. **19**(3): p. 163-168.
4. Lee, P.M., et al., *The extraction and tribological investigation of top ring zone oil from a gasoline engine.* Proc. Instn. Mech. Eng. Part J: J Engineering Tribology, 2005. **220**(J3): p. 171-180.
5. Lee, P.M., et al. *The tribological and chemical analysis of top ring zone samples of fully formulated oil taken from a four stroke gasoline engine.* in *World Tribology Congress III*. 2005. Washington DC.
6. Coates, J.P. and L. Setti, *Infrared Spectroscopy as a Tool for Monitoring Oil Degradation.* ASTM, 1984. **STP 916**: p. 57-78.
7. Stark, M. *Oil Flow in Gasoline Engines.* in *STLE Annual Meeting*. 2002. Houston.
8. Stark, M.S., et al. *The Degradation of Lubricants in Gasoline Engines: Lubricant Flow and Degradation in the Piston Assembly.* in *31st Leeds-Lyon. Life Cycle Tribology*. 2004. Leeds, UK: Elsevier.
9. Stark, M.S., et al. *The Degradation of Lubricants in Gasoline Engines: Part 1: A Combined Tribological, Rheological and Chemical Model for Lubrication Degradation in the Piston Ring Pack.* in *STLE annual conference*. 2003. Toronto.
10. Lee, P.M., et al. *Lubricant Degradation Studies using a Single Cylinder Research Engine.* in *World Tribology Congress III*. 2005. Washington DC.
11. Lee, P.M., et al. *The Study of the Residence Time and Flow Rate of Lubricating Oil through a Piston Ring Pack.* in *IMechE Mission of Tribology 13*. 2004. Birdcage Walk, London: IMechE.
12. Lee, P.M., *Lubricant Degradation, Transport and the Link to Piston Assembly Tribology*, in *School of Mechanical Engineering*. 2006, University of Leeds: Leeds.
13. Mufti, R.A. and M. Priest, *Experimental and theoretical study of instantaneous engine valve train friction.* Journal of Tribology, ASME, 2003. **125**(3): p. 628-637.
14. Mufti, R.A. and M. Priest, *Experimental evaluation of piston assembly friction under motored and fired conditions in a gasoline engine.* Journal of Tribology, ASME, 2005. **127**(4): p. 826-836.
15. Smith, O., et al. *In-cylinder fuel and lubricant effects on gasoline engine friction.* in *Proc. World Tribology Congress III*. 2005. Washington DC, USA.

Determining the effect of lubricating oil properties on diesel engine fuel economy

C.D. Bannister, J.G. Hawley, C.J. Brace
Department of Mechanical Engineering, University of Bath, UK

I. Pegg
European P/T Research & Advanced, The Ford Motor Company Limited, UK

J.C. Dumenil, A. Brown
Global Lubricants Technology, BP plc, UK

ABSTRACT

Different formulations of lubricating oil have been evaluated in a Design of Experiment (DoE) approach to examine the effect of their viscometric properties on engine fuel economy. The performance of the oils has been evaluated on a transient test bed over the New European Drive Cycle (NEDC) using a production automotive Diesel engine. The main viscometric properties investigated were cold cranking shear (CCS), kinematic viscosity at 100°C (Kv100) and high temperature high shear (HTHS). Results have shown that CCS is dominant during the first phase of the NEDC (ECE15) whilst HTHS and Kv100 become more dominant during the second phase (EUDC). The viscometric changes for the aged oils evaluated showed a 3.5% fuel economy reduction over the NEDC when comparing the 0W30 grade with 10W40 oil. A DoE analysis confirmed the observed trends and facilitated the creation of a multi-dimensional response model which can be used to predict brake specific fuel consumption (BSFC) over the NEDC based on changes in oil viscometrics.

LIST OF SYMBOLS AND UNITS

ACEA	Association of European Car Manufacturers	dti	Department of Trade and Industry
BSFC	Brake Specific Fuel Consumption (g/kWh)	ECE15	Economic Commission for Europe regulation 15, or UDC – Urban Drive Cycle
CCS	Cold Cranking Shear (cP)		
CO	Carbon Monoxide	ECU	Engine Control Unit
CO_2	Carbon Dioxide	EU	European Union
DoE	Design of Experiments	EUDC	Extra Urban Drive Cycle
		HC	Hydrocarbons

© IMechE 2007

HTHS	High Temperature High Shear (cP)	NEDC	New European Drive Cycle
		NOx	Oxides of Nitrogen
Kv_{100}	Kinematic Viscosity measured at 100°C (cSt)	PM	Particulate Matter
		THC	Total Hydrocarbons

1. INTRODUCTION

In February 2007 the EU introduced new proposals to encourage car manufacturers to reduce fleet CO2 emissions to an average of 130g/km by 2012 [1] the current average is 163g/km. This came about after a voluntary target of 120g/km CO2, proposed by the association of European car manufacturers (ACEA) [2], was not met. The revised target amounts to a 22.2% reduction in fleet CO2 emissions. The expectation is that in the longer term the target will revert to the original EU proposed level of 120g/km[3]. Additionally, in the period 1990-2004, CO2 emissions from road transport increased by 26%, relying exclusively on fossil fuels, consuming 60 % of all the oil used within the EU[3]. It is also accepted that viable and cost effective fuel cells and the associated hydrogen infrastructure to support such a transport mechanism will not be available to the market until post 2020 at the earliest [4]. As such, minimising the level of CO2 produced by vehicle powertrains is a significant and demanding challenge. The work presented in this paper is aimed at reducing the carbon footprint of automotive diesel engines by focusing on the formulation of the lubricating oil. The question to be answered is:

"Can changes in lubricating oil formulation lead to a significant reduction in the CO2 emissions of future vehicle powertrains?"

This study was part of a larger industry/University collaborative project sponsored by the dti. The work conducted was focused on examining the effects of both the physical and chemical properties of the lubrication oil on fuel consumption when fresh, and after a period of ageing.

2. APPROACH

The three main lubricating oil viscometric properties investigated were cold cranking shear (CCS), kinematic viscosity at 100°C (Kv100) and high temperature high shear (HTHS). A brief description is given below.

Cold Cranking Shear (CCS)
CCS is measured using a Cold Cranking Simulator Test and provides a measure of the viscosity of the lubricating oil under conditions of low temperature and high shear rates. In general, a reduction in the CCS value of the lubrication oil lowers the viscosity under cold start conditions and results in reduced engine friction and, in turn, fuel consumption.

High Temperature High Shear (HTHS)
HTHS is a measure of the lubricant's minimum dynamic viscosity when subjected to high temperature and shear conditions similar to those typically experienced during use in-engine. A higher HTHS value would suggest a higher dynamic viscosity within engine bearings resulting in increased fuel consumption but potentially reduced wear.

Kinematic Viscosity at 100°C (Kv100)
Kv100 gives the lubricating oil's resistance to shear or flow caused by intermolecular friction exerted when layers of molecules within the fluid move relative to each other, measured at 100°C.

Specifically the kinematic viscosity is the ratio of dynamic viscosity to density. The greater the Kv100 value, the greater the fluid's resistance to flow resulting in increased engine friction and fuel consumption.

The study was broken down into a number of distinct phases and a design of experiments (DoE) approach was adopted for each phase. It was a requirement to measure very small changes in fuel consumption, so it was necessary to develop robust process control techniques to limit test-to-test variation and improve confidence in experimental results.

The experimental project phases were divided as follows:

- Phase 1 – Cold Cranking Shear (CCS) Study
- Phase 2 – Cold Cranking Shear (CCS), High Temperature High Shear (HTHS) and Kinematic Viscosity (Kv_{100}) Study

3. EXPERIMENTAL PROCEDURE

The experimental study was performed on a modern 2.4l Diesel engine mounted on a dynamic AC engine dynamometer. The New European Drive Cycle (NEDC) was the experimental datum. Each cycle was started from a thermally stable condition of approximately 20°C.

The experimental procedure was as follows and is shown schematically in Figure 1:

1. For each fresh fill of oil the engine was run over a set of three conditioned NEDC's to measure the characteristics of the fresh oil.
2. The engine was run over an ageing cycle for 10 hours.
3. A set of six drive cycles was performed. The first NEDC of the day (AM) was followed by a conditioning cycle and a cool-down period to allow the engine head, coolant and oil temperatures to return to initial values before a second NEDC (PM) and a conditioning cycle.
4. The engine was run over a second ageing cycle lasting for 20 hours
5. A final set of six drivecycles were undertaken, comprising of AM and PM cycles as previously discussed.

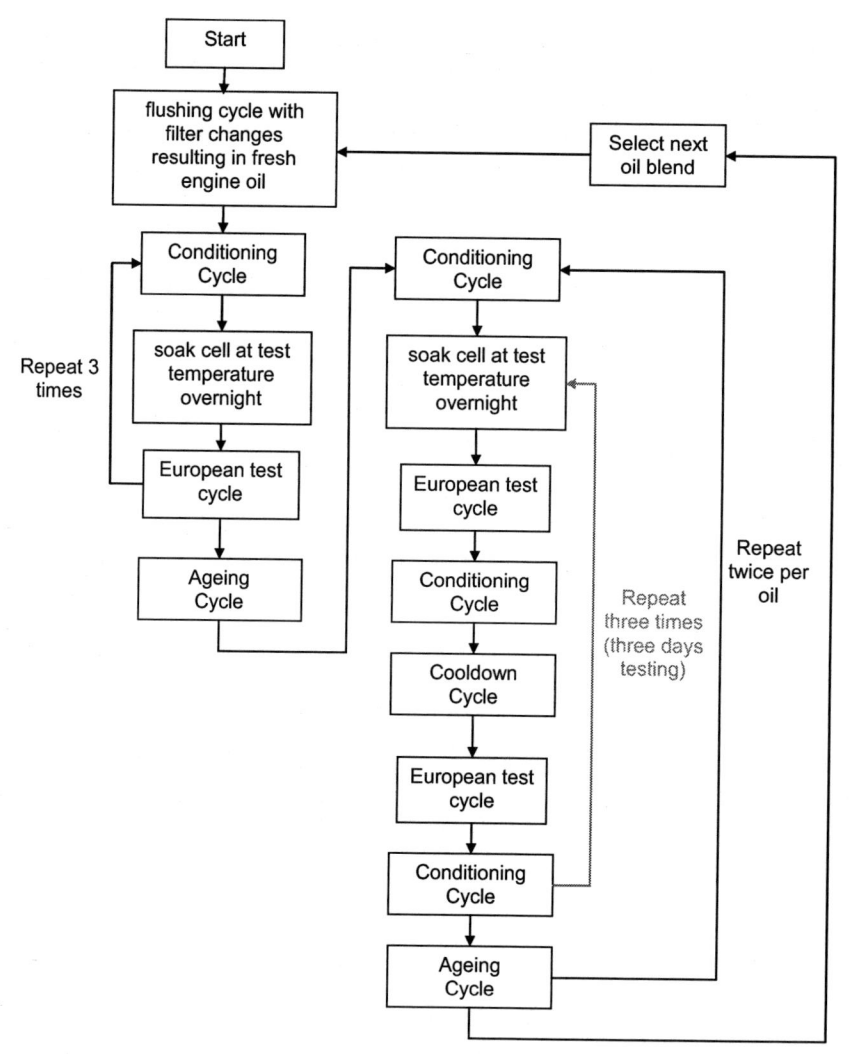

Figure 1 - Lubrication Study: Experimental Procedure

The aim of this procedure was to investigate the effects on fuel economy between fresh and aged oil.

Fuel consumption was measured by four discreet methods and the performance of each, as well as the trends observed, were compared:
- Carbon balance using Horiba MEXA 7000 exhaust gas analyser sampling post catalyst
- Carbon balance using Horiba MEXA 9000 exhaust gas analyser sampling pre catalyst
- Physical fuel mass determined by CP Engineering gravimetric fuel balance
- Inferred fuel usage from engine ECU fuel demand data

Figure 2 shows the four fuel consumption data sets for one of the test oils while Table 1 shows the mean, standard deviation (SD) and coefficient of variation (SD expressed as a ratio of the mean) achieved by each method. Taken across the entire test program, the four methods show good agreement with a maximum bias (taken here as the variation in mean between the different measurement methods) of 2%. The precision achieved by each method, which will include all sources of experimental scatter, can be seen from the coefficient of variation, which is, at worst, 1%, at best 0.7%. This is larger than would be seen typically for this number of repeat tests but the ageing effect on the oil contributes to this scatter. The carbon balance method, using the MEXA 7000 emissions analysers, appears to be the most precise and the gravimetric method the least. Data presented during the remainder of this discussion will, therefore, be that derived using the carbon balance method from data captured using the MEXA 7000 analyser.

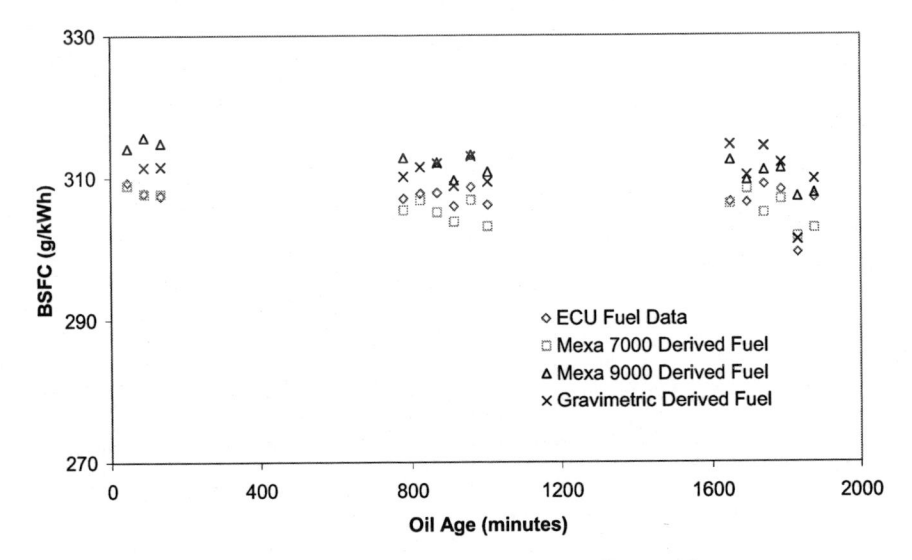

Figure 2 - Fuel Consumption Method Comparison

Table 1 - Fuel Consumption Method Comparison

Method	Mean (g/kWh)	Standard Deviation (g/kWh)	Coefficient of Variation (%)
MEXA 7000	305.81	2.15	0.70
MEXA 9000	311.63	2.48	0.80
Fuel Balance	310.74	3.21	1.03
ECU Fuelling	307.03	2.32	0.76

4. PHASE 1 – CCS STUDY

The approach selected was to systematically change the oil formulation to vary CCS while keeping all other factors, particularly the high temperature high shear (HTHS) performance, as constant as possible. This was achieved by varying the proportions of base oils in the mix. Four batches of lubricant were blended and their properties are shown in Table 2. It was expected that the oils would offer improving fuel consumption in the order 2,1,3,4.

Table 2 - Phase 1: Oil Properties

Oil Number	CCS@ -35°C (cP)	CCS@ -30°C (cP)	CCS@ -25°C (cP)	HTHS@150°C (cP)
1	2210	4000	7530	2.87
2	2510	4330	7910	2.89
3	2000	3290	6040	2.90
4	1780	2810	4900	2.90

5. PHASE 1 – RESULTS & DISCUSSION

The standard NEDC can be split into two sections, the ECE which accounts for the first 780 seconds, followed by the EUDC which encompasses the remaining 420 seconds. It was expected that CCS would play a role in determining engine parasitic losses and hence fuel consumption over the early part of the cycle but, as the oil warms up, the effect of CCS was expected to diminish while the impact of high temperature viscometrics would become more prevalent. Since the high temperature properties were held constant in all four blends, the differences between oils were expected to reduce during the hot phase of the tests.

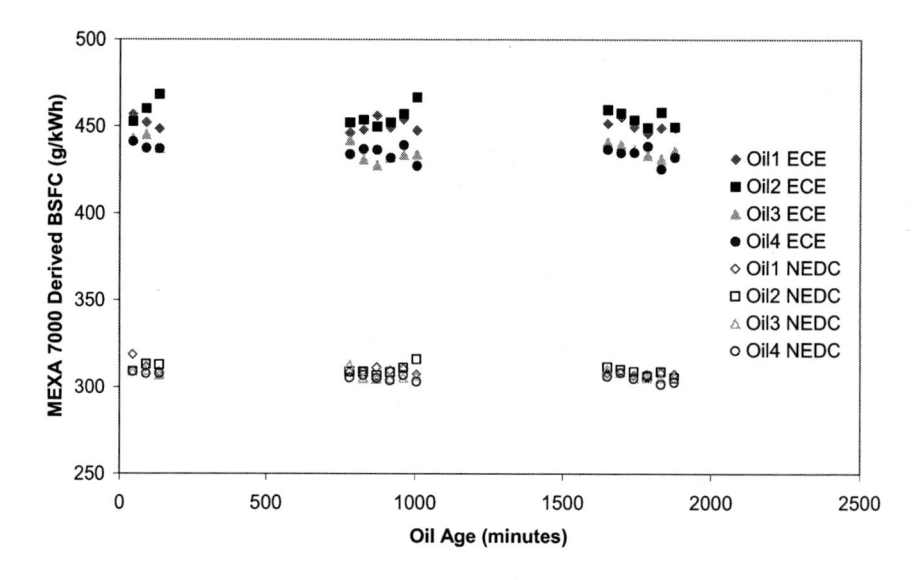

Figure 3 - Phase 1: ECE & NEDC Oil Performance Based on MEXA 7000 Derived BSFC

Figure 3 shows the performance of the different oils during the ECE and for the NEDC as a whole. It can be seen that, regardless of oil formulation, there is a slight decrease in measured BSFC with increasing oil age. This can be attributed to a gradual permanent shear thinning resulting in reduced oil viscosity. Examination of averaged data obtained from the individual results displayed in Figure 3 shows a rank order, from lowest to highest fuel consumption, of 4, 3, 1, 2 (for aged oil) with a maximum fuel consumption benefit between oils 4 and 2 of approximately 5.5%. The oil rank order determined from ECE data correlates with that expected based on oil viscometrics, namely the designed CCS values.

Although not presented here, experimental data during the hot, EUDC portion of the drive cycle showed a much reduced spread between the BSFC obtained with different test oils compared with those observed during the ECE. There is a less pronounced differentiation in BSFC and this reduces the confidence in any derived rank order for the different oils over the EUDC. As was seen in the ECE data, the gradual decrease in BSFC with oil age was also observed in the EUDC. The hypothesis that this trend was due to a reduction in oil viscosity due to permanent shear thinning over time was confirmed by analysis of oil samples taken periodically during the experimental program.

It can be seen that the effect of CCS on the NEDC as a whole accounts for a maximum improvement in fuel consumption of approximately 1.5% between oils 2 and 4. Although shorter in duration than the ECE, the EUDC accounts for a larger proportion of the total fuel burnt during the NEDC. As such, benefits of reduced CCS values on BSFC observed during the cold ECE do not equate to a substantial reduction in fuel consumption over the entire NEDC due to high temperature viscometrics being more dominant during the hot EUDC. For these reasons the second phase of work would examine the effects of high temperature viscometric properties of the oils as well as CCS with the identification of any possible interactions between them.

6. PHASE 2 – CCS, HTHS AND KV_{100} STUDY

After the completion of the Phase 1 CCS study it was decided that, in order to have a greater influence on total drive cycle fuel consumption, high temperature viscometric properties of the oils needed to be investigated in more detail. The two high temperature properties considered were high temperature, high shear (HTHS) and kinematic viscosity (Kv_{100} – measured at 100°C).

Table 3 - Phase 2: Oil Properties

Oil Number	Oil Grade	CCS @ -25°C (cP)	Kv_{100} (cSt)	HTHS @ 150°C (cP)
1	SAE 10W-40	6600	14.20	3.9
2	SAE 5W-40	4774	14.93	3.9
3	SAE 10W-20	6601	8.20	2.71
4	SAE 5W-20	4800	8.55	2.70
5	SAE 0W-30	196	9.57	2.69

An experimental design was proposed as shown in Table 3 and Figure 4. Initially it was intended to only vary oil properties within generally accepted ranges seen in production oils but an additional oil was formulated with exceptionally low CCS values in order to assess and quantify the impact this extreme would have on drive cycle fuel consumption.

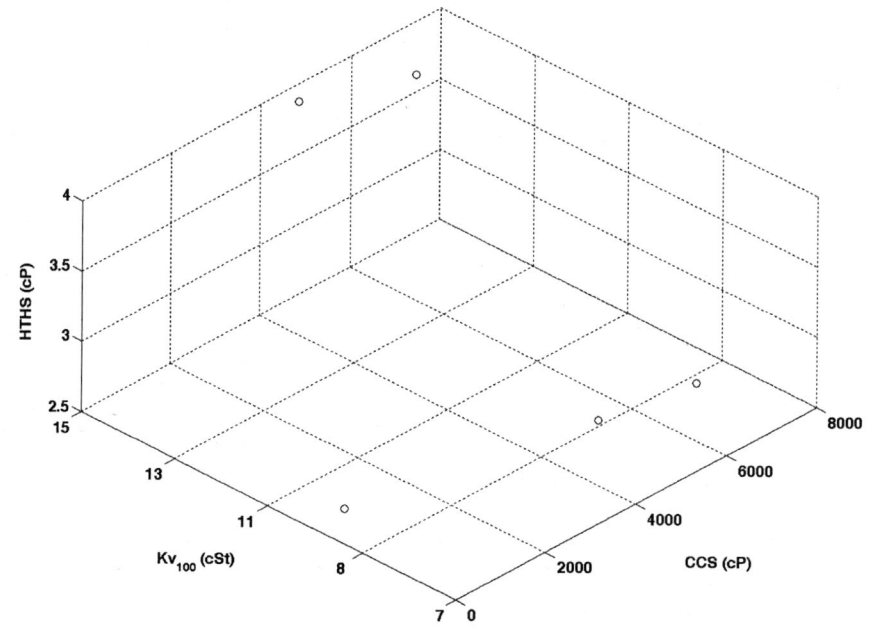

Figure 4 - Phase 2: Oil Formulation Design

7. PHASE 2 – RESULTS & DISCUSSION

For ease of viewing, data has been averaged into three groups, fresh, partially aged and fully aged oil. As with Phase 1 data Phase 2 results are shown for BSFC derived using the carbon balance method from data captured by the MEXA 7000 emissions analyser.

Figure 5 shows the BSFC for all 5 oil formulations over the whole NEDC drive cycle with error bars showing ±2 standard deviations (96% confidence). Of the 4 oils falling within generally accepted viscometrics (oils 1-4), the oil with low values of CCS, Kv_{100} and HTHS performed the best both when fresh and when aged. High values of CCS, Kv_{100} and HTHS seen in oil 1 resulted in approximately a 10g/kWh (3.2%) increase in BSFC compared with oil 4 regardless of oil age.

Oil 5 exhibited a much reduced BSFC when fresh compared with the other oils manifesting as a 3.2% reduction in BSFC over oil 4. However, oil 5 did not display the characteristic reduction in fuel consumption normally seen with increasing oil age, instead demonstrating a marked rise. Table 4 summarizes the results of analysis carried out on oil samples taken at the start and on completion of the experimental program for oils 2 and 5. Oil 2 was chosen as a comparison as it too did not display a reduction in fuel consumption with time, however fully aged data for oil 2 were subject to a larger-than-normal degree of experimental scatter resulting in reduced confidence in the averaged value. In Table 4, it can be seen that CCS, HTHS and Kv_{100} all decrease (9.3%, 8.6% and 6.4% respectively) as oil 2 is aged which would be expected to cause a reduction in fuel consumption thus implying that the slight increase in observed BSFC can be attributed to experimental error.

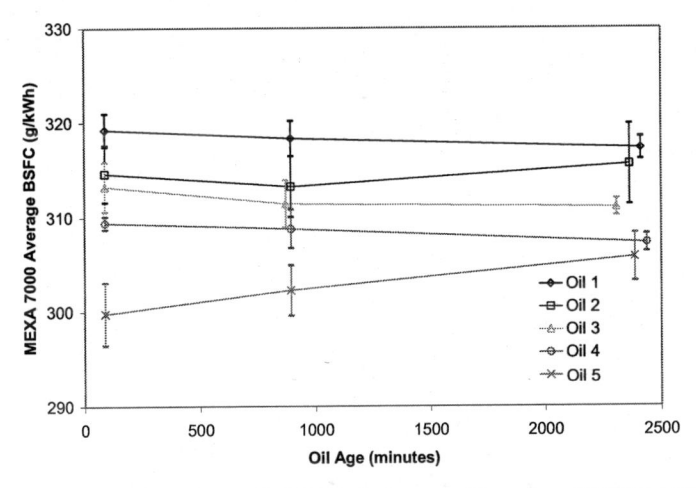

Figure 5 - Phase 2: NEDC Oil Performance Based on MEXA 7000 Derived BSFC

Table 4 - Phase 2: Oil Sample Analysis

Oil Property	Oil 1 Fresh	Oil 1 Aged	Oil 2 Fresh	Oil 2 Aged	Oil 3 Fresh	Oil 3 Aged	Oil 4 Fresh	Oil 4 Aged	Oil 5 Fresh	Oil 5 Aged
CCS @ -30°C (cP)	7180	7100	4720	4280	7340	7040	4720	4350	657	1330
HTHS @ 150°C (cP)	4.01	3.71	4.07	3.72	2.89	2.73	2.77	2.75	2.69	3.43
Kv_{100} @ 100°C (cSt)	14.53	13.61	15.33	14.32	8.12	8.28	8.45	8.41	9.57	12.18

CCS, HTHS and Kv_{100} values for oil 5 significantly increased with age by 102.4, 27.5 and 27.3% respectively in agreement with the observed increase in BSFC, thus adding confidence in this trend as a real result. This trend of increasing CCS, HTHS and Kv_{100} values with age was likely to be caused by lighter, more volatile organic compounds within the oil evaporating off over time and causing the oil to 'thicken'. Oil 5 would be particularly susceptible to this effect due to it containing a higher than normal proportion of these volatile fractions in order to achieve the much-reduced CCS value.

The effect of the differing oil formulations on BSFC during the ECE portion of the NEDC can be seen in Figure 6. As would be expected from the oil CCS values, oil 4 exhibits a 3.4% reduction in BSFC compared with oil 1 when aged, while oil 5 reduces the BSFC by an additional 4.1%.

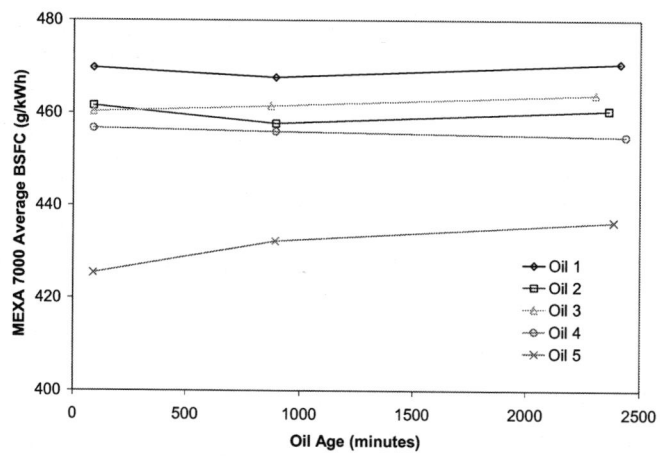

Figure 6 - Phase 2: ECE Oil Performance Based on MEXA 7000 Derived BSFC

Results for the EUDC correlate exceptionally well with the high temperature oil viscometrics, HTHS and Kv_{100} values, with the measured rank order of oils 1 and 2 having the highest fuel consumption with oils 3, 4 and 5 performing better. Once the oils are fully aged the increase in HTHS and Kv_{100} for oil 5 (due to evaporation of volatile fractions) causes a rise in BSFC to a level greater than oils 3 and 4 but still less than that of oil 1 and 2. This trend is as expected based on aged oil sample property analysis. EUDC data suggests a maximum BSFC improvement of 3.1% (between oils 2 and 4) can be obtained for aged oil by varying the high temperature viscometric properties.

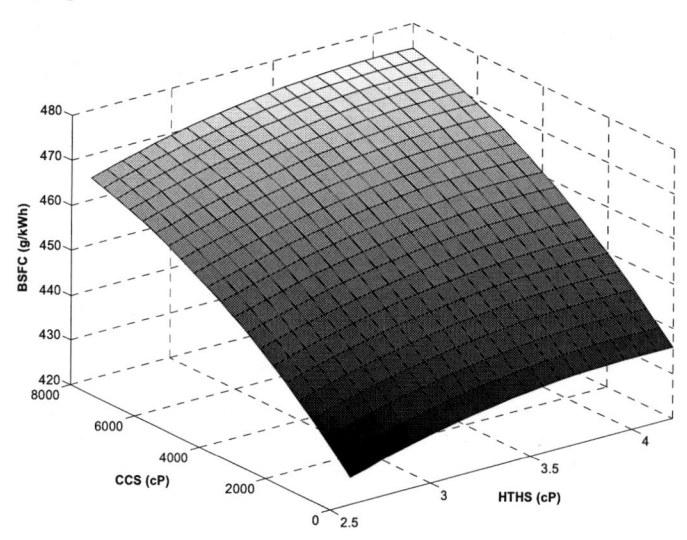

Figure 7 - Phase 2: ECE DoE Results

The effect of CCS, HTHS and Kv_{100} on BSFC for aged oil during the ECE, EUDC and NEDC as a whole was modelled using the Matlab model based calibration toolbox and the results are shown in Figure 7 and Figure 9 respectively. It is evident that CCS is by far the most significant factor during the cold ECE section of the cycle while HTHS and Kv_{100} become dominant during the EUDC. All three factors have a significant effect on total-cycle (NEDC) fuel consumption.

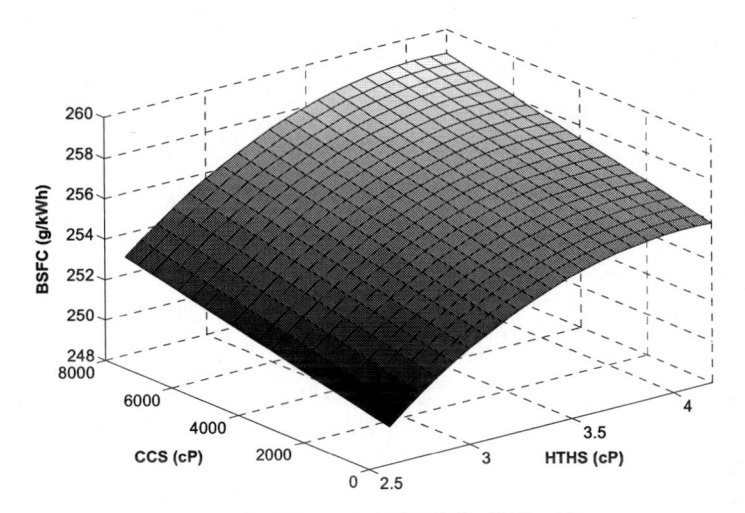

Figure 8 - Phase 2: EUDC DoE Results

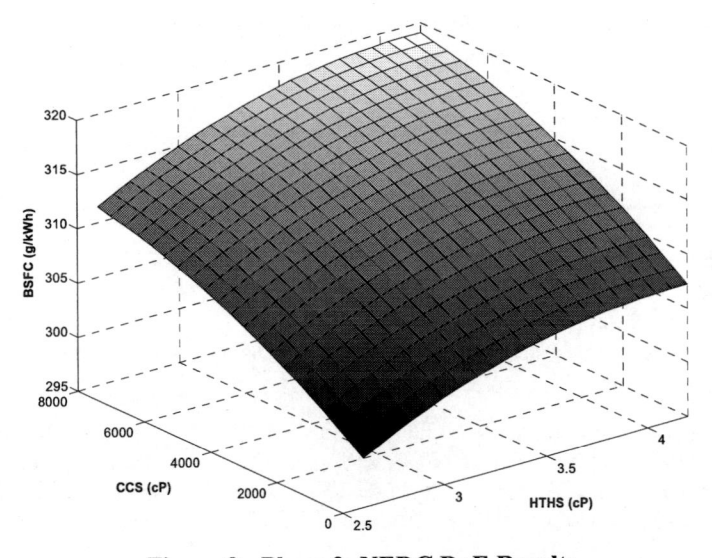

Figure 9 - Phase 2: NEDC DoE Results

8. CONCLUSIONS

The following conclusions can be drawn from the data presented and discussed in this paper:
- CCS is the dominant factor on fuel consumption during the early stages (ECE) of the NEDC drive cycle when the oil temperature is still low

- A reduction in CCS of around 30% resulted in a fuel economy improvement of around 5.5% during the ECE. This advantage is not seen during the hot portions of the test and the overall NEDC gain is slight at 1.5%
- HTHS and Kv_{100} become more dominant with increasing oil temperature as seen during the EUDC portion of the NEDC drive cycle. Reduced values of HTHS and Kv_{100} were observed to improve fuel consumption by 3.1% over the EUDC for a fully aged oil
- A combination of low CCS, HTHS and Kv_{100} values reduced total cycle BSFC, for fully aged oil, by a maximum of 3.2%
- Oil formulated with exceptionally low CCS values performed well when fresh but suffered from an increase in viscosity (and hence BSFC) with age as volatile compounds were lost via evaporation, negating potential benefits
- The derived response model can be used to accurately predict fuel consumption over an NEDC for given oil viscometric properties

ACKNOWLEDGEMENTS

This work has been conducted under the dti "Succeeding through innovation" programme (project number TP/2/5/10036) and the funding and support is acknowledged. Also, the funding from the collaborating partners, the Ford Motor Company and BP and the support received is also acknowledged as well as the permission to publish this paper. The work has been conducted at the Powertrain and Vehicle Research Centre in the Department of Mechanical Engineering at the University of Bath.

REFERENCES

1. Europa, 7th February 2007. *Commission plans legislative framework to ensure the EU meets its target for cutting CO_2 emissions from cars* [online]. IP/07/155. Brussels: Rapide. Available from: http://europa.eu/rapid/pressReleasesAction.do?reference=IP/07/155&format=HTML&aged=0&language=EN&guiLanguage=en [Accessed 20th June 2007].

2. European Automobile Manufacturers Association, 9th July 2002. *European Automotive Industry Further Reduces New Car CO_2 Emissions in 2001* [online]. Brussels. Available from: http://www.acea.be/files/20020709PressRelease.pdf [Accessed 20th June 2007].

3. Ricardo Consulting Engineers Ltd, 2005. *Diesel Passenger Car and Light Commercial Vehicle Markets in western Europe* [online]. Available from: http://www.ricardo.com/engineeringservices/technicalsupport.aspx?page=dieselreport [Accessed 20th June 2007]

4. Owen, N., Gordon, R. 2003. 'Carbon to Hydrogen' Roadmaps for Passenger Cars: Update of the Study for the Department for Transport and the Department of Trade and Industry [online], Available from: http://www.azuredynamics.com/pdf/RicardoUKStudyNov2003.pdf [Accessed 20th June 2007].

CALIBRATION

A novel approach to investigating advanced boosting strategies of future diesel engines

S. Akehurst, M. Piddock
University of Bath, Powertrain & Vehicle Research Centre, UK

ABSTRACT

This paper describes the modelling of advanced air handling configurations using cycle simulation and a "black box" engine, design of experiments approach. The interactions between manifold pressures and temperatures, compression ratio and valve timing events are investigated and described. Optimisation of engine operating conditions for maximum torque and improved fuel consumption is discussed. Finally the development of novel boost system emulation hardware is described. The paper describes ongoing work to develop real time models of air handling components to integrate with the emulating hardware such that future technology combinations of supercharging, e-boosting and sequential turbo charging configurations can be rapidly evaluated in terms of both steady state and transient engine performance.

NOMENCLATURE

K number of factors in experimental design
CAHU Combustion Air Handling Unit
DoE Design of Experiments
EGR Exhaust Gas Recirculation
HSDI High Speed Direct Injection
MBC Matlab Model Based Calibration Toolbox
CAGE Calibration Generation Toolbox
VCR Variable Compression Ratio
VGT Variable Geometry Turbocharger
VVA Variable Valve Actuation
RMSE Root Mean Squared Error

1 INTRODUCTION

Exhaust emissions targets for automotive vehicles are becoming ever more stringent, while at the same time considerable efforts are being made to reduce the fuel consumption (effectively CO_2 emissions), with the aim of reducing air pollution and the output of greenhouse gases.

© IMechE 2007

To realise the required reductions in vehicle CO_2 emissions there is a move to increase the specific power output and thus efficiency of the automotive diesel engine. To this end there is a move towards engine downsizing and the adoption of significantly higher boost levels from air handling systems fitted to the engine. This task is complex and can often result in multiple configurations and iterations of prototype air handling hardware before performance criteria are met.

In an effort to achieve the current and future emissions and CO_2 reduction targets whilst delivering the performance demanded by consumers a number of Diesel powertrain technologies are available or will be available in the future. Examples of such technologies are variable valve activation (VVA) [1, 2], variable compression ratio, sequential/ parallel and electrically assisted turbocharging and supercharging. Each of these technologies is proposed to offer potential benefits to engine performance. Unfortunately a by-product of this is the escalation of the degree of complexity of the powertrain functionality as well increasing the number of variables the engine strategy has to control. This paper describes a simulation based approach to investigate the multiple interactions that can occur within such a complex multidimensional system, such that a reverse engineering approach may be applied to assessing future technology selections and their required performance capabilities.

2   SIMULATION BASED APPROACH

A simulation based approach to this research program was adopted as part of an overriding and ongoing project sponsored by the Engineering and Physical Sciences Research Council entitled "Lean powertrain Development", which aims to develop an integrated approach to powertrain design, performance optimisation and rapid calibration, through a simulation model based philosophy.

The key objective of this work is to shift as much of the powertrain development task in to the virtual environment, thus reducing development time and costs of expensive experimental work. In parallel to the proposed shift towards simulation based development, the program also proposes the integration of both the virtual (simulation) and real world (experimental) techniques by the adoption of hardware in the loop (HIL) testing and emulation of prototype systems where appropriate. The techniques described here allow this to occur.

2.1 Baseline Engine
The engine being investigated in this program of work is a production standard Ford 2.0 litre, in line 4 cylinder, HSDI, diesel engine, fitted with common rail fuel injection, a variable geometry turbocharger and exhaust gas recirculation (EGR). This engine is used in both the Ford Mondeo and Ford Transit type vehicles. Specifications of the baseline engine are given in **Table 1**. The engine is controlled using a Delphi single box ECU system.

Table 1 Baseline Engine Specifications

Bore (mm)	86
Stroke (mm)	96
Con Rod Length (mm)	152
Compression Ratio	19:1
Max Power (kW) @ Rated Speed (rev/min)	95.6, 4000
Max Torque (Nm) @ Rated Speed (rev/min)	320, 1900

2.2 Air Handling Technologies

The air charge handling technologies investigated are highlighted below in some detail

Variable Valve Actuation

Variable valve actuation covers a considerable range of potential technologies and methods of valve motion modification. However the factors manipulated in this work were those easily identifiable to a valve train engineer and those that could be easily manipulated within the engine modelling software, namely:
1. Inlet and exhaust valve phasing with both positive and negative overlap
2. inlet and exhaust valve lift multiplication
3. inlet and exhaust valve duration multiplication

The available range of valve operating envelope is shown schematically in **Figure 1**.

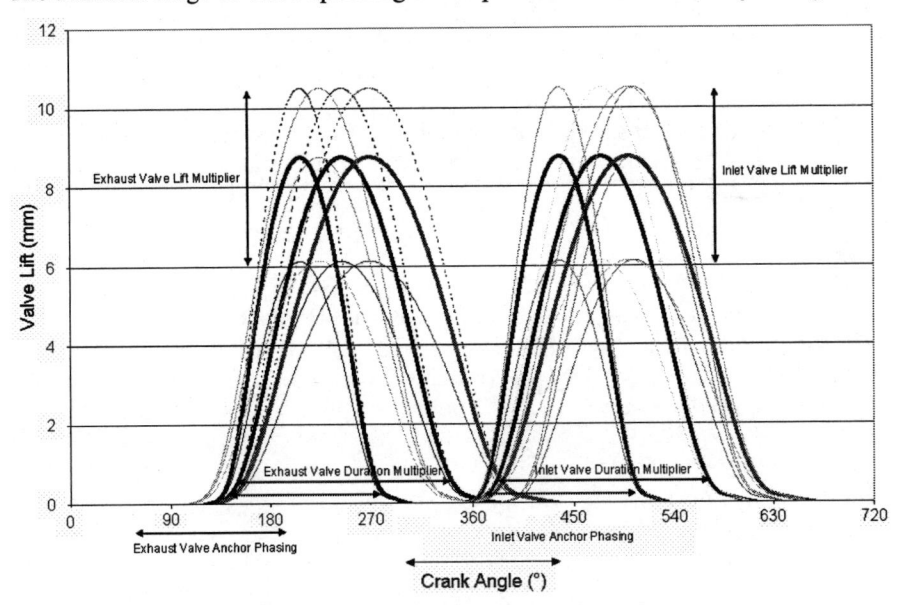

Figure 1 Scope of Variable Valve Actuation

Variable Compression Ratio

The compression ratio of a fixed compression ratio engine is generally arrived at from a set of compromises. At low loads and cranking speeds a high compression ratio is desirable to maximise combustion efficiency and starting quality. While as engine boost levels are increased compression ratio is generally reduced to maintain maximum combustion pressure within safety constraints that prevent damage to engine components. Compression ratio is consider here mainly due to the above stated interaction and such that the compression ratio boost trade offs may be investigated. Variable compression ratio can be achieved by a number of methods, but for brevity in this paper it is considered as a simple change in the clearance volume at TDC. As such the compression ratio may be directly input in to the simulation package. In reality VCR is typically achieved through

1. Tilting engine block design
2. Eccentric crank shaft mechanisms
3. Variable height pistons
4. Variable chamber volume at TDC

Advanced Boosting Concepts

Advanced boosting concepts is a generic term to cover future iterations of turbo charging, such as sequential (two stage) turbocharging, parallel turbocharger arrangements [3-6], electric boosting [7-11], supercharging[12 & 13] and exhaust gas energy recovery[13]. Therefore the engine simulation was tested over a range on operating conditions for inlet pressure and temperature and exhaust back pack pressure. The range 1-4 bar was chosen as this represented a realistic maximum achievable boost level for a two stage turbocharging system with typical compressor pressure ratios for each of the stages.

After Treatment Effects

The modelling work undertaken within this paper takes a simplified approach to after treatment effects on engine air flow. A brief review of literature on the subject [15] was undertaken to ascertain suitable levels of back pressure generated by after treatment devices such as oxidation catalysts, lean NOx traps, SCR catalysts and diesel particulate filters.

Exhaust Gas Recirculation

The work undertaken in this paper is primarily concerned with the maximum performance of the engine rather than the low loads experienced during drive cycle operation. Therefore for brevity EGR performance was not considered in detail, although it is planned to extend the techniques and degrees of freedom modelled in the work to include these effects in the future.

Engine Modelling

The baseline engine was modelled using Wave, an industry standard engine cycle simulation package. This was validated under full load operating conditions. Validation data for the model is highlighted in **Figure 2** showing experimental vs. simulation data for mass airflow, torque, power and intercooler performance. The results clearly show that all significant trends are modelled, while in general errors are due to lack of accuracy in the applied turbocharger maps and the accuracy of interpolation between turbine maps at different VGT settings. The concept described in the following section removes some of these inaccuracies from the simulation.

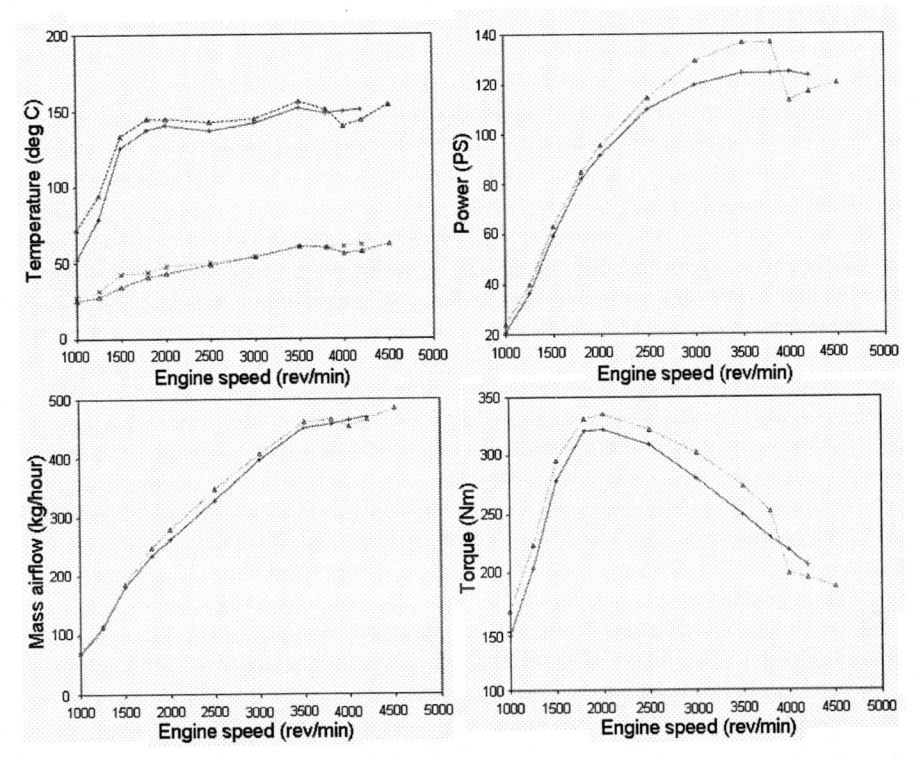

Figure 2 Validation of Limiting Torque Curve Model for Baseline Engine

2.3 Black Box Model concept

Once the baseline engine model was accepted as validated, the model was stripped back to a black box engine, all ancillary air flow components were removed from the simulation, such that the model would now represent the engine block with control over the engine boundary conditions. These conditions were set as
1. Inlet Manifold charge pressure
2. Inlet Manifold charge temperature
3. Exhaust manifold back pressure

Additionally the model was adapted to accept variable conditions for a range of other input parameters on actuators considered to influence the engine air flow. These were:
1. Inlet valve timing
2. Exhaust valve timing
3. Inlet valve opening duration
4. Exhaust valve opening duration
5. Inlet valve lift
6. Exhaust valve lift
7. Compression Ratio

Furthermore the model was set up to run at a range of engine speeds and fuelling rates. Fuelling injectors in the simulation were set to control to AFR (15:1 to 70:1) such that the model would adapt to the airflow through the engine and unrealistic over rich operating conditions would not be simulated. The range of design factors used in the simulation program are highlighted in **Table 2**.

Table 2 Design Factors and Associated Ranges

Variable #	Variable Description	Units	Range	Notes
1	Inlet valve timing	°CA	+/- 20°	Relative phasing to anchor point of 351°
2	Exhaust valve timing	°CA	+/- 20°	Relative phasing to anchor point of 131°
3	Inlet valve opening duration	#	0.8-1.2	Multiplier of standard range 262
4	Exhaust valve opening duration	#	0.8-1.2	Multiplier of standard range 262°
5	Inlet valve lift	#	0.7-1.2	Multiplier of standard lift 8.75mm
6	Exhaust valve lift	#	0.7-1.2	Multiplier of standard lift 8.75mm
7	Compression Ratio	#	15-22	Baseline compression ratio 19:1
8	Inlet Manifold charge pressure	Bar	1-4	
9	Inlet Manifold charge temperature	°K	323-423	Represent a range of Intercooler capabilities
10	Exhaust manifold back pressure	Bar	1-4	
11	Engine Speed	rpm	1000-4500	
12	Fuelling	Kg/hr	0.5-45	

To reduce computation time at each experiment point, settings within the simulation package were optimised to reduce the number of iterative cycles required, these included: Turning off re-initialisation between experiments since only relatively small perturbations of the various actuators occur; Setting a suitable convergence detection tolerance (typically 0.5%) to prevent excessive simulation cycles being computed

Experimental Design
A design of experiments tool exists within the cycle simulation package, however this only offers the potential to undertake Full/Half factorial (2 level) designs or central composite (3 level) designs, which results in rapid growth of experiment numbers as soon as design factors begin to increase considerably. It was therefore decided to adopt the Matlab Model based calibration (MBC) toolbox to undertake the experimental design and evaluation. This offered a range of advantages, although it also required the development of additional code for porting designs and results between the two software packages.

A range of experimental designs were considered from an early stage in the investigation and evaluated using the MBC toolbox. Due to the large number of design inputs adopting a full factorial design would result in an excessively large number of experiments being required for model evaluation, particularly if the aim of achieving accurate response surfaces was to be achieved, see **Table 3**. Partial factorial designs were also considered but resulted in poor potential model fit; therefore a space filling design based on Latin hypercube sampling was used. As the simulation results were built up in phases initial programs of simulation had been undertaken at fixed AFR rates and therefore multiple space filling designs were run at 6 different levels of AFR ranging from 15:1 to 70:1. This resulted in a total number of 3332 experiments, which took just over 35hours to run using the developed model on a high specification PC (dual core 4Gb RAM).

Table 3 Experimental Designs and Run Numbers

		\multicolumn{5}{c}{Number of Experimental Factors (k)}
Experimental Design	Design Formula	2	3	 10	11	12
Full Factorial 2 Level	2^k	4	8	 1024	2048	4096
Full Factorial 3 Level	3^k	9	27	 59049	177147	531441

Model Responses
Response models were developed using the model fitting tools in the Matlab MBC toolbox. A range of models were considered including polynomial surface fits, radial basis functions and neural networks. All of these are readily available and configurable within the toolbox. The use of radial basis functions was rapidly discounted due to the calculation complexity involved with the many dimensions considered.

Table 4 Model responses from Engine Simulation Package

Response #	Response Description	Neural Network Models Accuracy of Model R^2	RMSE	Quadratic RSM Accuracy of Model R^2	RMSE
1	Trapped AFR	0.999	0.3792	0.792	13.367
2	Air flow	0.998	7.6053	0.991	16.393
3	Brake Power	0.999	0.4818	0.997	1.5715
4	BSFC	0.999	0.0415	0.163	1.2001
5	Volumetric Efficiency	0.993	0.0091	0.910	0.0345
6	Max. rate of change of cylinder pressure	0.986	0.3123	0.868	0.9444
7	Brake Thermal Efficiency	0.998	0.4719	0.845	4.2888
8	Heat Transfer Rate %	0.959	2330.8	0.990	1119.3
9	Max. Cylinder pressure	0.999	1.5755	0.995	3.6312
10	Exhaust Temperature	0.998	9.0739	0.955	49.658
11	Engine Torque	0.999	1.5342	0.981	15.402

Table 4 shows the modelled responses and their respective accuracy of fit. Shaded areas indicate the final model selected. Initially it was planned to adopt multi-dimensional response surface models based on quadratic functions, due to their simplicity to evaluate and differentiate when seeking maxima and minima in the optimisation processes. However these were found in many circumstances to have relatively poor accuracy of fit (low R^2) values and hence models based on neural networks were also developed, which had considerably improved fit levels particularly in the highly non-linear or discontinuous areas of the response models.

Figure 3 shows a typical example of the accuracy of fit achieved using the trained neural network to predict torque relative to the observed results from the WAVE simulation code

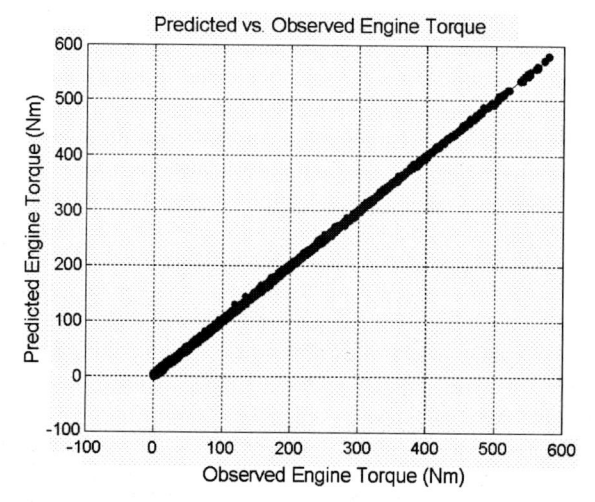

Figure 3 Accuracy of Fit for torque model predicted (Neural Network) vs. observed (WAVE simulation)

Once these models were developed and verified in the MBC toolbox they are readily available for evaluation and manipulation with the Matlab CAGE (calibration generation) component of MBC. This component allows the development of function models based on the imported RSMs and the use of optimisation tools to evaluate the potential tradeoffs between the various actuators discussed. **Table 5** shows some of the calculated function models derived in CAGE.

Table 5 Calculated responses derived from Model responses

Response #	Response Description	Response Equation
1	Exhaust Mass Air Flow	$\dot{m}_{exh} = \dot{m}_{in} + \dot{m}_{fuel}$
2	Compressor Power Requirement	$W_{Comp} = \dfrac{\dot{m}_{in} c_{p,in} T_{amb}}{\eta_{comp}} \left[\left(\dfrac{P_{boost}}{P_{amb}} \right)^{(\gamma-1)/\gamma} - 1 \right]$
3	Turbine Power Delivery	$W_{Turb} = \dot{m}_{exh} c_{p,exh} \eta_{turb} T_{exh} \left[1 - \left(\dfrac{P_{post}}{P_{pre}} \right)^{(\gamma_{exh}-1)/\gamma_{exh}} \right]$
4	Engine Pressure Gradient	$dP_{engine} = P_{boost} - P_{exhaust}$
5	After Treatment Back Pressure	$dP_{aftertreat} = P_1 + k\dot{m}_{exh}$
6	Turbocharger Power Balance	$W_{Turb} - W_{Comp} \geq 0$
7	Supercharger Torque from Engine	$Tq_{Schgr} = \dfrac{W_{Comp}}{\omega_{engine}}$

Model Constraints

Constraint conditions were set up on the model to limit the range of some operating conditions. These were

1. Maximum Combustion pressure (160 bar)
2. Maximum Boost pressure (4 bar)
3. Maximum Back Pressure (4 bar)
4. AFR limit (17:1) for combustion with no visible smoke
5. Maximum exhaust gas temperature pre turbine (1073 °K)

Additionally the effects of various air handling systems can be investigated by the application of some simple constraints. For example turbocharging systems can be represented by the turbine power delivery being constrained to equal the compressor power requirement, while a supercharged system may be represented by removing the compressor power requirement from the crankshaft torque RSM. Further modelling fidelity can be implemented with respect to the efficiency values for the turbine and compressor. These can either be set as arbitrary constants for first pass assessments or assigned there own functional models based on pressure ratio and mass air flow, the data for which can be readily derived from compressor and turbine performance maps. Similarly Specific heat capacities and ratios may be developed as functions of gas temperatures.

3 RESULTS AND ANALYSIS

Figure 4 & **Figure 5** show some typical sections through the response surface models for airflow and torque respectively. For each plot all actuators not displayed are set to the mid point of their respective ranges. Clearly Air flow responses are dominated by engine speed and boost pressure **Figure 4d**, in a fairly linear relationship, while valve timing and lift events have a rather higher order effect, although significant and some considerably non-linear effects on engine air flow. The graphs here indicate the complexity of the multi-dimensional interactions that occur as only 2 variables may be represented in any one graph. Taking **Figure 4a** as an example, this shows that increasing the compression ratio, increases the suction effect on intake stroke and therefore there is a more dense air charge and hence air flow. The Exhaust Valve Anchor (EVA) effect is more complex to explain, but air flow clearly tails off at late EVA and high compression ratios which may be due to increased levels of internal EGR, since less volume for the fresh air charge is available.

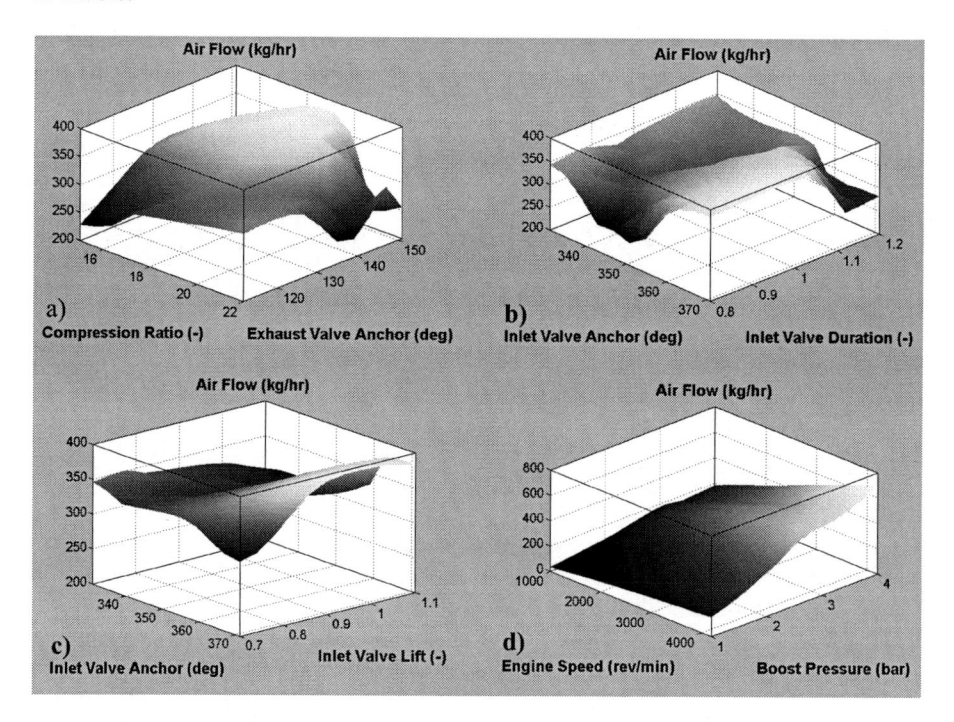

Figure 4 Air flow response models with respect to various model inputs

Similarly, the torque response, **Figure 5**, is dominated by fuelling and engine speed **Figure 5d**, and to a lesser extent boost pressure and exhaust back pressure **Figure 5b** and engine compression ratio and exhaust valve timing **Figure 5a**. Similar figures for other response surfaces may be generated, but are not shown here for brevity and since they are unable to show all the interactions occurring across all actuator ranges.

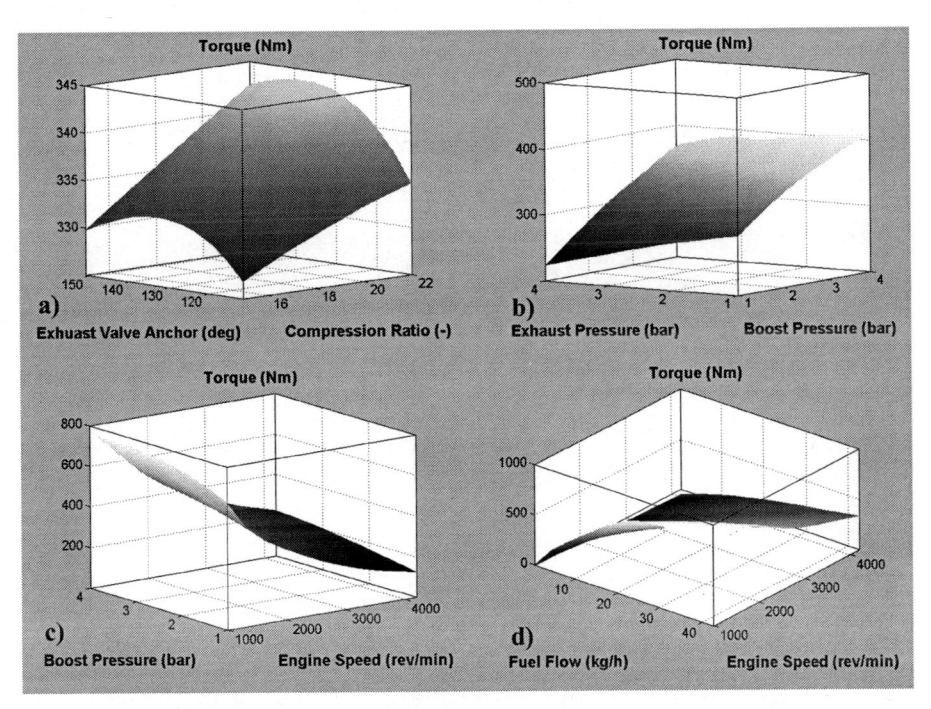

Figure 5 Torque response models with respect to various model inputs

3.1 Limiting Torque Curve Performance

To investigate the performance of the engine at limiting torque curve the models are constrained at a number of fixed engine speeds, while the other model inputs are allowed to vary. Optimisation routines are then implemented to maximise torque for the given speed with respect to the other constraints described. Furthermore the optimisation can investigate the tradeoffs between maximising torque and minimising BSFC. It must be noted that the turbocharger operating envelope is assumed limitless (no surge or choke boundaries). This for the purposes of this paper is valid, since the CAHU will be able to operate unaffected in this way. It is however anticipated that in the latter practical phases of the project a more realistic turbocharger operating envelope will be adopted. As it is currently the simulation provides a 'first cut' and directional trend of which area the practical testing should be aimed towards. **Figure 6** shows some typical Limiting torque curve performance responses, simulating a turbocharged engine (turbocharger power balance constraint) with an assumed efficiency of 50% for both turbine and compressor. **Figure 6a** shows the power curves of the baseline engine compared to the ultimate power curve (maximising power) and a target power curve derived from the objective of raising the torque curve, **Figure 6b**, nominally by 25% and then filling in the low speed torque curve to improve vehicle driveability. **Figure 6c** shows the comparable BSFC lines for the 3 configurations both 'concept' power curves improved considerably on the baseline engine primarily by increasing specific power relative to the nominally constant parasitic losses in the engine. **Figure 6d** shows the exhaust gas temperatures of each configuration indicating that these have been elevated in the 'concept' configurations to approach the prescribed exhaust gas temperature limit.

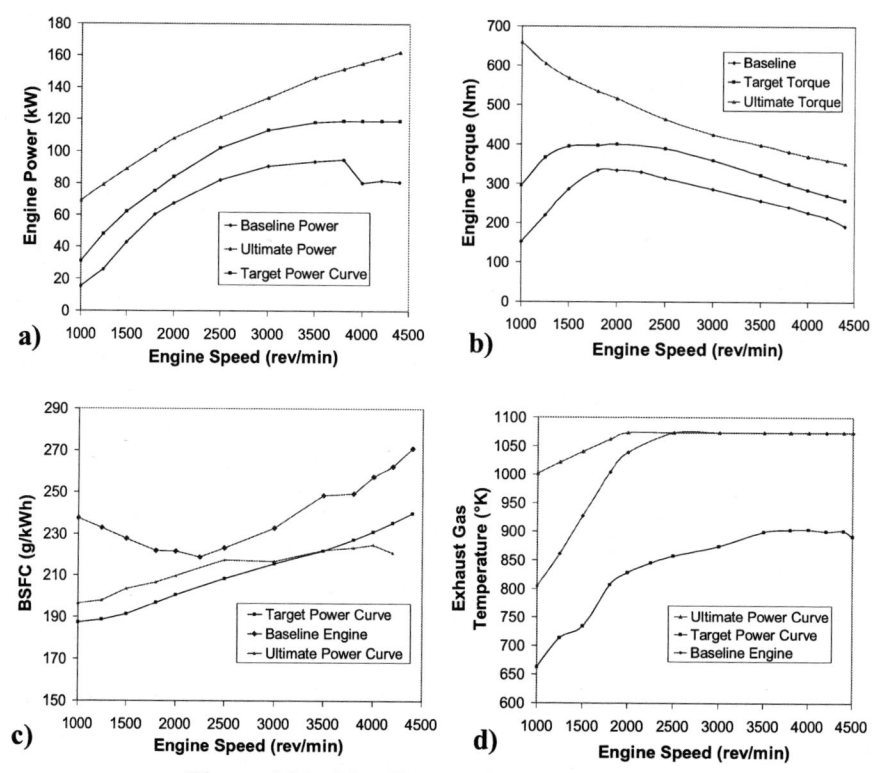

Figure 6 Limiting Torque Curve Performance

BSFC conditions at low speed may be optimistic due to the assumed efficiency of the turbocharger components at low mass flows, it is the aim of the ongoing work to assess this with more fidelity and propose boost systems that may be able to develop high efficiencies across the engine speed range.

Table 6 shows the optimised actuator setting and performance characteristics derived for the target power curve developed above, when optimising to minimise BSFC. At limiting torque curve performance, inlet and exhaust valve phase was found to be relatively well optimised for the baseline engine and thus these were fixed at the default values. It is however expected that these setting may offer some tradeoffs at lower loads. Compression ratio is driven to the lower limit of 15:1 to achieve the desired constraint on maximum cylinder pressure, although it begins to increase from 3500 rev/min onwards. As expected optimum BSFC is achieved at the minimum inlet temperature condition of 320°K indicating high levels of inter-cooling. Exhaust back pressure was as expected near or at maximum to extract maximum exhaust gas energy. Interestingly it appears beneficial for the engine at high boost to operate with lower valve lifts, and a variable inlet duration shortening at higher engine speeds. These influences can be investigated in more detail by further interrogation of the cycle simulation results. Exhaust valve duration was found to change slightly extending at higher engine speeds, although this would need to be carefully considered due to the possibility of valve to piston contact around TDC. Maximum Cylinder pressures became a limiting constraint from around the peak torque point. Air fuel ratio values were significantly high compared to the smoke limiting constraint normally adhered to for limiting torque performance.

Table 6 Performance and Actuator Settings for Target Power Curve

RPM	PIN	TIN	CR	PBP	IVL	EVL	IVD	EVD	IVA	EVA	AFR	BSFC	Power	TEXH	PMAXSI
rev/min	bar	°K	#	bar	#	#	#	#	°CA	°CA	#	g/kWhr	kW	°K	bar
1000	2.71	320	15	4	0.7	0.7	1.2	1	351	131	28	187.3	31	804.9	109
1250	2.88	320	15	4	0.7	0.7	1.02	0.99	351	131	25.8	188.8	48	861.7	145
1500	3.09	320	15	4	0.7	0.7	0.97	0.99	351	131	23.7	191.5	62	927.4	159
1800	3.36	320	15	4	0.7	0.7	0.82	0.99	351	131	21.9	197	75	1004.2	160
2000	3.48	320	15	4	0.7	0.7	0.8	1	351	131	21.4	200.7	84	1038.1	160
2500	3.61	320	15	4	0.7	0.7	0.8	1.07	351	131	21.3	208.4	102	1073	160
3000	3.6	320	15	4	0.7	0.7	0.86	1.1	351	131	22.1	215.7	113	1073	160
3500	3.59	320	15.8	4	0.7	0.7	0.86	1.05	351	131	22.7	221.9	118	1073	160
3800	3.59	320	16.9	4	0.7	0.7	0.83	1.06	351	131	22.9	227.2	119	1073	160
4000	3.59	320	17.9	4	0.7	0.7	0.8	1.06	351	131	23	231.1	119	1073	160
4200	3.55	320	18.2	3.94	0.7	0.7	0.8	1.07	351	131	23.2	235.4	119	1073	156
4400	3.51	320	18.8	3.86	0.7	0.7	0.8	1.08	351	131	23.4	239.9	119	1073	155

4 COMBUSTION AIR HANDLING EMULATION

A further novel approach in this research program is the adoption of hardware that enables the boost and back pressure conditions of the test engine in to be manipulated to accurately represents prototype air handling systems without the requirement to develop the prototype hardware. This is achieved by the use of a combustion air handling system, which supplies a controlled air flow to the engine to desired temperature and pressure (or Mass air flow) set points.

Compressed air is supplied from and industrial compressor system at a steady 7bar the CAHU can then manipulate inlet manifold pressure or mass air flow by rapidly controlling the amount of air that is dumped back to atmosphere by a by-pass system. Additional control exists to manipulate the CAHU exit temperature such that compressor heating effects and inter-cooling may be simulated. Turbine simulation is achieved by manipulating a fast response throttle valve at the exit of the exhaust manifold to generate appropriate levels of exhaust back pressure. The system is controlled by the CP engineering control and data acquisition system to include safety trips which also communicates with the main test cell host system. As such superchargers may be simulated by applying the required torque load of the supercharger to the engine dynamometer. In its entirety it is proposed that the system will allow real time models of air handling systems to be developed in Simulink, executed on a dSPACE real time controller and implemented by the CAHU.

5 CONCLUSIONS

The techniques described in the paper allow for the rapid development of engine models with a large number of degrees of freedom. This has been achieved using the Matlab model based calibration toolbox and a design of experiments approach based on engine cycle simulation. These models are then readily available to undertake optimisation and trade-off investigations at a very early stage in powertrain development. This technique has many advantages over traditional experimental investigations. Primarily these are: reduced cost; reduced time; improved interrogation of engine operation from simulation results and the capability to test beyond real world limits, these later capabilities allow the building of more robust models. Furthermore the techniques developed allow the response models to be exported as Simulink blocks and thus may be used in existing engine model structures. This enables control system development and real time engine modelling to be performed early in the powertrain development task.

6 FUTURE WORK

The work detailed in this paper investigates only the performance of the engine over the limiting torque curve operating area, i.e. the ultimate power potential of applying these devices. An additional program of work is being undertaken to investigate engine performance requirements at partial loads equivalent to those seen in standard drive cycles such as the NEDC and the FTP75.

Additionally a phase of work is proposed to validate the simulation findings to date. This is being undertaken by the adoption of the combustion air handling unit for emulating and thus prototyping advanced air handling systems. This emulating hardware will be used in conjunction with real time models of the proposed air handling systems in a Hardware-in-the-loop (HIL) configuration.

Supplementary data extracted from these experimental phases of the work will be imported in to the model design. This will also be use for the development of emissions response models, particularly with respect to NOx so that the interactions of these systems with EGR can be more rigorously investigated.

Furthermore the complex interactions of these advanced air handling systems are difficult to model physically in a real time environment. The methods proposed here allow the rapid generation of response surface models that may then be easily ported into existing Simulink based real time engine model structure to represent engine performance criteria such as volumetric efficiency.

7 ACKNOWLEDGEMENTS

The authors would like to acknowledge the Engineering and Physical Sciences Research Council for funding this research under projects EP/C540883/1 and EP/C540891/1 and also the Ford Motor Company for assisting with the research in terms of initial model development and advice throughout the project.

REFERENCES

1. H. Fessler et al.- 2004 – 'An Electro-Hydraulic "Lost Motion" VVA System for a 3.0-Liter Diesel Engine' - SAE Technical paper 2004-01-3018
2. T. Lancefield – 2003 – 'The influence of variable valve actuation on the part-load fuel economy of a modern light-duty Diesel engine' - SAE Technical paper 2003-01-0028
3. X. Tauzia, J.F. Hetet, P. Chesse, G. Grosshans, and L. Mouillard – 1998 – 'Computer aided study of the transient performance of a highly rated sequentially turbocharged marine diesel engine' - Proc IMechE Vol 212 Part A, pg 185, 1998.
4. Z. Ren,, T. Campbell,, J. Yang - 1998 - 'Investigation on a computer controlled sequential turbocharging system for medium speed diesel engines' - SAE paper 981480
5. M. Mario – 2000 – 'Some problems of sequential turbocharging of diesel engines for racing boats' - SAE paper 2000-01-0528
6. R.S. Wijetunge, M. Criddle, J. Dixon and G. Morris – 2004 – 'Comparative Performance of Boosting Systems for a High Output, Small Capacity Diesel Engine' - Fisita 2004 World Automotive Congress, F2004F195, 2004
7. Y. Yamashita, S. Ibaraki and H. Ogita – 2006 – 'Development of electrically assisted turbocharger for diesel engine' - 8th International conference on Turbochargers and Turbocharging IMechE Combustion Engines & Fuels Group p 147 – 155, ISBN 1 84569 174 1
8. O. Ryder, H. Sutter and L. Jaeger – 2006 – 'The design and testing of an electrically assisted turbocharger for heavy duty diesel engines' - 8th international conference on Turbochargers and Turbocharging IMechE Combustion Engines & Fuels Group p 157 – 166, ISBN 1 84569 174 1.
9. T. Kattwinkel, R. Weiss, and J. Boeschlin – 2003 – 'Mechatronic Solution for electric turbocharger' - SAE paper 2003-01-0712
10. K. Fieweger, H. Paffrath and N. Schorn – 2002 – 'Drivability assessment of an HSDI Diesel Engine with electrically assisted boosting systems' – IMechE, Conference on Turbochargers and Turbocharging paper C602/009/2002
11. S.M. Shahed – 2006 – 'An Analysis of Assisted Turbocharging with light hybrid powertrain' - SAE paper 2006-01-0019
12. N. Ueda, N. Matsuda, M. Kamata, H. Sakai. and H. Kanesaka – 2001 - 'Proposal of New Supercharging System for Heavy Duty Vehicular Diesel and Simulation Results of Transient Characteristic' - SAE paper 2001-01-0277
13. G. Cantore, E. Mattarelli and S. Fontanesi – 2001 – 'A new concept of supercharging applied to high speed DI diesel engine' - SAE paper 2001-01-2485

14. T.E. Darlington and A.A. Frank – 2003 – 'Exhaust Gas Driven Generator with Altitude Compensation for Battery Dominant Hybrid Electric Vehicles' - SAE paper 2003-01-3276
15. M. Masoudi, A. Heibel, P.M. Then – 2000 - 'Predicting Pressure Drop of Wall-Flow Diesel Particulate Filters - Theory and Experiment' - SAE Paper 2000-01-0184

Smart calibration – turbocharger speed limitation – an example

M. Wellers, B. Carnochan, F. Ewen, U. Genc, Eike Martini
AVL Powertrain UK Ltd., UK

ABSTRACT

A model-based turbocharger protection function is proposed and discussed. Turbocharger speed estimation is required to prevent turbocharger over speeding. A model of the compressor behaviour is calculated from the measured data to identify a compressor map and a detailed explanation starting from the modelling up to the calibration is provided. Turbocharger speed limitation is an example of the AVL Smart Calibration ^{TM} concept (1). This paper describes a novel approach to vehicle optimization of turbocharger speed limitation. It shows that improved processes for function development and calibration can be developed based on the know-how of the engine, combustion and electronics in conjunction with improved and adjusted tools.

1. INTRODUCTION

In recent years modern vehicles have achieved high levels of customer satisfaction with respect to performance, quietness and fuel economy. However, customer demands require further improvements especially in performance and driveability. With evermore tighter emission standards to meet, the calibration tasks are becoming more onerous. In order to cope with these demands a smart calibration process has been developed. The targets are set to improve efficiency, create structured and traceable calibrations, whilst also improving quality and permitting reusability. While most established engine calibration processes are divided into two major steps: steady state and transient, the aim of the smart calibration activities are to combine the power of steady state testing with new concepts of transient testing to create new powertrain calibration processes.

There have been several publications on mean value modelling of diesel engines with turbochargers and EGR (2, 3, 4). Such models lead model-based functionalities in modern ECU's and open doors to new calibration processes. It is essential that already during the development phase of these functionalities the calibration of the values and maps are kept in mind. Modelling and identification processes have to go

© AVL Powertrain UK Ltd., 2007

hand in hand. An example of such a way of development is shown in this paper using the turbocharger speed limitation. Starting with modelling of the compressor and the turbocharger speed estimation, the calibration of the map is discussed and finally the limitation of the turbo speed introduced.

2. TURBOCHARGER SPEED ESTIMATION

Modern Diesel engines combine external exhaust gas recirculation and turbo supercharging to meet emission restrictions as well as performance targets. In order to achieve good performance of turbocharged engines the turbocharger itself is used up to its limits. Especially critical is the turbocharger speed. The turbocharger speed limit is especially critical at altitude. Therefore the usual process is to test the turbocharger behaviour in high altitudes on vehicle test drives in order to ensure the overspeed protection is obtained. This experimental calibration method is time consuming.

Starting from a steady state assumption of the compressor a model based limitation can be achieved. Already in 2001, the basis of this functionality was patented (5). Nowadays similar functionalities can be found in modern ECU's.

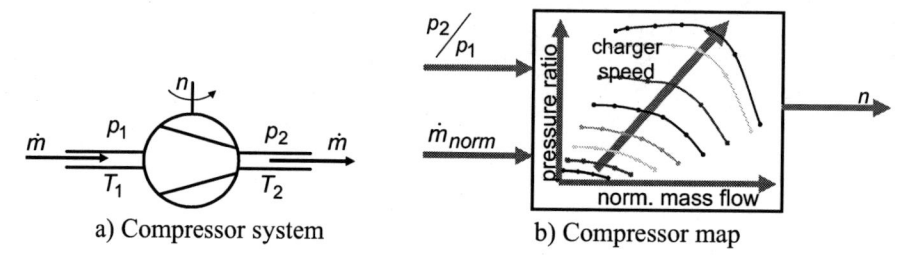

a) Compressor system

b) Compressor map

Figure 1: Compressor

In figure 1a the compressor system is shown including the turbocharger speed n, the compressor inlet conditions pressure p_1, temperature T_1 and the corresponding compressor outlet conditions of p_2 and T_2. The air mass flow going in and out of the compressor $\dot{m}$ is assumed to be the same. The compressor works at steady state and the behaviour can be described by the compressor map shown in figure 1b. The normalised air mass flow

$$\dot{m}_{norm} = \dot{m} \cdot \sqrt{\frac{T_1}{T_{ref}}} \cdot \frac{p_{ref}}{p_1} \qquad (2.1)$$

is calculated using the reference values for intake temperature T_{ref} and intake pressure p_{ref}.

Using the calculation of the normalised air mass flow and the pressure ratio of the compressor p_2/p_1 the turbocharger speed can be calculated with the compressor map (shown in figure 1b).

3. TURBOCHARGER SPEED LIMITATION

In order to limit the turbocharger speed it is possible to use the turbocharger map. Assuming a maximum speed (n = max) of the turbocharger a maximum pressure ratio is determined for each normalised air mass flow, see figure 2.

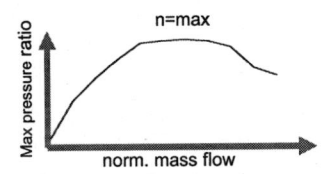

Figure 2: Maximum pressure ratio for a given maximum turbocharger speed

The maximum pressure ratio is transformed by

$$p_{max} = \frac{p_{max}}{p_1} \cdot p_1 \qquad (3.1)$$

into a maximum pressure after the compressor p_{max}.

This maximum pressure is limiting the boost pressure setpoint in this model based approach. An alternative is to limit the fuel set point. For the calculation of maximum boost pressure the measured signals p_1, T_1, and $\dot{m}$ are necessary as well as the maximum pressure ratio curve shown in figure 2. The signals p_1, and T_1 can be taken from the ambient pressure and the ambient temperature. Some vehicles have installed sensors after the air filter – in the position before the compressor. The mass air flow is measured in modern vehicles, or can be calculated using the engine speed, boost pressure and temperature as well as other signals like valve timing. The introduction of this functionality does not require additional sensors in the vehicle.

The benefit of this functionality is the consideration of changing ambient conditions (like ambient pressure and temperature) within the layout of the functionality. By following the described approach the guarantee of the limitation of the turbocharger speed even in high altitude is given. Once the map is calibrated at sea level or other reference conditions, in theory no additional calibration has to be done. Nevertheless, a verification of the functionality is essential.

4. IDENTIFICATION OF THE COMPRESSOR MAP

The calibration task of the turbocharger speed limitation functionality is simplified to the calibration of the curve shown in figure 2. In the first step the compressor map of the turbocharger supplier can be used. However, the compressor map provided by the supplier is often not appropriate for this functionality due to the installation of turbocharger in the vehicle. Furthermore, the installation of the mass air flow, pressure and temperature sensors have to be taken into account. If the pressure sensors and the temperature sensor are not near by the compressor, additional maps can be used to take account of losses in the air path. Nevertheless, the maximum pressure ratio curve can incorporate these losses by using the following identification method.

Considering the calibration task of this functionality a simplified identification is introduced. Measured signals from the test bench or the vehicle are used for the identification of the map including a turbocharger speed sensor. These sensors are often available on some turbochargers during the calibration process.

Figure 3 shows the preferred approach for the identification of the curve using the measured data. The aim is to create a curve with defined breakpoints $x_A, x_B, \ldots x_E$. For each of these breakpoints the corresponding values $y_A, y_B, \ldots y_E$ have to be identified.

For example, in order to define the value y_C of the breakpoint x_C all the measured data are considered with the condition $x_B \leq x_i \leq x_D$. Where x_B and x_D are the breakpoints to the left and right of the considered breakpoint x_C.

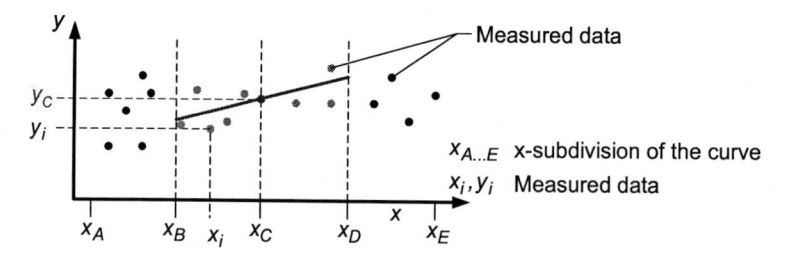

Figure 3: Identification of a curve using measured data

A best fit straight line using the Least Squares method has to be calculated in the region $x_B \leq x_i \leq x_D$. The value y_C calculates as the value of the best fit straight line at the breakpoint x_C.

When this method is extended to two independent dimensions, such as air mass flow and pressure ratio, and one dependent dimension such as turbocharger speed, a whole compressor map can be identified.

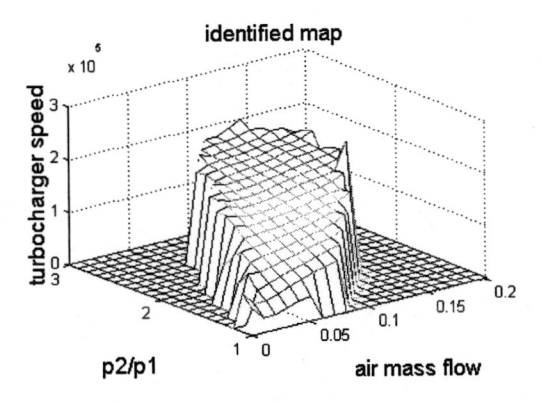

Figure 4: Identified compressor map

Figure 4 shows an example of the identified compressor map. Sufficient turbocharger speed, pressure ratio and air mass flow data have to be measured in order to create a complete map. To achieve high quality results, several repeated data sets have to be collected covering all areas of the map. These data files can be collected very quickly in the vehicle by transient acceleration in all gears. The large amount of data ensures a good fit. For the discussed functionality of a turbocharger speed limitation the turbocharger has to reach these limits otherwise the data only covers the uncritical areas of the compressor map.

On the basis of the identified map a horizontal plane at the maximum turbocharger speed provides a relationship between the maximum pressure ratio and air mass flow. The result is typically represented as shown in figure 2. This calibrated curve in conjunction with formula 3.1 is key for producing the model-based approach to limit the maximum turbocharger speed by using maximum boost pressure

5. VERIFICATION AND DISCUSSION OF RESULTS

From the above it has been described how the identified compressor map is used to limit the turbocharger speed and the boost pressure in the model-based approach.

Following this separate data files are collected in vehicle to verify the model of the turbocharger speed. Figure 5 plots a section of simulated and measured verification turbocharger speed data. Especially a good match at high speed is required for the satisfactory performance of the functionality.

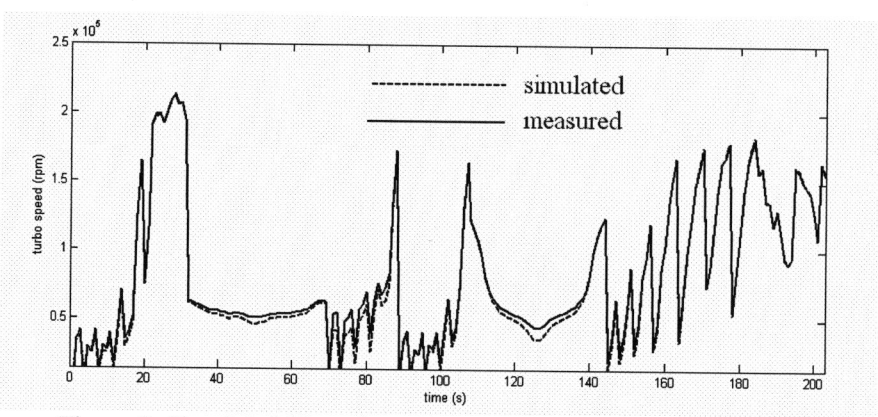

Figure 5: Simulated (dotted) and measured (solid) turbo speed using validation data

Simulated value tracks very well the measured value for most of the operating range especially at high speeds. At low speed the differences between the simulation and the measured data are more significant due to high sensitivity to pressure and air mass flow changes.

Using the identified and validated estimation of the turbocharger speed in the next step the limitation is discussed. In figure 6 is shown the unlimited and limited turbocharger speed. The measured unlimited turbocharger speed is shown as a solid line. The speed exceeds the limit of 190.000rpm. With the limitation functionality applied the turbocharger speed stay within the 190.000rpm limit as shown as a dotted line. In several applications this method was tested and verified in vehicles at altitude.

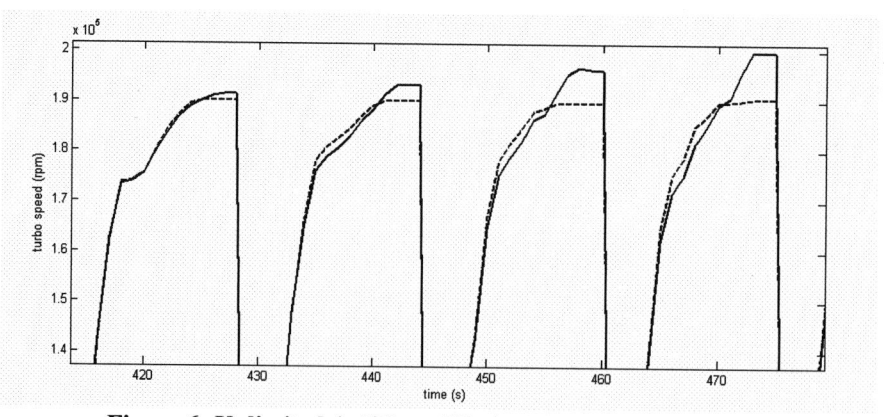

Figure 6: Unlimited (solid) and limited (dotted) turbo speed

6. SUMMARY AND CONCLUSION

In this paper a model-based approach for developing and calibrating a turbocharger speed limiter is discussed. Also this study is an example of AVL Smart Calibration ^{TM} concept (1). Smart Calibration approach not only allows model based ECU function development but also includes calibration and validation of the functions in the model based paradigm. This paper demonstrates by starting with first principles, model based functions can be developed, calibrated and validated for series production.

REFERENCE LIST

(1) Martini. E. et. al. (2007) Smart Calibration - Successful combination of process, know-how and tools. MTZ, 02/2007. ISSN 0024-8525 10814, Vieweg Verlag.
(2) M. Kao, J.J. Moskwa (1995) Turbocharged Diesel Engine Modelling for Nonlinear Engine Control and State Estimation. ASME Journal of Dynamics Systems, Measurement and Control, 1995, Vol.117.
(3) P. E. Moraal, I.V. Kolmanovsky (1999) Martini.- Turbo Charger Modelling for Automotive Applications. SAE, 1999, Paper 1999-01-0906
(4) J. P. Jensen, A. F. Kristensen, S.C. Sorensen, N. Houbak, E. Hendricks (1991) Mean Value Modelling of a Small Turbocharged Diesel Engine. SAE, 1991, Paper 910070.
(5) Wellers, M. and Elicker. M. (2001). Verfahren zur Begrenzung eines abgasturboladers fuer eine Brennkraftmaschine. DaimlerChrysler Patent Amtl. Aktenzeichen 10160469.6.

The effect of multiple fuel-injections on emissions of NOx and smoke with partially-premixed diesel combustion in a common-rail diesel engine

P. Eastwood[1], Y. Hardalupas[2], T. Morris[1], A.M.K.P. Taylor[2], K. Tufail[1], T. Winstanley[1]

(1) Diesel Powertrain Development and Integration Group, Ford Motor Company Ltd., UK
(2) Mechanical Engineering Department, Imperial College London, UK

ABSTRACT

The aim of this study is to quantify, using partially-premixed combustion, the effect of fuel-injection schedules on the NOx-smoke trade-off. Steady-state tests were conducted, between 1200 - 2000rpm and 3bar BMEP with conventional (*two*-injections) and partially-premixed combustion (*three*-, *four*- and *five*-injections). Compared with conventional combustion, partially-premixed achieved up to 61% reduction in NOx, for similar smoke. However, *four*-injections, compared with *three*- and *five*-injections, gave the lowest NOx-smoke trade-off and 19.5% increase in Specific Fuel Consumption (SFC) compared with conventional combustion. The most effective fuel distribution was achieved when high quantities were injected earlier and smaller ones later *i.e.* closer to TDC. The last fuel-injection quantity made the greatest contribution to combustion noise. The general level of *three*-injection partially-premixed combustion noise, compared with conventional at 2000rpm and 2bar BMEP, was more spectrally uniform with frequency.

1. INTRODUCTION

Conventional diesel combustion consists of two sequential phases. First, combustion is governed by the rate at which fuel and air react *i.e.*, the rate is "kinetically controlled". Secondly, combustion is governed by the rate at which fuel and air mix *i.e.*, the rate is "mixing controlled" [1-3]. (We choose to distinguish carefully here, between a "phase" and a "flame"; because the "diffusion" phase nevertheless comprises a premixed flame [4]). In both of these phases, a flame is established, but the implications for emissions of NOx and smoke (soot) are different. Generally speaking, NOx is held to be a strong function of flame temperature (among other variables), whereas smoke forms predominantly in the second phase (diffusion), wherever the Air-Fuel Ratio (AFR) is locally rich [1]. The question is: how can we control – *i.e.* reduce - both of these emissions simultaneously?

© *Ford Motor Company Ltd, 2007*

A potential method of simultaneous control is to promote a kinetically-controlled combustion *i.e.*, minimizing the diffusion phase [5-7]. Strictly speaking, kinetically-controlled combustion can exist in many desired and undesired forms *e.g.* misfire, flame quenching, premix, partially-premixed, Homogeneous Charge Compression Ignition (HCCI) [7]. The particular form of interest for us here is to achieve a desired HCCI-like combustion *e.g.* partially-premixed, which results in low emissions of NOx and smoke.

The present study used a conventional diesel engine designed for a medium sized modern passenger car / light-duty truck *i.e.* High Pressure Common Rail (HPCR), Fuel Injection Equipment (FIE), Direction Injection (DI), Variable Geometry Turbocharger (VGT) and Exhaust Gas Recirculation (EGR). The aim was to achieve partially-premixed combustion under steady-state operation, without modification to any of the engine hardware. The objective of this study, using multiple fuel-injection events *i.e. three, four* and *five* separate injections, was to quantify the effects of the number of injection events on emissions of NOx and smoke, and on specific fuel consumption (SFC), *via* modifications made to engine calibration. A potential operating range (engine speed) for partially-premixed combustion was then examined where injection quantity distribution of each event was to be investigated. Analysis of combustion noise and frequency content was also to be conducted whilst operating under partially-premixed combustion utilizing *three* injection events. This paper is divided into four sections: namely, a review of previous work; a description of experimental facilities / tests conducted; and, finally, a discussion of test results, highlighting our conclusions.

2. KINETICALLY-CONTROLLED COMBUSTION LITERATURE REVIEW

This section briefly describes the differences in types of kinetically-controlled combustion and then evaluates some aspects of diesel engine fuelling control on partially-premixed combustion and formation of emissions.

2.1 Conventional and non-conventional diesel combustion

In a *conventional* diesel engine, where injection is made close to piston Top Dead Centre (TDC), the combustion invariably takes place in *two* phases, commonly referred to as "premix" and "diffusion". Extensive research has been conducted in the field of conventional combustion, as reflected in ref. [4, 8-10]. *Non-conventional* diesel combustion is developing rapidly at the moment and takes many forms [5, 11-27]: here, the present paper is concerned with early injection *i.e.*, "significantly" before TDC. In this case, the fuel injection is completed before Start Of Combustion (SOC), giving rise to a number of ways in which the fuel may burn. In so-called HCCI combustion, all the fuel is thought to be fully mixed on molecular scales, and the charge strength is - spatially - almost homogeneous and lean, before combustion. But cases exist in which the charge strength is not spatially uniform but nevertheless everywhere lean, and these may result in combustion designated "premixed". A further possibility is where the fuel is not only insufficiently mixed with air, and not fully spatially homogenized, but also rich in places: in this case, although there is a "premix phase", there may also be a diffusion flame; and hence we choose to describe this combustion as partially-premixed. The difference relative to conventional diesel combustion is that this diffusion flame takes place *after* all the fuel has been injected. In these situations, the distinction between what is mixing and kinetically controlled phase may become difficult to define – certainly in the absence of optical measurements. We use "kinetic control" as an umbrella term to describe premixed and partially-premixed combustion, as well as HCCI.

2.1.1 Kinetically-controlled combustion
Kinetically-controlled combustion is brought about by early injections; which lead to long ignition delays, short combustion durations, and rapid heat releases. Usually, less smoke is emitted than in conventional combustion: this is the expected result due to greater premixing of the reactants and gives rise to either a formation period that is shorter than in conventional combustion, or because higher temperatures enhance soot oxidation rates [24]. In order, therefore, to minimize the formation of NOx and smoke during partially-premixed combustion, early injection (relative to conventional combustion) *e.g.* 90°bTDC and high EGR rates are scheduled with a view to prolonging the Ignition Delay (ID): enough time is then allowed for fuel and oxygen to mix before SOC, and formation of NOx is minimised, by avoiding high charge temperatures [7, 28-30].

Therefore, to attain partially-premixed combustion (under lean operation) the main parameters to adjust are the local AFR (distribution of equivalence ratio) [17], the local mixture temperature (as partly determined by fuel injection schedule) and the heat capacity of the reactants (through EGR) [13, 20]. The challenge, in the field of kinetically controlled combustion research, is to produce low emissions of NOx and smoke without degradation in SFC and combustion noise [31]. This challenge can be explored by experimenting with injection quantity scheduling.

2.2 Injection scheduling and premix combustion
In order to improve mixing, *i.e.*, to allow ample homogenisation time, the fuel is injected early, significantly before TDC, an action which exploits, additionally, the prolongation this bestows on the ignition delay. In achieving partial premix, the ignition delay must be longer than or equal to the homogenization time of the mixture [13, 20]. Thus, early injection meets the criterion of minimizing smoke. It should be noted that, by invoking early injection, fuel is injected into a low-temperature charge: this is different from conventional combustion, wherein – after SOC – liquid fuel is usually injected towards a premix flame [14]. Although the local AFR becomes spatially more uniform with earlier injection (due to long Ignition Delay, ID), inhomogeneities may remain which give rise to locally high temperatures and thus to conditions that enhance emissions of NOx. One method of avoiding this is to promote further mixing by introducing the fuel in multiple events [18-19, 26, 32-33]. Multiple injection events may result in a "less inhomogeneous" distribution of fuel, as opposed to a single injection event, whereby incoming fuel continually replenishes a rich-burning, soot-producing region - even though, globally, conditions remain comfortably lean.

Therefore the number of injections, their fuel mass (distribution) and SOI are variables which provide degrees of freedom to shape and control local AFR and Rate Of Heat Release (ROHR) and thus to determine conditions before and during combustion [10]. More importantly, the above variables give rise to the challenge of striking a balance between achieving low temperature combustion with minimum emissions and controllability, with a desire to achieve a specific type of kinetically-controlled combustion *i.e.* partially-premixed.

2.3 Final remark

Most of the reported investigations in the literature have used custom-built engine hardware. Multiple-fuel injection investigations have been conducted and reported extensively in the open literature. However, fuel quantity distribution between multiple injections during partial premix requires further research. The present investigation uses current-production hardware, for which the fuel injection events are the only means of realizing partially-premixed combustion. With this approach, we will demonstrate the limitations in contemporary engine hardware for premixed combustion, and try to establish optimum trends for multiple fuel-injection quantity distribution.

3. TEST FACILITY AND EXPERIMENTS

This section documents the test-facility, powertrain configuration used and experimental tests conducted. The measurements were recorded by a computer data-logging system and subsequent processing.

3.1 Test facility and powertrain configuration

3.1.1 Transient test-cell dynamometer

AVL EMCON series 302 dynamometer controllers with an Elin-AVL APA 208/3.5 series AC dynamometer were employed in conjunction with AVL Puma v5.6 (ISAC) software to conduct steady state test-points. The Puma system recorded data at 100Hz and Engine Control Unit (ECU) recorded engine management data at 20ms resolution. Emissions were measured using an AVL 415 smoke meter and gaseous emissions were measured using Horiba Mexa 7000 gas analysers. All emissions were sampled at engine-out location, (rather than at the tail-pipe exit). Test room equipment also consisted of a portable AVL-Indiwin unit, to conduct combustion analysis, by employing Kistler in-cylinder pressure transducers.

3.1.2 Powertrain configuration

The diesel engine used was typical of the contemporary passenger-car market: Direct Injection (DI) High Pressure Common Rail (HPCR), 16-valve I4 configuration. The engine featured a VGT, cooled EGR and air-to-air inter-cooling. EGR and VGT actuators were controlled by electric and vacuum motors, respectively. Fuel Injection Equipment (FIE) employed a common rail system capable of delivering 1600bar maximum rail pressure through piezo-electric controlled injectors. Steady-state tests simulated a vehicle with the inertia class of a typical passenger car *i.e.*, C/CD inertia class.

3.2 Experiments conducted

Four main test cases, each represented by steady-state engine speed and load, were conducted according to typical operating conditions of the New European Drive Cycle (NEDC). All tests employed two modes of combustion, namely *conventional* i.e. two injections (baseline), and *partially-premixed*, using multiple injections i.e. three, four and five injections. Separate and combined variations in multiple injections were evaluated *via* a single-factor testing matrix. It should be noted that all of the above injection events occur before TDC. In the case of multiple-injection events, the injection closest to TDC is designated *late*; that furthest from TDC is designated *early*. Details of all four tests conducted are listed below:

1. Determining number of injections (*three-*, *four-* and *five-*injections) for optimum partially premixed combustion and for comparison with conventional combustion (*two-*injection) at 1200rpm and 3bar BMEP.
2. Determining four injection (partially-premixed) part-load operating range at 1200, 1500 and 2000rpm and 3bar BMEP.
3. Determining four injection fuel quantity distribution for optimum partially-premixed combustion at 1500rpm and 3bar BMEP (see Table 1 below)
4. Comparison between conventional (two injections) and partially-premixed (*three-*injections) combustion noise and frequency at 2000rpm and 2bar BMEP; Case A: conventional combustion with 400bar Rail Pressure, Case B: partially premixed combustion with 500bar Rail Pressure and Case C: partially premixed with 400bar Rail Pressure

Table 1 - Test cases considered

Case Partially-premixed	Injection 1 Quantity [mg/st] (SOI CA °bTDC)	Injection 2 Quantity [mg/st] (SOI CA °bTDC)	Injection 3 Quantity [mg/st] (SOI CA °bTDC)	Injection 4 Quantity [mg/st] (SOI CA °bTDC)	Total Quantity [mg/st]
I	4.6 (39.5)	3.5 (29.5)	2.5 (19.5)	1.5 (9.5)	12.3
II	4.3 (39.5)	3.5 (29.5)	2.8 (19.5)	2.3 (9.5)	13
III	3.7 (39.5)	2.4 (29.5)	3.1 (19.5)	2.9 (9.5)	13
IV	3.2 (39.5)	3.2 (29.5)	3.2 (19.5)	3.2 (9.5)	13
V	2.0 (39.5)	2.7 (29.5)	3.3 (19.5)	4.3 (9.5)	13
VI	4.6 (39.5)	3.5 (29.5)	2.5 (19.5)	1.5 (9.5)	12.3
Conventional					
bl			1.1 (7)	9 (-2)	10

4. RESULTS AND DISCUSSION

Figure 1 compares performance and emissions (*i.e.* BMEP and NOx / smoke), at engine speed of 1200rpm and 3bar BMEP load, between conventional- *(two-injections)* and partially-premixed combustion with varying number of fuel injections, *i.e. three, four and five*. Figure 1(a) presents BMEP and measured fuel quantity as a function of injection quantity. Recall that the intended BMEP is 3bar, as highlighted on the graph, and the "baseline" injection schedule – with conventional combustion, that is to say pilot and main – is also plotted for comparison. It can be seen, from Figure 1(a), that as number of injections is increased, a larger deviation from the target BMEP is observed *i.e. five*-injections show a 5% reduction (from target of 3bar BMEP). Deviations in the total fuel-quantity (*i.e.* delivered/measured), resulting from multiple-injections, suggests that a variation in combustion efficiency (*i.e.* SFC) exists.

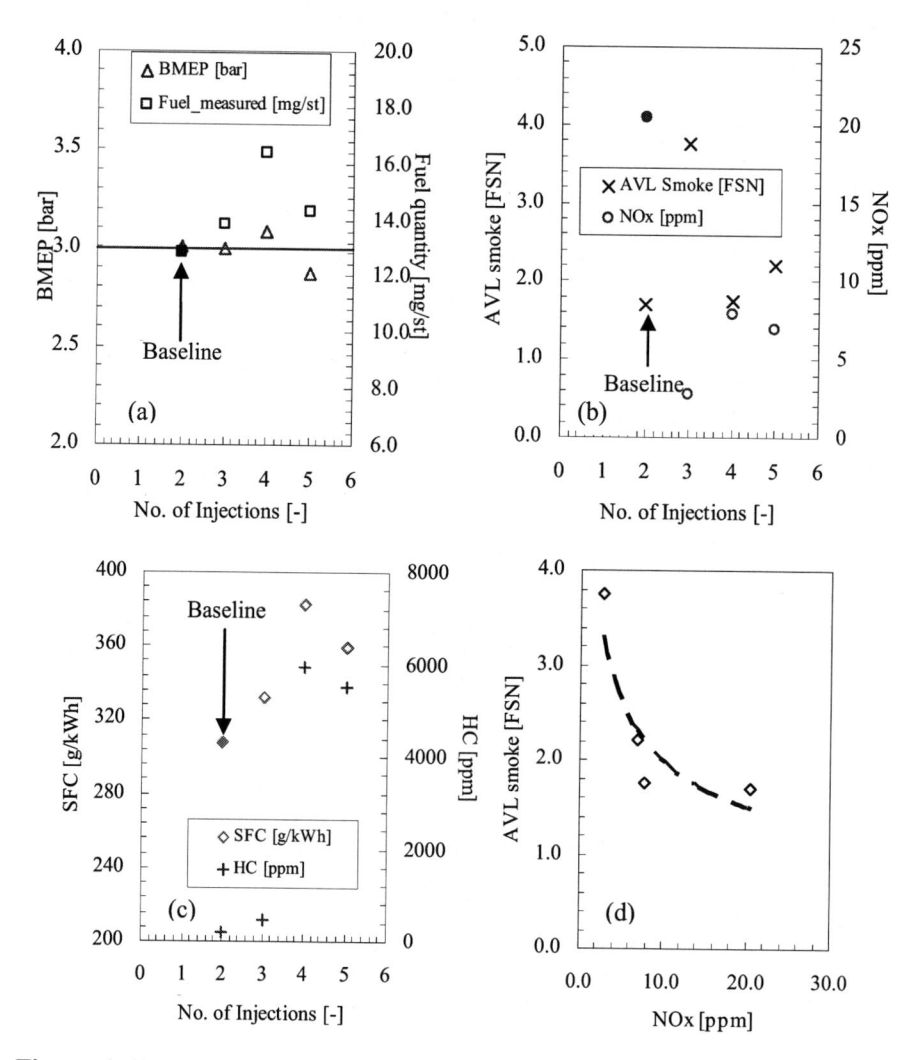

Figure 1 The effect of number of injections on engine performance and emissions *i.e. two*-injections (baseline), *three*-, *four*- and *five*-injections (partially-premixed) at 1200 rpm and 3bar demanded BMEP, (a) Brake Mean Effective Pressure, BMEP [bar] and measured fuel quantity [mg/st] *vs.* number of injections, (b) AVL smoke [FSN] and NOx [ppm] *vs.* number of injections, (c) Specific Fuel Consumption, [g/kWh] and Hydrocarbons [ppm] *vs.* number of injections, (d) AVL smoke [FSN] *vs.* NOx [ppm], for all injection configurations.

Figure 1(b) presents emissions of NOx and smoke as a function of number of injections. It can be seen from Figure 1(b) that largest reduction of NOx *i.e.* 86% and an increase in smoke *i.e.* 55 % from baseline is observed with *three*-injections; *four*-injections maintain baseline (conventional combustion) levels of smoke and reduce NOx by 61%, while *five*-injection partially-premixed combustion exhibits similar levels of NOx and smoke (compared with *four*). Therefore, it can be concluded from comparison between *three*-, *four*- and *five*- injections that the lowest NOx-smoke emissions are produced with *four*-injection partially-premixed combustion.

Figure 1(c) illustrates that, with partially-premixed combustion, as the number of injections are increased from *three* to *four*, SFC rises *e.g.* +13% from *three*- to *four*-injections and +19.5% from conventional to *four*-injections. This is probably because, due to early SOI, spray penetration occurs outside the combustion bowl. As the fuel mass per injection is lowered, with the increase in the number of injections (*i.e. five*-injections), a reduction in SFC is observed. A similar trend in emissions of HC is also observed, however, *three*-injection partially-premixed combustion, relative to conventional combustion (*two*-injections), does not exhibit a significant deviation in HC. Finally, Figure 1(d) illustrates that partially-premixed is susceptible to the same NOx-smoke trade-off as is conventional diesel combustion.

Figure 2(a) presents in-cylinder cycle analysis *e.g.* rate of cylinder-pressure rise-rate, cumulative Heat Release (cHR), Rate Of Heat Release (ROHR) and in-cylinder pressure trace) of conventional- (*two*-injection) and multiple-injection (*three*-, *four*- and *five*) partially-premixed combustion events. Figure 2(b) is an expanded view of Figure 2(a), to highlight the effect of fuel injection activation on SOC and consequently on in-cylinder pressure rise rate, to infer combustion noise from the latter. It can be seen from Figure 2(a) that a considerable reduction in combustion duration occurs between *two*-injection (baseline) and *three*-, *four*- and *five*-injections, indicating the occurrence of partially-premixed combustion.

Figure 2(b) illustrates that ignition delay is prolonged as the number of injections is increased. This trend is in agreement with earlier results observed in Figure 1(b) where highest smoke is achieved with *three*-injections (*i.e.* shortest ignition-delay), and thus possibly poorer mixing, compared with lower smoke emissions produced with *four*- and *five*- injections (*i.e.* with longer ignition-delays) and thus better mixing. It is also noticed from Figure 2(b) that *four*-injection combustion results in a smooth ROHR compared with *three*-injection combustion, and thus may contain the least amount of diffusion-phase burning (*i.e.* shortest injection duration) and consequently results in the lowest NOx values, as seen in Figure 1(b), at comparable smoke to conventional combustion (*two*-injections).

An increase of 80% in the rate of cylinder pressure rise rate, compared with conventional combustion (*two*-injections), (Figure 2b), indicates that partially-premixed combustion noise, compared with conventional, is significantly louder. Partially-premixed combustion with *five*-injections lowers the peak value of cylinder pressure rise-rate, indicating that combustion noise can be reduced by activating an additional injection. However, the flexibility between injection quantity and combustion stability is compromised as the earliest injection is fixed by FIE hardware suppliers *i.e.* a limit exists in the maximum quantity that could be scheduled for the fifth injection. Finally, from the analysis of Figures 1 and 2 it can be concluded that *four*-injections gives optimum emission (NOx-smoke) results and thus should be the subject of further study in this paper. The next section attempts to define a successful engine speed region, at constant engine load of 3bar BMEP utilizing *four*-injection partially-premixed combustion.

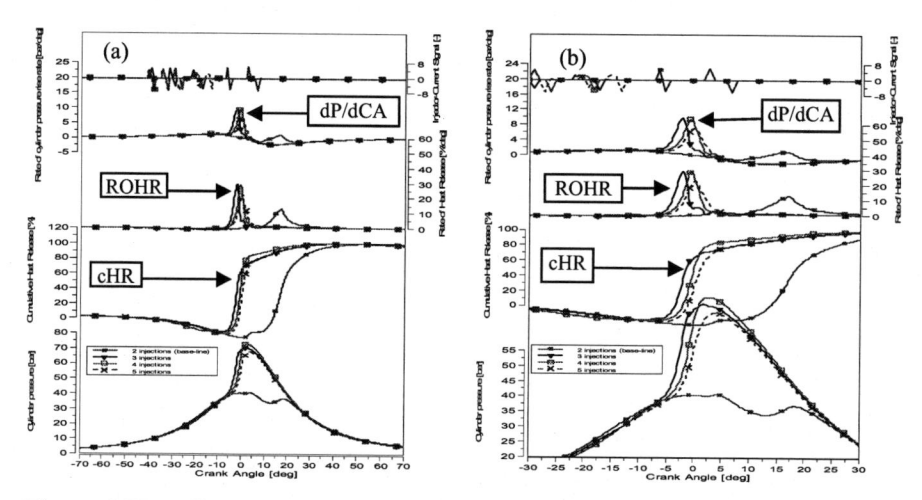

Figure 2 The effect of number of injections in-cylinder heat-release analysis at 1200rpm and 3bar demanded BMEP *e.g. two*-injections (baseline), *three-, four-* and *five*-injections (partially-premixed); (a) Cylinder pressure [bar], Injector current-clamp signal [-], Rate Of Heat Release, ROHR [%], Cumulative Heat Release [%] and Rate of cylinder pressure rise-rate [bar/CA], (b) Repeat of (a) for CA range -30 to 30°bTDC for all injection configurations.

Figure 3 presents a comparison of performance and emissions (*i.e.* NOx and smoke) between conventional- *(two*-injections*)* and partially-premixed combustion *(four*-injections*)* with varying engine speeds *i.e.* 1200, 1500, 2000rpm and constant load of 3bar BMEP. It can be seen from Figure 3(a) that degradation of 22% in BMEP (from the target of 3bar) is recorded at 2000rpm, suggesting that maximum range of operation lies up to 1500rpm engine speed. Figure 3(b) illustrates that smoke emission increases by 61% (compared with 1500rpm) at 2000rpm, suggesting that the "diffusion" content of the overall combustion has increased. Figure 3(c) shows that SFC is reduced as engine speed is increased, possibly due to the lower BMEPs achieved and longer ignition-delays resulting in SOC closer to TDC. Hence it can be concluded, from analysis of Figure 3, that the maximum range of *four*-injections partially premixed combustion lies up to 1500rpm engine speed and 3bar BMEP. The next section attempts to study, (with *four*-injections at 1500rpm and 3bar BMEP), the influence of injection quantity distribution between each multi-injection event.

Figure 3 The effect of four injections on engine performance and emissions of partially-premixed combustion at 3bar demanded BMEP; (a) Brake Mean Effective Pressure, BMEP [bar] and fuel quantity [mg/st] - (b) AVL smoke [FSN] and NOx [ppm] - (c) Specific Fuel Consumption, [g/kWh] and unburned Hydrocarbons [ppm] vs Engine speed [rpm].

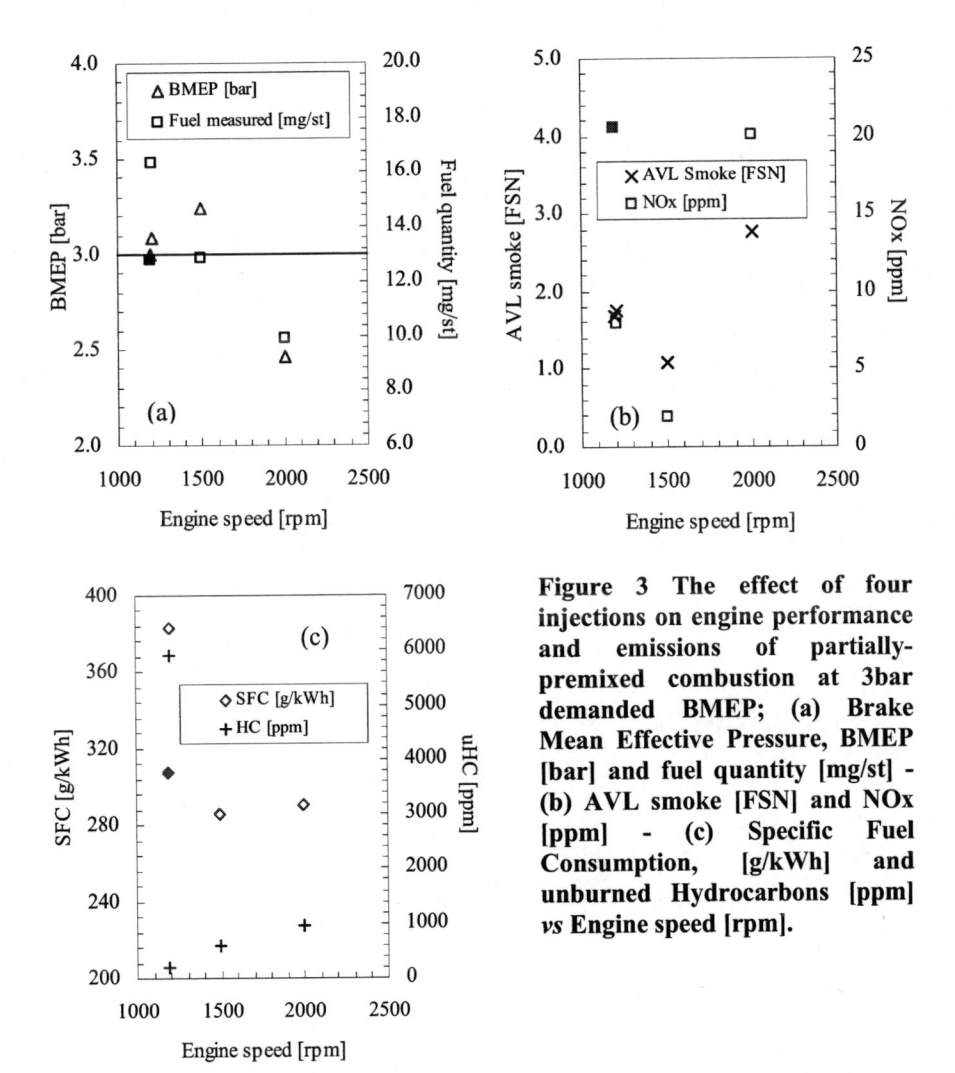

Figures 4-5 present results of the seven test cases described in Table 1. It can be seen from Figure 4(a) that all test cases achieved the targeted 3bar BMEP operating point (within ±4% error margin). Variability in measured total fuel injection quantity suggests that combustion efficiency may be varying as fuel injection schedules are transitioned from *high-* (early) to *low-* (early) quantity (*i.e.* test cases I-VI). Figure 4(b) presents emissions of AVL smoke and NOx for all seven test cases. It can be seen that test-case (I) produce relatively low smoke and NOx, compared with the baseline (designated as case-bl). As injection quantity scheduling is reduced from *first*-injection, smoke started to increase (as indicated by arrows on the plot). Highest smoke was achieved with case (V) that was operating with the highest injection quantity at late SOI *i.e.* close to TDC. An acceptable level of test-test repeatability is also demonstrated, as test-case (VI) reduces the smoke and NOx back to their approximate original values (test-case (I)).

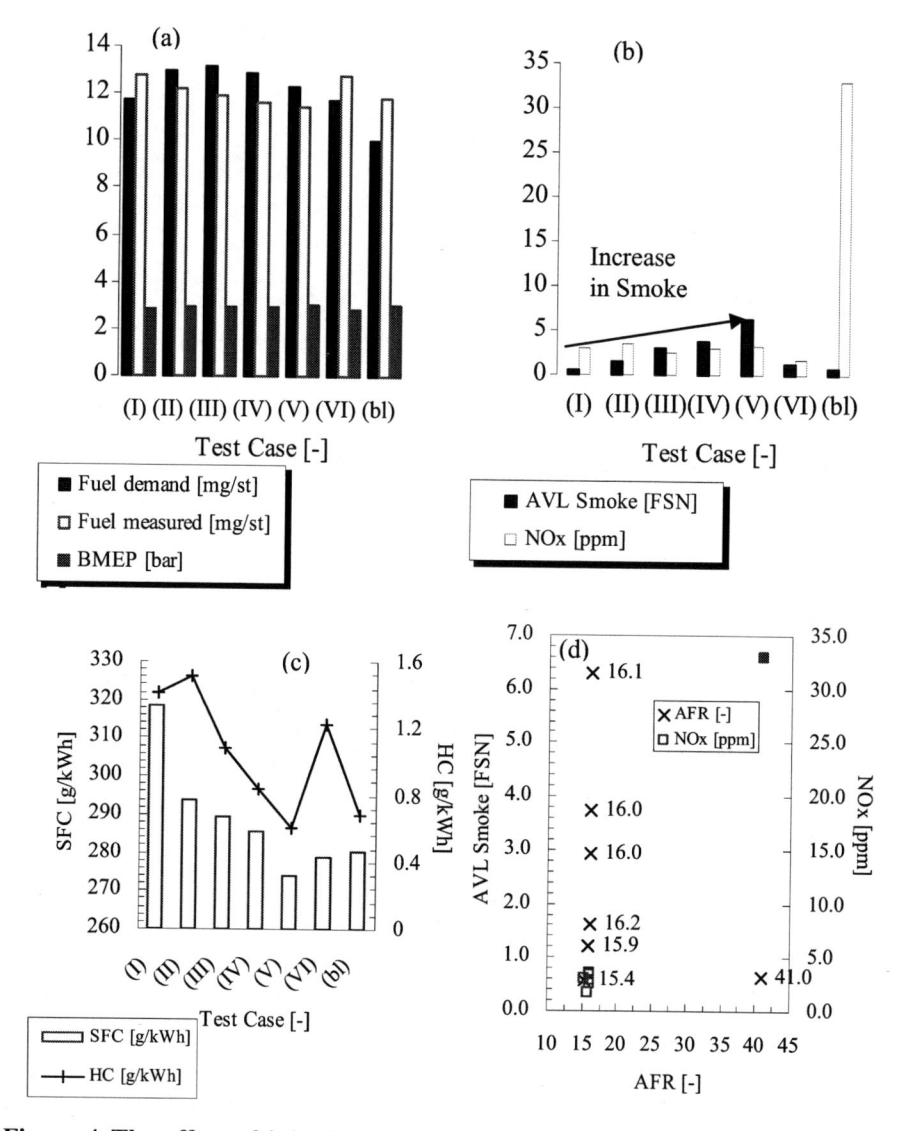

Figure 4 The effect of injection fuel quantity scheduling (*four*-injections) on engine performance and emissions during partially-premixed combustion at 1500rpm and 3bar demanded BMEP; (a) BMEP [bar] and Fuel quantity [mg/st] *vs.* reference Injection schedule [-]; (b) AVL smoke [FSN] and NOx [ppm] *vs.* reference injection schedule [-]; (c) Specific Fuel Consumption, [g/kWh] and unburned Hydrocarbons [ppm] vs. reference Injection schedule [-]; (d) AVL smoke [FSN] and NOx [ppm] *vs.* AFR [-], for all injection configurations; cases (I – VI) correspond to conditions given in Table 1.

Figure 4(c) presents SFC and HC emissions for all seven test cases. It can be seen that although smoke is increased with moving injection scheduling earlier, SFC is reduced indicating that combustion efficiency improves as higher fuelling quantities are injected closer to TDC. The discrepancy between test cases (I) and (V) SFC values cannot be explained. Figure 4(d) illustrates the same smoke and NOx results but this time as a function of global AFR. It can be seen (from Figure 4(d)) that although no significant change in AFR is observed between different test cases, a significant increase in smoke is observed. This indicates that although no significant difference is observed in *global* AFR, AFR *local* variation must be changing – given that a variation in smoke is being produced. This indicates that the mixture is not fully premixed but, rather, partially premixed.

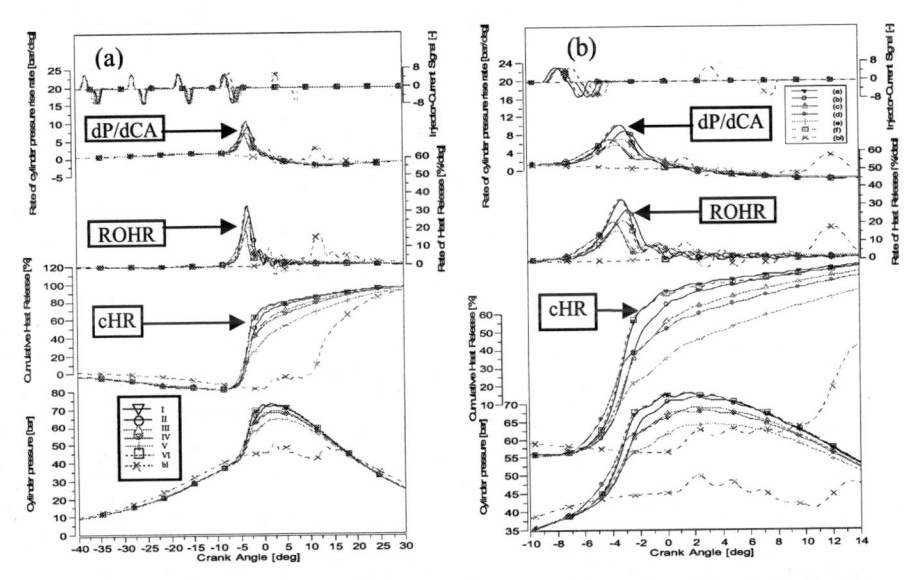

Figure 5 The effect of injection fuel quantity scheduling (*four*-injections) on engine emissions and corresponding in-cylinder heat-release analysis at 1500rpm and 3bar BMEP; (a) Cylinder pressure [bar], Injector current clamp signal [-], Rate Of Heat Release, ROHR [%], Cumulative Heat Release [%] and Rate of cylinder pressure rise-rate [bar/CA], (b) Repeat of (a) for CA range -10 to 14°bTDC for all injection configurations.

In-cylinder cycle analysis (*e.g.* rate of cylinder-pressure rise-rate, cumulative Heat Release (cHR), Rate Of Heat Release (ROHR) and in-cylinder pressure trace) of all seven test-cases is presented in Figure 5. Figure 5(a) shows that, as fuel injection quantity is shifted towards TDC, the rate of combustion decreases: *i.e.* the 50% cumulative heat release rate is reached at later CA. This would indicate that the larger quantities do not have enough time to premix fully before combustion starts and therefore the diffusion "proportion" of the overall combustion is increasing as larger quantities are injected closer to TDC. The above could be a plausible explanation for the smoke increasing as larger injection quantities are moved towards TDC. Figure 5(b) shows that SOC is not changed significantly as injection schedules are changed. It can also be seen that the initial rise rate of cumulative heat release is faster with later injected fuel compared with earlier ones. This would, once again, indicate that mixture properties before the start of combustion may have not influenced greatly the ignition delay but affect combustion duration after ignition.

The absence of any significant effect in NOx emissions (Figure 4b) also indicates that in-cylinder combustion temperatures are not greatly varied by injection scheduling; however, a variation in smoke resulting from different fuel injection quantities - without overall AFR variation, indicates that the mixture is not fully premixed but is more partially premixed. Varying levels of smoke could be hence due to varying levels of the diffusion content in the overall combustion.

Finally, a popular C/CD inertial class NEDC part-load operating point of 2000rpm and 2bar BMEP was selected for conducting Noise Vibration and Harshness (NVH) analysis with varying injection rail pressure (Section 3.2, Case 4). Partially premixed combustion with *three*-injections is utilized since it gives an added flexibility in SOI timing over *four*-injections (due to earliest possible SOI hardware limitations). Figure 6 compares, at 2000rpm and 2bar BMEP, emissions of smoke and NOx from *two*-injection conventional with *three*-injection partial premixed combustion; varying injection pressure from 500bar to 400bar respectively. It can be seen that reduction of 90% in NOx, at similar smoke, is observed between conventional and partially-premixed combustion (Test-case B). A further reduction in smoke (34%) is observed by reducing injection rail-pressure to 400bar (Test-case C).

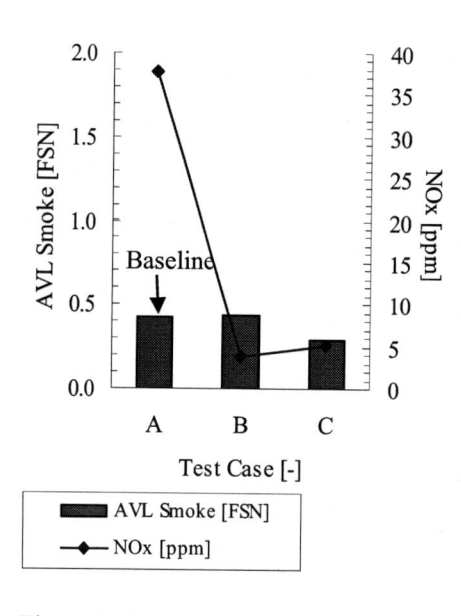

Figure 6 The effect of injection fuel quantity scheduling (*three*-injections) and fuel injection rail pressure on engine emissions and corresponding in-cylinder heat-release analysis at 2000 rpm and 2bar BMEP; AVL smoke [FSN] and NOx [ppm] vs. reference Injection schedule [-] according to Section 3.2 Case 4.

Figure 7 presents in-cylinder pressure trace analysis for the above three test cases. It can be seen that before the start of combustion, the partially-premixed in-cylinder pressure trace is significantly different from the conventional combustion and this can be explained by the large amount of EGR being introduced in to the cylinders. Large amounts of EGR dilute the in-cylinder charge and thus reduce overall boost pressure. Once the combustion is initiated (Figure 7b), it can be seen that the rate of cylinder pressure rise rate with partially-premixed combustion is 5bar/deg compared to 1.5bar/deg with conventional combustion. Furthermore, no significant difference in rate of cylinder pressure rise rate is observed when injection pressure is decreased by 100bar (Test-case C). The above trend indicates that a significant increase in overall combustion noise is expected with partially-premixed combustion, as compared with conventional operation.

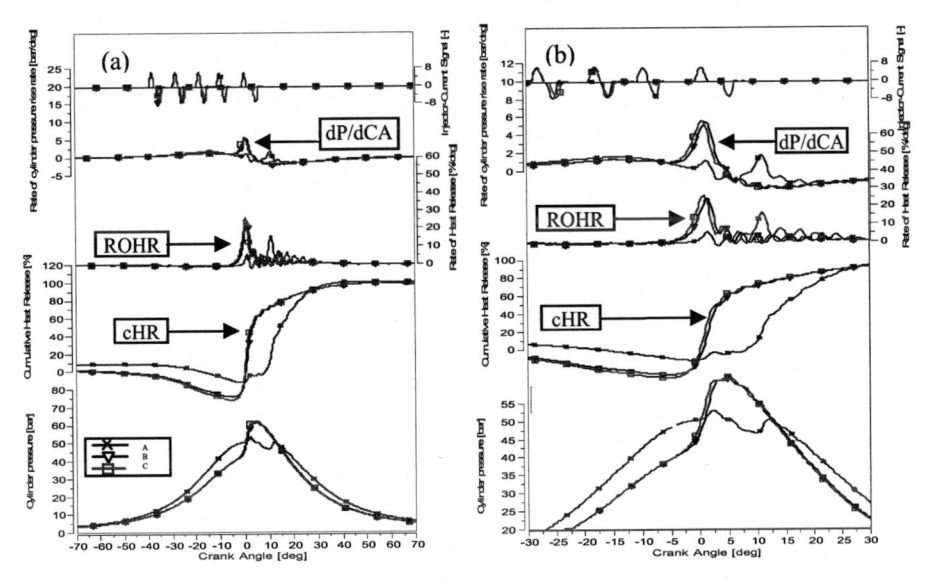

Figure 7 The effect of injection fuel quantity scheduling (*three* injections) and fuel injection rail pressure on engine emissions and corresponding in-cylinder heat-release analysis at 2000rpm and 2bar BMEP; (a) Cylinder pressure [bar], Injector current clamp signal [-], Rate Of Heat Release, ROHR [%], Cumulative Heat Release [%] and Rate of cylinder pressure rise-rate [bar/CA], (d)) Repeat of (a) for CA range -30 to 30 °bTDC, for all injection configurations

Figure 8 presents combustion noise and frequency analysis corresponding to the three pressure traces (Figure 7); a maximum of 8dbA increase in combustion noise was observed (in the range of human audible frequency; as indicated by dotted lines). However, the general level of noise is more uniform with frequency in the case of partially-premixed combustion.

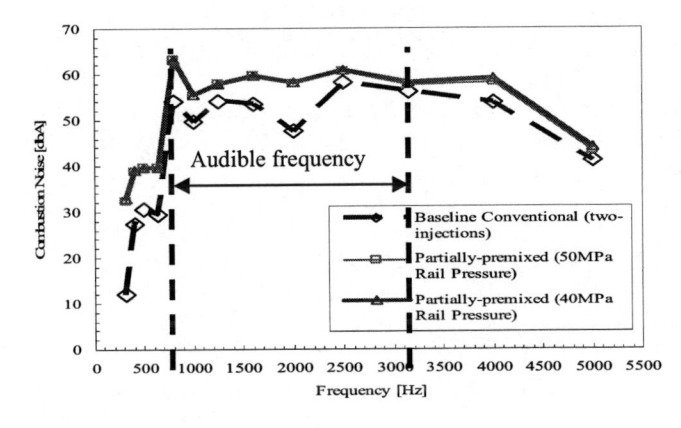

Figure 8 The effect of injection fuel quantity scheduling (*three* injections) and fuel injection rail pressure on engine combustion noise and corresponding frequency analysis at 2000 rpm and 2bar BMEP

5. CONCLUSIONS

This paper investigated, solely through engine calibration, benefits in emissions of NO_X and smoke, while operating with *multiple*-injection partially-premixed combustion on a conventional diesel engine for the contemporary passenger car market. The effect of fuel injection quantity scheduling on engine load, partially-premixed combustion and emissions of NOx and smoke were quantified.

Comparison between *two*-injection conventional operation and *three- four-* and *five*-injection partially premixed-combustion, at steady-state operating conditions of 1200rpm and 3bar BMEP, reduce NOx, by 61% with *four*-injections, at similar levels of smoke, to conventional combustion. However, a 13% increase in SFC was observed with *four*- as compared with *three*- injections and +19.5% with *four*-injections as compared with conventional combustion. The increase in SFC can be mainly attributed due to early SOC (5°bTDC) compared with late SOC (*i.e.* 5 °aTDC) with conventional combustion. All three partially-premixed *multiple*-injections, compared with conventional combustion, exhibit an 80% rise in dP/dCA. It was also noticed that partially-premixed is susceptible to the same NOx-smoke trade-off as conventional diesel combustion. Overall, *four*-injection partially premixed combustion, compared with *three*- and *five*-injections, resulted in the best NOx-smoke results.

Operating with *four*-injections at 3bar BMEP and varying engine speed we found that the maximum range of engine speed operation lay between 1500 and 2000rpm (*e.g.* a reduction in 22% torque was observed at 2000rpm and 3bar BMEP). Rescheduling fuel quantity distribution in *four*-injection mode (keeping total fuel quantity approximately constant) at 1500rpm and 3bar BMEP, we found that:

1. Larger fuel quantities injected earlier give the lowest emissions of smoke
2. Emissions of NOx were not significantly affected by variation in the distribution of injection quantities
3. An 85% reduction in NOx was achieved with partial premix combustion whilst similar levels of smoke emissions were produced as conventional operation.

Finally, operating at 2000rpm and 2bar BMEP with *three*-injection partially-premixed combustion gave a 90% reduction in NOx at similar levels of smoke. Comparison between *two*-injection conventional operation and *three*-injection partially-premixed combustion noise indicated a rise of approximately 5dbA value: however, the overall level was more uniform in frequency space as compared with conventional combustion.

5.1 Further Work
In this study, only three, four and five injections events were studied: it is recommended that further work be conducted to understand the influence of *single*- and *dual*- injections, and the distribution of fuel mass between these (for a minimum of two injections), by measurement of fuel-mass corresponding to each injection event. It would also be interesting to conduct a study of the combustion noise using

such schedules. This is because combustion noise is determined principally by the rate of heat release, which is sharper in partially-premixed than in conventional diesel combustion. Our work should be repeated with a narrow angle injector, rather that the wide-angle spray used here. Finally, we acknowledge that our work investigated combustion only as a "global" phenomenon and that, through future optical experiments in a single-cylinder engine, more confident assertions could be made about the amount and distribution of partially-premixed combustion as a "local" phenomenon.

REFERENCES

1. Oppenheim, A.K., (2004) "Combustion in piston engines: Technology, Evolution, Diagnosis and Control", Springer-Verlag Berlin Heidelberg 2004
2. Hsu, B.D., (2002) "Practical Diesel-Engine Combustion Analysis", Society of Automotive Engineers, Inc., PA, USA
3. Heywood, J.B., (1988) "Internal Combustion Engine Fundamentals", McGraw-Hill Book Company 1988
4. Dec, J.E., "A conceptual model of DI diesel combustion based on laser-sheet imaging", SAE Paper No. 970873 (1997)
5. Bianchi, G.M., Cazzoli, G., Pelloni, P. and Corcione, F.E., "Numerical study towards smoke-less and NOx-less HSDI diesel engine combustion", SAE Paper No. 2002-01-1115 (2002)
6. Yokota, H., Kudo, Y., Nakajima, H., Kakegawa, T. and Suzuki, T., "A new concept for low emission diesel combustion", SAE Paper No. 970891 (1997)
7. Najt, P.M. and Foster, D.E., "Compression-Ignited homogeneous Charge Combustion", SAE Paper No. 830264 (1983)
8. Carlucci, P., Ficarella, A. and Laforgia, D., "Effects of pilot injection parameters on combustion for common rail diesel engines", SAE Paper No. 2003-01-0700 (2003)
9. Ishikawa, N., Uekusa, T., Nakada, T. and Hariyoshi, R., "DI Diesel emission control by optimized fuel injection", SAE Paper No. 2004-01-0117 (2004)
10. Liu, Y. and Reitz, R.D., "Optimising HSDI diesel combustion and emissions using multiple injection strategies", SAE Paper No. 2005-01-0212 (2005)
11. Jacobs, T.J. and Assanis, D.N., (2007) "The attainment of premixed compression ignition low-temperature combustion in a compression ignition direct injection engine", Proceedings of the Combustion Institute **31** (2007) 2913-2920
12. Pickett, L.M., (2005) "Low flame temperature limits for mixing-controlled Diesel combustion", Proceedings of the Combustion Institute 30 (2005) 2727-2735
13. Buchwald, R., Brauer, M., Blechstein, A., Sommer, A. and Kahrstedt, J., "Adaption of injection system parameters to homogeneous diesel combustion", SAE Paper No. 2004-01-0936

14. Fang, T., Coverdill, R.E., Lee, C.F. and White, R.A., "Combustion and soot visualization of low temperature combustion within an HSDI Diesel engine using multiple injection strategy", SAE Paper No. 2006-01-0078 (2006)
15. Kook, S., Bae, C., Miles, P.C., Choi, D., Bergin, M. and Reitz, R.D., "The effect of swirl ratio and fuel injection parameters on CO emission and fuel conversion efficiency for high-dilution, low-temperature combustion in an automotive diesel engine", SAE Paper No. 2006-01-0197 (2006)
16. Harada, A., Shimazaki, N., Sasaki, S., Miyamoto, T., Akagawa, H. and Tsujimura, K., "The effects of mixture formation on premixed lean diesel combustion engine", SAE Paper No. 980533 (1998)
17. Lee, S., Gonzalez D.M.A., Reitz, R.D., "Effects of engine operating parameters on near stoichiometric diesel combustion characteristics", SAE Paper No. 2007-01-0121 (2007)
18. Simescu, S., Ryan III, T.W., Neely, G.D., Matheaus, A.C. and Surampudi, B., "Partial Pre-mixed combustion with cooled and uncooled EGR in a heavy-duty diesel engine", SAE Paper No. 2002-01-0963 (2002)
19. Simescu, S., Fiveland, S.B., Dodge, L.G., "An experimental investigation of PCCI-DI combustion and emissions in a heavy-duty diesel engine", SAE Paper No. 2003-01-0345 (2003)
20. Aiyoshizawa, E., Hasegawa, M., Kawashima, J. and Muranaka, S., "Combustion characteristics of a small DI Diesel engine", JSAE Review **21** (2000) 241-263
21. Nandha, K.P. and Abraham, J., "Dependence of fuel-air mixing characteristics on injection timing in an early-injection diesel engine", SAE Paper No. 2002-01-0944 (2002)
22. Ogawa, H., Kimura, S., Koike, M. and Enomoto, Y., "A study of heat rejection and combustion characteristics of a low-temperature and pre-mixed combustion concepts based on measurement of instantaneous heat flux in a direct-injection diesel engine", SAE Paper No. 2000-01-2792 (2000)
23. Hasegawa, R. and Yanagihara, H., "HCCI Combustion in DI diesel engine", SAE Paper No. 2003-01-0745 (2003)
24. Hashizume, T., Muyamoto, T., Akagawa, H. and Tsujimura, K., "Combustion and emission characteristics of multiple stage diesel combustion", SAE Paper No. 980505 (1998)
25. Singh, S., Reitz, R.D. and Musculus, M.P.B., "2-Colour thermometry experiments and high-speed imaging of multi-mode diesel engine combustion", SAE Paper No. 2005-01-3842 (2005)
26. Su, W., Lin, T. and Pei, Y., "A compound technology for HCCI combustion in a DI Diesel engine based on the multi-pulse injection and the BUMP combustion chamber", SAE Paper No. 2003-01-0741 (2003)
27. Neely, G.D., Shizuo, S. and Leet, J.A., "Experimental investigation of PCCI-DI combustion on emissions in a light-duty diesel engine", SAE Paper No. 2004-01-0121 (2004)

28. Dubreuil, A., Foucher, F., Mounaim-Rousselle, C.M., Dayma, G. and Dagaut, P., (2007) "HCCI combustion: Effect of NO in EGR", Proceedings of the Combustion Institute 31 (2007) 2879-2886
29. Jacobs, T.J., Knafi, A., Bohac, S.V., Assanis, D.N. and Szymkowicz, P.G., "The development of throttled and unthrottled PCI combustion in a light-duty engine", SAE Paper No. 2006-01-0202 (2006)
30. Klingbeil, A.E., Juneja, H., Ra, Y and Reitz, R.D., "Premixed diesel combustion analysis in a heavy-duty diesel engine", SAE Paper No. 2003-01-0341 (2003)
31. Kondo, M., Kimura, S., Hirano, I., Uraki, Y. and Maeda, R., (2000) "Development of noise reduction technologies for a small direct injection diesel engine", JSAE Review **21** (2000) 327-333
32. Kastner, O., Atzler, F., Muller, A., Weigand, A., Wenzlawski, K., Zellbeck, H., (2006) "Multiple injection strategies and their effect on pollutant emission in passenger car diesel engines", Thiesel 2006 Conference on Thermo- and Fluid Dynamic Processes in Diesel Engines
33. Dronniou, N., Lejeune, M., Balloul, I. and Higelin, P., "Combination of high EGR rates and multiple injection strategies to reduce pollutant emissions", SAE Paper No. 2005-01-3726 (2005)

Effects of extended exhaust and intake duration on CAI combustion in a multi-cylinder DI gasoline engine

Navin Kalian, Hua Zhao
Brunel University, Department of Engineering and Design, UK

ABSTRACT

CAI-combustion had been achieved in a 4-cylinder four-stroke gasoline DI engine using the method of early exhaust valve closure and trapping exhaust residual. This method uses a set of exhaust and intake camshafts with shorter-than-normal valve duration and lower-than-normal valve lift set in a Negative Valve Overlap configuration. It is known that valve opening duration and lift would affect the range of CAI operation using this method. It is expected that extended exhaust and intake duration may benefit the high speed CAI operation.

Using two sets of re-profiled camshafts, one with a 20 CA extended duration than the other; the effects of valve duration were studied. Testing was carried out at all permissible Exhaust Valve Closing and Intake Valve Opening times and the effects of valve opening duration on NIMEP were investigated for CAI combustion at an injection timing of -30 CA deg ATDC during the valve overlap period.

The engine speed was maintained at 1500 rpm and lambda was kept constant at 1.0. It was found that with longer duration camshafts, NIMEP values were lower compared with shorter duration camshafts. Pressure data was then studied to better understand this phenomenon.

1 INTRODUCTION

A new type of technology which offers a method for the vast reduction of emissions and improved efficiency is Homogeneous Charge Compression Ignition (HCCI) also known as Controlled Auto-Ignition (CAI). NOx emissions can be reduced by 90-98% [1-4], compared with SI engines. Fuel consumption savings of as much as 20% can be achieved within a 4-cylinder gasoline engine [5]. These engines are not radically different from spark-ignition engines found in production cars today. It is the actual combustion method which separates HCCI/CAI engines from standard SI engines. HCCI/CAI combustion is achieved by subjecting part or the entire cylinder

© *Brunel University 2007*

contents of premixed fuel/air mixture to temperatures above those required for auto-ignition of the reacting species.

The initial approach of achieving HCCI/CAI combustion utilized intake charge heating, this involved the use of an electrical heater to heat the intake air and provide the thermal energy needed for the charge to auto-ignite. Najt and Foster [6] used this method to highly dilute charge compositions to control the subsequent heat release. Even though initiating HCCI/CAI combustion through intake charge heating has been used in many subsequent studies, this method has limited automotive applications. Owing to the highly transient nature of automotive application, the large thermal inertias associated with heating the intake charge will make control of HCCI/CAI combustion with this method very hard to achieve. Najt and Foster's work established the potential of HCCI/CAI combustion to reduce fuel consumption in four-stroke engines.

Various other methods have been used to achieve HCCI/CAI combustion, these principally are: 1] Intake Charge Heating by exhaust gas through a heat exchanger, 2] Higher Compression Ratios, 3] More Auto-Ignitable fuel, 4] Recycling of burnt gases. A method which offers more flexibility in achieving HCCI/CAI combustion is the use of residual gas trapping through the use of negative valve overlap (Figure 1). With this method, the exhaust valve is closed early and a volume of exhaust gas is trapped, the intake valve is opened late preventing blow-back in the intake manifold [7]. The more advanced EVC, the greater the volume of trapped residual.

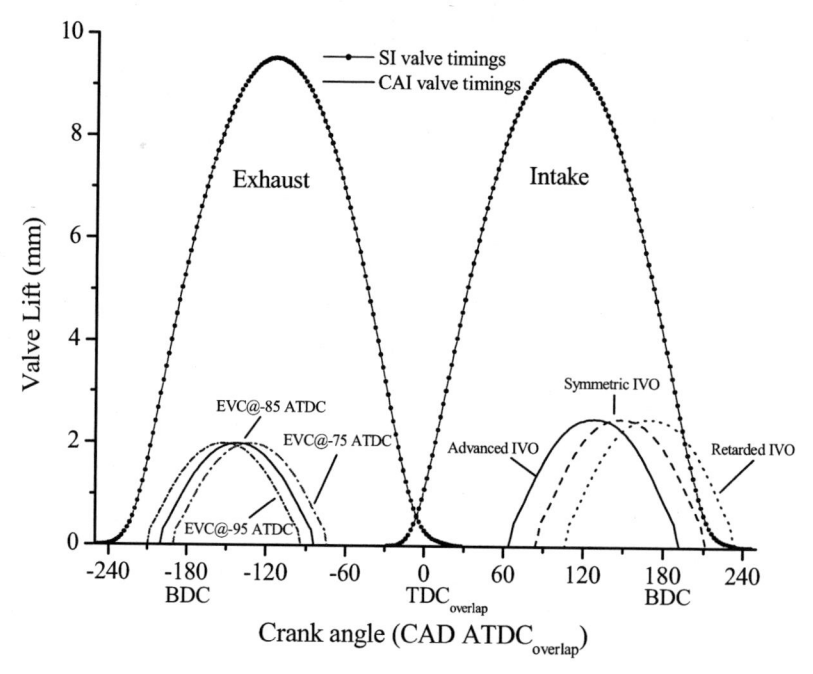

Figure 1 Indication of typical SI and CAI valve configuration

Investigators involved in the 4-SPACE [8] project realized HCCI/CAI combustion of gasoline at the part load, low speed range of the engine using fixed camshafts, but with shorter than normal valve opening duration and lower than normal valve lift. They coined the term Controlled Auto-Ignition (CAI) which has become synonyms for HCCI combustion generated by trapping residuals. Due to this reason the author will refer to HCCI combustion as CAI combustion within the results and discussion section.

HCCI/CAI combustion also presents several hurdles and challenges which must be overcome if this combustion is to be considered for application to commercial engines. HCCI/CAI combustion is governed by chemical kinetics and has no flame propagation. HCCI/CAI combustion is achieved by controlling the temperature, pressure and composition of the air and the fuel mixture so that it spontaneously ignites in the engine [9]. Therefore determining the phasing and rate of combustion presents many difficulties. Various approaches have been proposed to overcome these difficulties [10-12].

Another challenge with HCCI/CAI combustion is that CO and uHC emission levels of CAI combustion can reach or even surpass those of SI engines [13]. Probably, the most major hurdle blocking progression to commercial production of HCCI/CAI engines is the limited operating boundary compared with traditional SI operation. Due to the trapping of exhaust gas residual, limited volume of fresh charge is inducted; ultimately limiting high load operation [14]. Furthermore, the upper boundary of the HCCI/CAI operating regime suffers from knocking [15]. Low load operation is restricted by lack of thermal energy for auto-ignition of the mixture. Yang et al. [16] also reports that when load decreases, the demand of thermal energy for mixture auto-ignition increases, but the available thermal energy decreases. The lack of adequate thermal energy for mixture auto-ignition eventually leads to engine misfire.

This paper investigates how the valve duration and lift for a HCCI/CAI engine with a negative valve configuration affect the load output.

2 EXPERIMENTAL SET UP

A 1.6 Litre Ford prototype DI gasoline engine with four cylinders, 16 valves and DOHC was used to achieve CAI (Figure 2). The cylinder block was based on a standard Ford 1.6 L Zetec-SE engine. The cylinder head was modified to incorporate a VCT system on both the intake and exhaust side. The fuel was supplied via a low pressure fuel pump at 4 bar, before it was pressurized at 100 bar by a high pressure fuel pump and directly injected into the cylinders. Speed and load control was achieved by coupling the engine to an eddy-current water-cooled dynamometer. A Kistler 6061B pressure transducer was used to acquire in-cylinder pressure measurements. A LabVIEW™ based data acquisition system was used to calculate the Net Indicated Mean Effective Pressure (NIMEP), heat release data and 10%, 50% and 90% Mass Fraction Burned (MFB). The cooling water temperature was carefully controlled at 80°C.

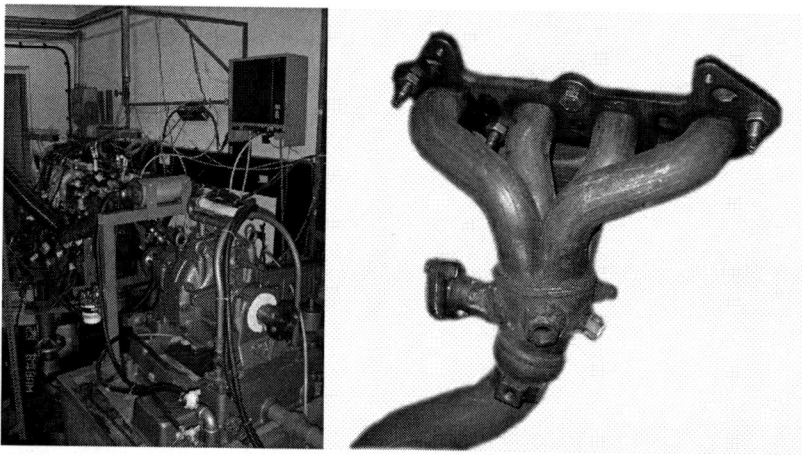

Figure 2: The Ford 1.6 L Zetec experimental engine test bed (right) and four-branch exhaust manifold with central downpipe used on the test engine

The engine was started through SI combustion and allowed to warm up. A heat exchanger was used to cool the coolant water, in addition a thermostat was used to regulate the coolant water temperature. It was possible to achieve CAI when the coolant temperature reached 80°C and the valve timing was adjusted to trap considerable amounts of exhaust residual. The air fuel ratio along with spark ignition, direct injection timing and valve timing were controlled using the Bosch ETAS tools VS-100 engine management system. The throttle was kept at wide open position and lambda was maintained at 1.0 through the variation of fuel injection amount by the engine management system using closed loop control. Where a test for typical CAI was carried out, the spark was turned off.

Table 1: Engine specifications

Engine Type	Inline 4-cylinder
Bore (mm)	79
Stroke (mm)	81.4
Displacement (cm^3)	1596
Compression ratio	11.5
Fuel Supply	Direct Injection
Fuel Injector (Spray cone angle)	Swirl Injector (60°)
Fuel Rail Pressure	10MPa
Fuel	Gasoline 95 RON

Table 2: Shorter and longer CAI camshafts specifications

Shorter CAI camshafts		Opening Duration	Lift
	Intake Valve	120 CA deg	2mm
	Exhaust Valve	110 CA deg	2mm
Longer CAI camshafts		**Opening Duration**	**Lift**
	Intake Valve	140 CA	3mm
	Exhaust Valve	130 CA	3mm

Two pairs of camshafts were machined which are referred to as the shorter CAI camshaft and the longer CAI camshaft. Table 2 details the specifications of the two pairs of camshafts. The valve lift between the shorter and longer CAI camshaft varies by 1mm for both the intake and exhaust valve. This was due to a machining restriction rather than an experimental requisite.

3 RESULTS

3.1 Use of EVC-IVO as control parameters for HCCI combustion

Conventional representation of SI data utilizes load-speed maps. In order to investigate the effect of EVC-IVO, charts have been produced which show EVC timing against IVO timing plotted for load and an indicated specific emission value at 1500 rpm, lambda 1.0. EVC-IVO maps allow a complete representation of the operational range of the experimental engine taking into consideration SI and CAI combustion. The Spark-Ignition and the CAI regions are identified on the EVC-IVO plots. In order to characterize SI and CAI combustion, burn duration and ISNOx values were analyzed (Table 3 and Table 4). A long burn duration and high ISNOx values indicated SI combustion; whereas CAI combustion has a short burn duration and low ISNOx values. There also exists an operational region where combustion occurs partly through spark-ignition and partly by auto-ignition and has been labeled the spark-assisted CAI region [17]. For this region, the burn duration and ISNOx values are lower than SI combustion but higher than CAI combustion. From Table 3, it can be observed that at EVC 85 CA deg BTDC, IVO 110 CA deg ATDC for the shorter CAI camshaft, the burn duration is 13 CA deg and the ISNOx value is 1.61 g/kW.h. The burn duration and ISNOx value is approximately half that of typical SI combustion and almost double that of typical CAI combustion. Therefore, although EVC 75 CA deg BTDC, IVO 100 CA deg ATDC and EVC 85 CA deg BTDC, IVO 110 CA deg ATDC for the shorter CAI camshaft have been indicated as SI combustion, they have characteristics of Spark-Assisted CAI combustion.

Valve timing can be used as a control parameter to vary load as demonstrated in previous work [18]. Figure 3 shows a typical EVC-IVO plot for an engine with low-lift, short duration valve events and the set-up with a Negative Valve Overlap configuration. As EVC is advanced, a greater volume of residual is trapped and hence load decreases. The EVC-IVO plot shows this gradual variation as valve timing is varied in increments. The effect of IVO on combustion is dependent on a number of factors.

Advancing IVO causes the effective compression ratio to increase which may be beneficial in increasing NIMEP. In the case of Figure 3, as IVO is advanced there is a general increase in NIMEP. However, at the retarded intake timing, IVO 105 CA deg ATDC and EVC 105 CA deg BTDC, NIMEP values are high due to symmetrical valve timings which experience lower pumping losses. Therefore, as highlighted, due to the condition-dependent nature of IVO timings, it is considered that EVC-IVO maps provide a holistic representation of HCCI data.

Table 3: Burn duration and ISNOx values for the shorter CAI camshaft at SOI -30 CA deg ATDC

SI combustion - Ignition Timing 20 CA deg BTDC							CAI combustion			
EVC (CA deg BTDC)	75	75	75	75	75	75	85	95	105	105	105
IVO (CA deg ATDC)	50	60	70	80	90	100	110	110	80	90	100
Burn Duration (CA)	21	50	25	21	20	11	13	12	5	7	8
ISNOx (g/kW.h)	3.12	3.74	4.53	4.55	4.53	2.40	1.61	0.75	0.84	0.75	0.73

Table 4: Burn duration and ISNOx values for the longer CAI camshaft at SOI -30 CA deg ATDC

SI combustion- Ignition Timing 20 CA deg BTDC								
EVC (CA deg BTDC)	55	55	55	55	55	65	65	65
IVO (CA deg ATDC)	70	80	90	100	110	80	90	100
Burn Duration (CA)	13	21	13	26	21	31	41	39
ISNOx (g/kW.h)	4.52	4.51	4.51	4.53	4.56	2.42	2.64	2.67
CAI combustion								
EVC (CA deg BTDC)	75	75	75	75	75	85	85	85	95
IVO (CA deg ATDC)	60	70	80	90	100	60	70	80	90	90
Burn Duration (CA)	16	21	22	21	14	11	12	20	19	14
ISNOx (g/kW.h)	0.66	0.99	1.05	1.05	0.6	0.38	0.48	0.58	0.2	0.2

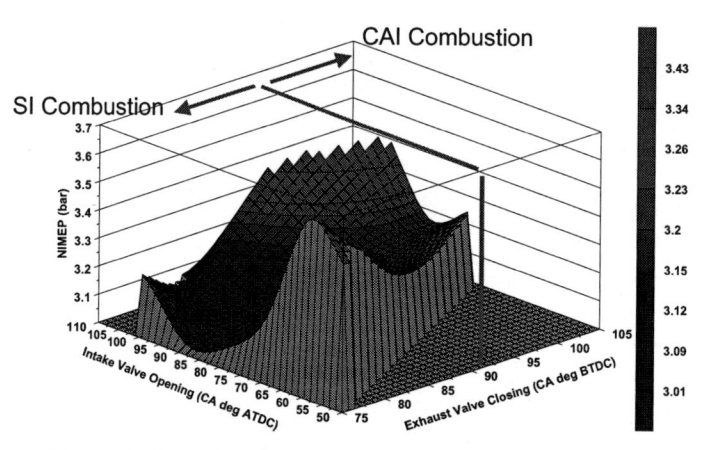

Figure 3: 3-D map depicting EVC-IVO for NIMEP values for the shorter camshaft

3.2 Effect of extended duration on NIMEP

Figure 4 shows a pressure against crank angle diagram for both the shorter and longer CAI camshaft, pressure data was inspected for each case presented in this paper in order to ensure the validity of experimental results. A comparison of NIMEP values at Start of Injection (SOI) -30 CA deg ATDC during the negative valve overlap period between shorter and longer CAI camshafts (Figure 5) shows that for the entire valve timing range, the shorter CAI camshafts have a higher NIMEP. For example at EVC 95 CA deg BTDC, IVO 90 CA deg ATDC and SOI -30 CA deg ATDC, the NIMEP for the shorter CAI camshafts is 3.16 bar, whereas for the longer CAI camshafts the NIMEP is 2.27 bar; almost a 1 bar drop. It would normally be expected that for a longer intake and exhaust duration that a greater volume of fresh charge would be inducted and NIMEP and ISNOx values would increase. The reason for the lower NIMEP values for the longer CAI camshaft is due to a few reasons. Firstly, EVO for the longer CAI camshaft occurs 20 CA deg before EVO for the shorter CAI camshaft. Meaning that the exhaust valve opens earlier during the power stroke, taking away positive work generated during this stroke. Pumping losses are much higher for the longer CAI camshaft over the valve timing range (Figure 6). Furthermore, the effective compression ratio is lower for the longer CAI camshaft as IVC is retarded by 20 CA deg.

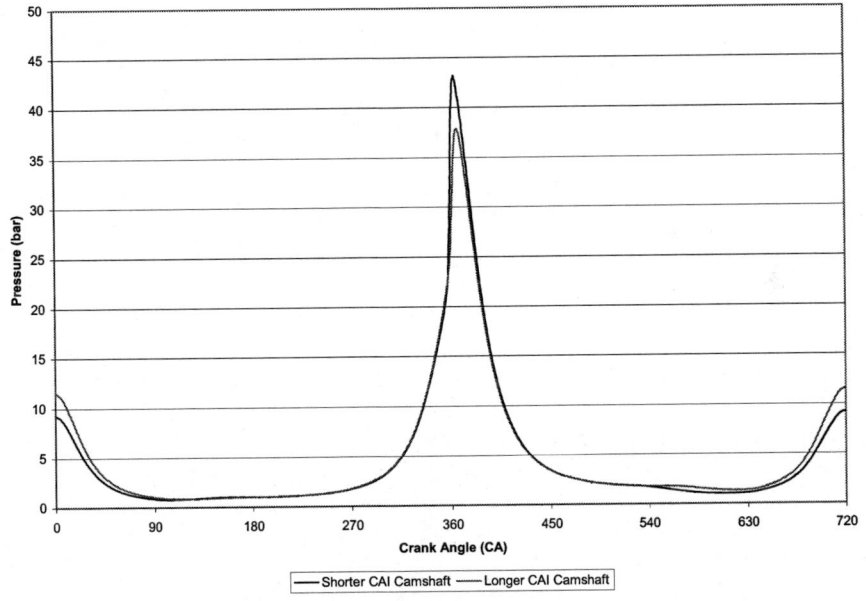

Figure 4: Pressure versus crank angle at EVC 75 CA deg BTDC, IVO 90 CA deg ATDC, speed 1500, lambda = 1.0, SOI -30 CA deg ATDC

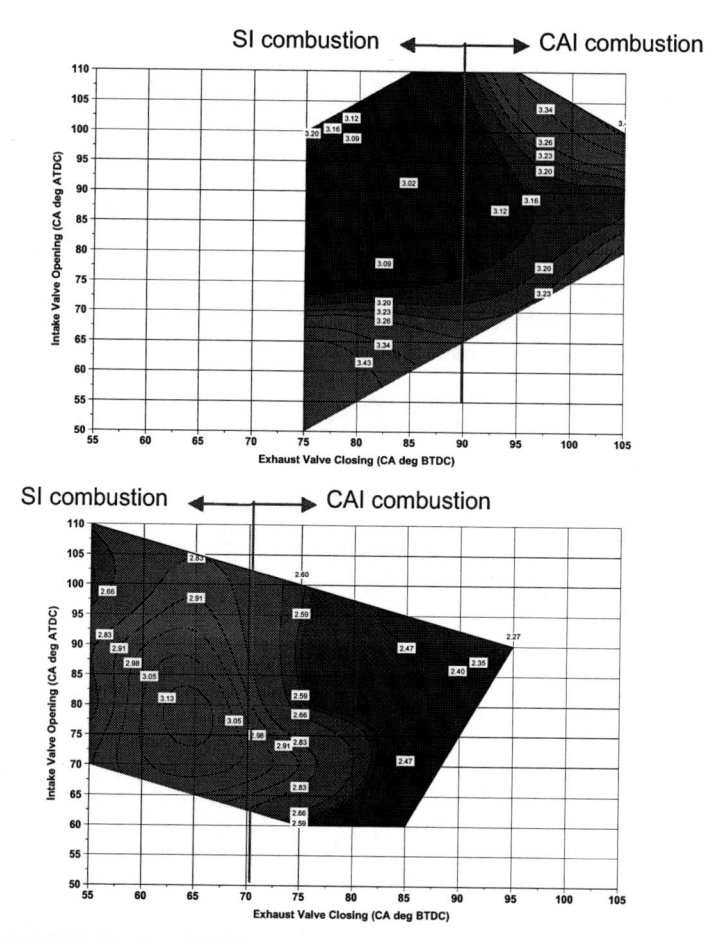

Figure 5: NIMEP (bar) at EVC versus IVO timings for the shorter (above) and the longer (below) CAI camshaft at lambda 1.0, SOI -30 CA deg ATDC

Another main reason for lower NIMEP values for the longer CAI camshaft at SOI - 30 CA deg ATDC compared with the shorter CAI camshaft is due to the higher percentage of trapped residual at the same valve timings (Figure 7). The percentage of exhaust residual was calculated using the ideal gas law, at EVC the mass of residual was calculated. The percentage of trapped residual was then calculated by dividing the mass of trapped residual by the total in-cylinder charge. For example at EVC 95 CA deg BTDC, IVO 90 CA deg ATDC, the percentage of trapped residual for the shorter CAI camshaft is 45.6%, for the longer CAI camshaft the percentage is 52.7%. This trend is true for all valve timings, on average the difference is approximately 7% higher residual for the longer CAI camshaft compared with the shorter CAI camshaft at a given valve timing. The higher percentage of trapped residual results in less inducted fresh charge, hence lower NIMEP values. The reason for higher trapped residual for the longer CAI camshaft is due to the fact that at EVC, the in-cylinder pressure is higher for the longer duration camshaft compared with the shorter duration camshaft (Figure 8), therefore at IVO there will be less inducted fresh charge. It appears that for the longer CAI camshafts, the in-cylinder pressure is higher during the exhaust stroke compared with the shorter CAI camshafts. On Figure 8 – longer CAI camshaft, there is an indication of the expected pressure trace for the longer CAI camshaft. It is usually expected that the in-cylinder pressure should drop at EVO and reach atmospheric pressure and allow complete exhausting of gases. There appears to be a reduction of the volume of exhaust gas exhausted out from the engine. A possible explanation is due to the interaction of exhaust gas within the exhaust manifold after it has been expelled from each individual cylinder. With individual exhaust ports, the gases from each cylinder will be discharged into the manifold branch pipe under identical conditions but the shortness of branches can still produce exhaust gas interference between cylinders as the gas back-pressure builds up in the down-pipe. As with high-lift, long-duration SI combustion, low-lift, short-duration HCCI combustion is also dependent on manifold design and length and should be taken into consideration when valve duration is altered.

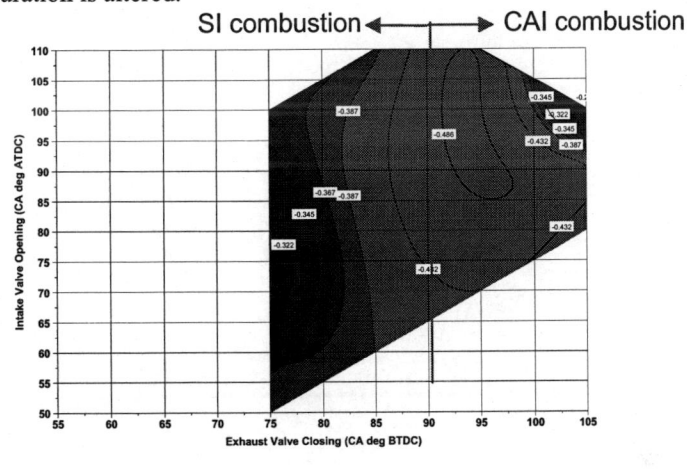

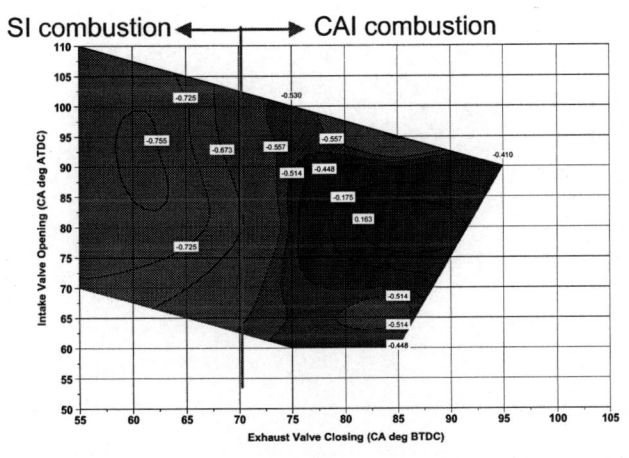

Figure 6: Pumping losses (bar) at EVC versus IVO timings for the shorter (above) and the longer (below) CAI camshaft at lambda 1.0, SOI -30 CA deg ATDC

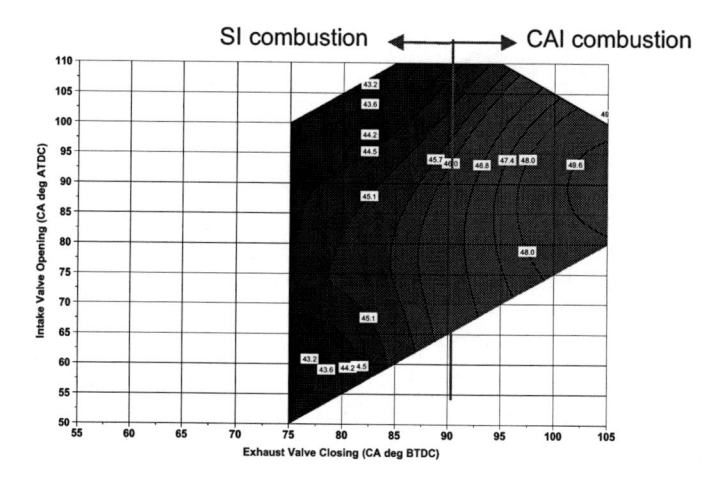

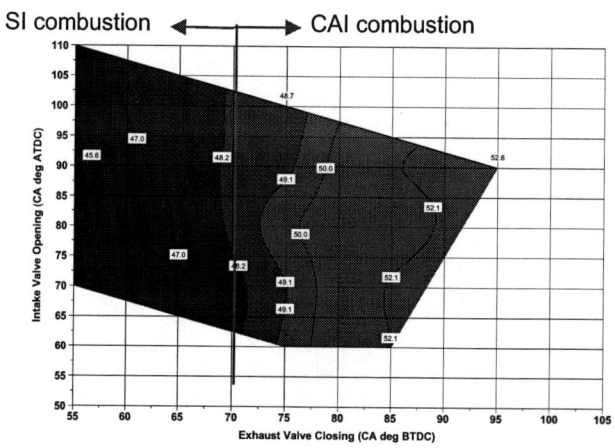

Figure 7: Percentage of trapped residual at EVC versus IVO timings for the shorter (above) and longer (below) CAI camshaft at lambda 1.0, SOI -30 CA deg ATDC

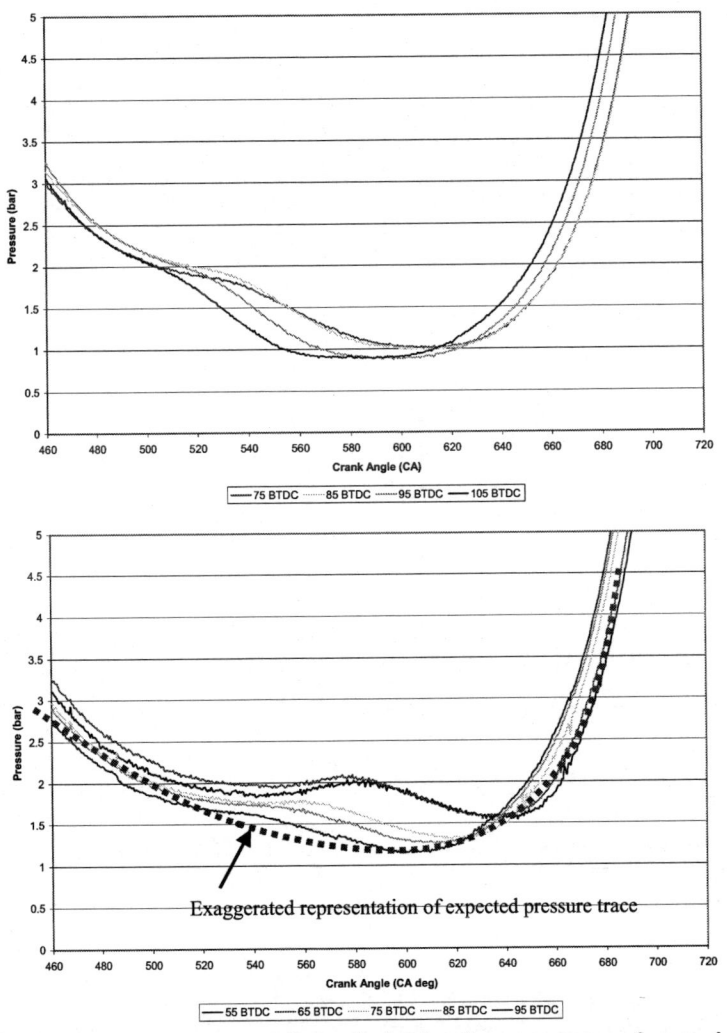

Figure 8: Pressure versus crank angle during the re-compression period at varying EVC, IVO 90 CA deg ATDC, 1500 rpm, lambda 1.0 for the shorter (above) and the longer (below) camshaft

4 SUMMARY

It was found that for the lambda 1.0 operation, the shorter CAI camshafts could lead to CAI operation at higher NIMEP values compared with the longer CAI camshafts. There are two reasons for the difference in NIMEP values. The first is that the exhaust camshaft with the 20 CA deg longer opening is experiencing Exhaust Valve Opening 20 CA deg before the shorter exhaust camshaft for a given EVC leading to reduced expansion work. The lower effective compression ratio associated with the longer CAI camshaft contributed to lower NIMEP values. The second reason for the difference in NIMEP values is due to the fact that at similar EVC timing, exhaust residuals are approximately 7% higher for the longer CAI camshaft as compared with the shorter CAI camshaft. The extra residual results in less inducted fresh charge and lower NIMEP values. The reason for the higher percentage of trapped residual for the longer CAI camshafts was due to higher in-cylinder pressure at EVC, hence less expelled exhaust gas. With hindsight, it is realized that the interaction of exhaust gas within the exhaust manifold for CAI combustion plays an important role in determining engine performance and hence any alteration of valve duration and lift should involve re-design of the exhaust and/or intake manifold.

ACKNOWLEDGMENTS

This work is supported by the Engineering and Physical Science Research Council (EPSRC) of UK. Additional support is provided with the engine and its controller by Ford Motor Company and partial funding for a studentship by Jaguar Cars Ltd.

REFERENCES

1. **Jian Li, Hua Zhao and Nicos Ladommatos,** Research and Development of Controlled Auto-ignition (CAI) Combustion in a 4-stroke Multi-cylinder Gasoline Engine, SAE Paper 2001-01-3608, 2001.
2. **Aoyama, T., Hattori, Y., Mizuta, J., (Toyota Central Research and Development Labs., Inc.) Sato, Y., (Toyota Motor Corp.)** An Experimental Study on Premixed-Charge Compression Ignition Gasoline Engine SAE Paper 960081, 1996.
3. **Willand, J., Nieberding, R.G., Vent, G., Enderle, C.,** The Knocking Syndrome – Its Cure and Its Potential SAE Paper 982483, 1998.
4. **Urushihara, T., Hiraya, K., Kakuhou, A., Itoh, T.,** Expansion of HCCI Operating Region by the Combination of Direct Fuel Injection, Negative Valve Overlap and Internal Fuel Reformation, SAE Paper 2003-01-0749, 2003.
5. **Yamaoka, S., Kakuya, H., Nakagawa, S., Nogi, T., (Hitachi Research Laboratory, Hitachi, Ltd) Shimada, A., Kihara, Y.,** A Study of Controlling the Auto-Ignition and Combustion in a Gasoline HCCI Engine SAE Paper 2004-01-0942, 2004.
6. **Najt, P. M. and Foster, D.E.**, Compression-ignited homogeneous charge combustion, SAE paper 830264, 1983.
7. **Yap, D., A. Megaritis, M.L. Wyszynski, and H. Xu**, Effect of inlet valve timing on boosted gasoline HCCI with residual gas trapping, SAE 2005-01-2136, SAE Fuels and Lubricants Meeting, Rio de Janeiro, May 2005, 2005.
8. **Lavy, J., Dabadie, J.C., Angelberger, C., Duret, P., Willand, J., Juretzka, A., Schaflein, J., Ma, T., Lendresse, Y., Satre, A., Schulz, C., Kramer, H., Zhao, H., and Damiano, L.**, Innovative Ultra-low NOx controlled auto-ignition combustion process for gasoline engines: the 4-SPACE project, SAE paper 2000-01-1873, 2000.

9. Kimura, S., Osamu, A., Yasuhisa, K., and Eiji, A., New combustion concept for ultra-clean and high-efficiency small DI diesel engines, SAE paper 1999-01-3681, 1999.
10. **Hyvonen, J.,(Fiat-GM Powertrain Sweden) Haraldson, G., Johansson, B., (Division of Combustion Engines, Lund Institute of Technology)** Operating Conditions Using Spark Assisted HCCI Combustion during Combustion Mode Transfer to SI in a Multi-Cylinder VCR-HCCI engine, SAE Paper 2005-01-0109, 2005.
11. **Milovanovic, M., Blundell, D., Pearson, R., Turner, J., (Lotus Engineers), Chen, R., (Department of Aeronautical and Automotive Engineering, Loughborough University)**, Enlarging the Operational Range of a Gasoline HCCI Engine by Controlling the Coolant Temperature, SAE Paper 2005-01-0157, 2005.
12. **Aroonsrisopon, T., Werner, P., Waldman, J.O., Sohm, V., Foster, D.E., Morikawa, T., and Lida, M.**, Expanding the HCCI Operation With the Charge Stratification, SAE Paper 2004-01-1756, 2004.
13. Aoyama, T., Hattori, Y., Mizuta, J., (Toyota Central Research and Development Labs., Inc.) Sato, Y., (Toyota Motor Corp.) An Experimental Study on Premixed-Charge Compression Ignition Gasoline Engine, SAE Paper 960081, 1996.
14. **Jian Li, Hua Zhao and Nicos Ladommatos,** Research and Development of Controlled Auto-ignition (CAI) Combustion in a 4-stroke Multi-cylinder Gasoline Engine, SAE Paper 2001-01-3608, 2001.
15. **Willand, J., Nieberding, R.G., Vent, G., Enderle, C.,** The Knocking Syndrome – Its Cure and Its Potential, SAE Paper 982483, 1998.
16. **Yang, J.,** Expanding the operating range of homogeneous charge compression ignition-spark ignition dual-mode engines in the homogeneous charge compression ignition mode, Int. J. Engine Research., 2006, 6(4), 279-288
17. **Kalian, N., Standing, R.H., Zhao, H., Effects of Ignition Timing on CAI Combustion in a Multi-Cylinder DI Gasoline Engine, SAE Paper, 2005-01-3720, 2005**
18. **Standing, R., Kalian, N., Ma, T., Zhao, H.,(Brunel University) Wirth, M., Schamel, A., (Ford Motor Company),** Effects of Injection Timing and Valve Timings on CAI Operation in A Multi-Cylinder DI Gasoline Engine SAE Paper 2005-01-0132, 2005.
19. **Stone, R.,** Introduction to internal combustion engines, Third Edition, Macmillan Press, ISBN 0-333-74013 0, 2000, pp99.
20. **Dec, J.E., Sjoberg, M., A.,** Parametric Study of HCCI Combustion – the Sources of Emissions at Low Loads and the Effects of GDI Fuel Injection, SAE Paper 2003-01-0752, 2003.
21. **Zhao, F., Asmus, T.W., Assanis, D.N., Dec, J.E., Eng, J.A., Najt, P.M.,** Homogeneous Charge Compression Ignition (HCCI) Engines: Key Research and Development Issues, 2003, pp.338.

LISTS OF ABBREVIATIONS

AI	Auto-Ignition	**ISFC**	Indicated Specific Fuel Consumption
ATDC	After Top Dead Center		
BDC	Bottom Dead Center	**ISHC**	Indicated Specific Hydrocarbon
BTDC	Before Top Dead Center		
CAD	Crank Angle Degree	**ISNOx**	Indicated Specific Oxides of Nitrogen
CAI	Controlled Auto-Ignition		
CO	Carbon Monoxide	**IVO**	Intake Valve Opening
DI	Direct Injection	**MFB**	Mass Fraction Burned
DOHC	Double Over-Head Cams	**NIMEP**	Net Mean Effective Pressure
ECU	Engine Control Unit		
EGR	Exhaust Gas Recirculation	**NOx**	Oxides of Nitrogen
		PFI	Port Fuel Injection
EVC	Exhaust Valve Closing	**PMEP**	Pumping Mean Effective Pressure
HC	Hydrocarbon		
HCCI	Homogeneous Charge Compression Ignition	**SC**	Stratify Charge
		SI	Spark Ignition
IMEP	Indicated Mean Effective Pressure	**TDC**	Top Dead Center
		VCT	Variable Camshaft Timing
IVC	Intake Valve Closing		

AUTHOR INDEX

Adolph, D. 49
Akehurst, S. 261
Aleiferis, P.G. 3, 223
Assanis, D. 27

Baert, R.S.G. 131
Banks, A. 71
Bannister, C.D. 247
Beasley, M. 71
Benjamin, S.F. 143
Blaxill, H. 223
Brace, C.J. 111, 247
Brown, A. 247
Busch, H. 49
Bushe, W.K. 213

Cairns, A. 223
Carnochan, B. 277
Cho, K. 27
Choi, B. 123
Christ, A. 87
Cracknell, R.F. 3, 15

Dumenil, J.C. 247

Eastwood, P. 285
Ewart, P. 15
Ewen, F. 277

Filipi, Z. 27

Genc, U. 277
Genz, T. 49
Guggenberger, B. 161

Hardalupas, Y. 285
Hathout, J.-P. 87
Hawley, J.G. 247
Hill, P.G. 213
Hoffman, H. 223
Hume, A. 223

Jeong, J.-W. 123
Jia, M. 99
Jorach, R.W. 61

Kalian, N. 303
Kawanabe, H. 199
Kolbeck, A. 49
Körfer, T. 49
Kulzer, A. 87

Lamping, M. 49
Lee, P.M. 235
Leverton, T. 71
Lim, M.-T. 123

Ma, H. 15
Malcolm, J.S. 223
Martini, E. 277
McTaggart-Cowan, G.P. ... 213

Morris, T.	285	Shioji, M.	199
Müller, J.	161	Skipton Carter, A.	71
Munshi, S.R.	213	Stevens, R.	15
		Stone, R.	15
Najt, P.	27	Such, C.	173
Nakagawa, K.	199	Szekely, G.	27
Nevard, R.T.	61		
Nicol, A.J.	173	Taylor, A.M.K.P.	285
		Thornthwaite, I.R.	61
Ohira, T.	199	Tily, R.	111
Otte, R.	161	Todd, A.R.	223
		Tolley, A.	71
Pegg, I.	247	Tufail, K.	285
Peng, Z.	99	Twigg, M.V.	185
Piddock, M.	261		
Pischinger, S.	49	van Romunde, Z.	3
Priest, M.	235		
		Wallace, S.	3, 15
Rask, R.	27	Walmsley, H.L.	3, 15
Richardson, D.	3, 15	Wang, X.	15
Roberts, C.A.	143	Weberbauer, F.	161
Rogak, S.N.	213	Wellers, M.	277
Rueckauf, J.	223	Williams, B.	15
		Wilson, N.D.	61
Sarnbratt, U.	173	Winstanley, T.	285
Sauer, C.	87		
Schoeppe, D.	61	Yamane, K.	199
Serras-Pereira, J.	3		
Seykens, X.L.J.	131	Zhao, H.	303